NANOMETER FREQUENCY SYNTHESIS BEYOND THE PHASE-LOCKED LOOP

IEEE Press
445 Hoes Lane
Piscataway, NJ 08854

Technical Reviewers
Prof. Michael Peter Kennedy, University College Cork
Associate Prof. Woogeun Rhee, Tsinghua University

Books in the IEEE Press Series on Microelectronic System: A complete list of the titles in this series appears at the end of this volume.

NANOMETER FREQUENCY SYNTHESIS BEYOND THE PHASE-LOCKED LOOP

LIMING XIU

IEEE PRESS

WILEY

A JOHN WILEY & SONS, INC., PUBLICATION

Published by John Wiley & Sons, Inc., Hoboken, New Jersey.
Published simultaneously in Canada

For general information on our other products and services or for technical support, please contact our Customer Care Department within the United States at (800) 762-2974, outside the United States at (317) 572-3993 or fax (317) 572-4002.

Wiley also publishes its books in a variety of electronic formats. Some content that appears in print may not be available in electronic formats. For more information about Wiley products, visit our web site at www.wiley.com.

Library of Congress Cataloging-in-Publication Data:
Xiu, Liming.
Nanometer frequency synthesis beyond the phase-locked loop / Liming Xiu.
p. cm.
ISBN 978-1-118-16263-7 (cloth)
1. Timing circuits. 2. Frequency synthesizers. 3. Very high speed integrated circuits. I. Title.
TK7868.T5X83 2012
621.381'32–dc23

2012001531

10 9 8 7 6 5 4 3 2 1

CONTENTS

PREFACE

I have no special talents. I am only passionately curious.

—Albert Einstein

In the great Einstein's view, passion, desire—and above all curiosity—are the very ignition switches to spark discovery and creation. More than two decades ago, when I was studying physics in Tsinghua University (Beijing, China), this confession seemed counterfactual. After 20 years of involvement in scientific and engineering work, it is gradually starting to make sense to me. Nowadays, there are 7 billion people living on this planet. If all the people who ever lived on Earth were included, this enormous number would be exponentially larger. Among this gigantic population there are countless gifted people who are born with talent. However, history shows that only a tiny handful of people have made paramount contributions to the understanding of the world we all live in. The force that separates these all-time greats from the exceptional group of the talented is the passion to ask *what* and *why*, sincerely and unyieldingly.

FREQUENCY IS CHANGED

I am neither the great nor the gifted. But this force of curiosity does have its hold over me. In my career as a very-large-scale integration (VLSI) circuit design professional, I have had the fortune to work in many different areas (please see my other book: *VLSI Circuit Design Methodology Demystified: A Conceptual Taxonomy*, 2007). This unique experience provides me with the opportunity to observe everything from a broader viewpoint, the ability to see things in the bigger picture. In the meantime, it engages my curiosity. It often drives me to challenge the conventional way of doing things. One particular example is the clock signal used in the VLSI circuit. As both a circuit level phase-locked loop (PLL) designer and a system-on-a-chip integration level PLL user, I have seen the story from both sides. I distinctly remember one afternoon in the summer of 2003, after spending a long time explaining the flying-adder architecture (invented in the late 1990s) to one of my colleagues, a question suddenly occurred to me: What is frequency? Why must all the cycles have equal lengths in time? In common sense, this question looks foolish and dangerous for anyone to ask. Curiosity about this issue has intrigued me for several years (secretly, for fear of being treated as an illiterate). In 2008, after a long period of serious investigation from both theoretical and experimental perspectives, I had built up enough nerve to formally introduce the concept of "time-average-frequency." It removes the constraint that all clock cycles must have the same length-in-time. This seemingly ridiculous or insignificant step is a bold move philosophically. Its aim is the two long-lasting problems in this field: arbitrary frequency generation and fast response in frequency switching. It will have profound influence in VLSI circuit design since clock signal is used in every chip. Along the running history of our progressive understanding of this world, it is shown that all the great advancements originate at the concept level. The greatest example is provided by Einstein. By changing our view of the two fundamental concepts of time and space, he brought us one giant step closer to the ultimate understanding of the universe. This has forever changed the way we live. In this book, the most important message that I want to share with reader is: *the concept of clock frequency is changed.*

> Your time is limited, so don't waste it living someone else's life. Don't be trapped by dogma—which is living with the results of other people's thinking. Don't let the noise of others' opinions drown out your own inner voice. And most important, have the courage to follow your heart and intuition.
>
> —Steve Jobs

The spirit behind this excerpt from Steve Jobs's famous speech (Stanford University, 2005) is not unfamiliar. Similar wisdom has been expressed in the past by great philosophers and pioneers. But Mr. Jobs's testimony is more touching and real to us as individuals because he lived in our time. He noticeably changed the face of technology and the modern way of life, and he preached his passion in a way that was pleasantly contagious. During the pursuit of time-average-frequency, I sometimes felt frustrated because this new thinking contradicts conventional wisdom. On several occasions, a painful price had to be paid to uphold what I believe. Today, whenever Jobs's remark is replayed, I feel a bit of warmth and encouragement. Looking at his journey, it is confirmed again that all the greats have their own obstacles. The key to success is not superior intellect or powerful financial muscle. Instead, it is the intrinsic drive to believe, to achieve, and to change. This book is my case of this testimony.

SIMPLE AND ELEGANT

Coupled with curiosity, the other important part of my mindset is the tenacious desire to pursue simplicity and elegance in almost everything. I admire beautiful things in life: beautiful music, beautiful art, beautiful literature, beautiful sportsmanship, a beautiful soul—the list goes on and on. During the creation of the flying-adder circuit, simplicity drove me to search unrelentingly for the simplest structure that required the minimum number of transistors possible. Elegance compelled me to ensure that there is a sophisticated and yet beautiful mechanism behind the simple circuit. I am a passionate believer of the "Principle of Least Action" (Pierre-Louis Maupertuis, 1774). I apply it to my circuit design whenever I can. I hope that I can convey this attitude to readers throughout this book.

TIME, NUMBER, AND THE BEAUTY OF MATHEMATICS

The key focus of this book—frequency—is closely related to the thing that we called *time*. Time is a major subject of religion, philosophy, and science. Among great thinkers, there are two distinct standpoints on time. One view is that time is part of the fundamental structure of the universe, a dimension in which events occur in sequence. The opposing view is that time does not refer to any kind of physical container that events and objects move through. Instead, time is part of a fundamental intellectual structure (made of *space*, *number*, and *time*) within which humans sequence and compare events. In this second view, time is a virtual subject, neither an event nor a thing, and thus is not itself measurable.

Another mysterious product from human brain is the *number*. The world is virtually made of numbers. Numbers were invented to fulfill the need to organize our life quantitatively, beyond just qualitatively. It is generally believed that this is one of the major reasons why humans and all other species have followed different evolutionary paths (language is among the others). In our daily life, time and number are connected though an entity called the atomic clock: the definition of *second*. In VLSI circuit design, time and number are related by a special signal called *clock*. In this engineering practice, however, the relationship between time (frequency) and number has not reached the harmonization achieved in our daily life. In this book, one of the goals is to see if something can be done to improve the situation (digital-to-frequency converter, the counterpart of digital-to-analog Converter). In this effort, two important mathematical tools are used: *Number Theory* and *Fourier Analysis*. During this process of reasoning and learning from several "beautiful minds," I am amazed at the power and the striking beauty of mathematics. I am deeply touched by the mysterious harmony rooted in our number system. In this book, I want to share this joy with reader.

PLAY TIME AS WE PLAY LEVEL

The entire VLSI circuit design business is built on the fact that we use *level* (voltage or current level) to represent information. In analog processing, level is organized in multiple elevations. In the digital domain, it is in binary fashion. As process technology advances, some momentous changes emerge: the transistor is switching faster and faster, and the supply voltage is reduced lower and lower. Consequently, *time* (or rate-of-switching) becomes an attractive option to represent information. This will unquestionably influence the way that we design circuits. In this book, a million-dollar question is asked: "Can we play time as we play level?"

This book is organized in the following way:

Chapter 1 discusses how the clock signal is used in all electronic applications. The aim of this chapter is to understand our targeting problem in depth. Chapter 2 briefly reviews the existing clock generation techniques. This chapter focuses on the explanation of how this problem is conventionally dealt with. Chapter 3 looks at the root of the clock problem. It investigates the very concept of frequency and introduces the breakthrough viewpoint that leads us on an entirely new path. Chapter 4 presents the supporting technology, flying-adder architecture, which implements this new concept into circuitry. This is the hardware implementation of this novel approach introduced in chapter three. Based on the time-average-frequency concept and the flying-adder circuit, Chapter 5 coins a new device: the digital-to-frequency converter. Chapter 6 shows some examples of using this innovative technology to build cheaper, faster, and better systems. It illustrates the strength of this new technology. Chapter 7 is the visionary discussion of using "time" for signal processing. It brings forth new directions for future chip design. Its goal is to inspire the next generation researcher and engineer with new opportunities.

This book was inspired by *Stay Hungry, Stay Foolish*, which I second from the bottom of my heart. This mindset is the invisible hand that has created our magnificent civilization out of the void. It will serve as the lighthouse to guide us in the journey of seeking the ultimate paradise. It is my wish that this book can play a role in achieving the goal of designing "cheaper, faster, and better" electronic products that will ultimately make for a more enjoyable life.

I would like to thank my dear wife, Zhihong You, for supporting me in the completion of this book. Without her selfless effort, this book would never have been published. She has always stood beside me through both "thick and thin." As a fellow professional who works in similar area and was trained in the same schools, her gifted mental might is highly respected by me. Fortunately, it appears that her exceptional competence has been passed to our lovely daughters Katherine and Helen. I also want to thank Katherine Xiu for helping me in English proofreading and in creating the index.

LIMING XIU

CHAPTER 1

CLOCK SIGNAL IN ELECTRONIC SYSTEMS

1.1 THE SIGNIFICANCE OF CLOCK SIGNAL

1.1.1 Clock Signal

In modern electronic-driven society, our everyday lives are supported by various kinds of electronic devices. At home, TV, computer, audio system, game machine, and digital camera are indispensable for our entertainment and relaxation. Away from home, mobile phones keep us connected with the world all the time. On the road, automobiles and airplanes with countless built-in electronic devices make them safe to be driven/flown and comfortable to ride in. At work, we spend most of our time dealing with the computer, fax machine, copier, printer, projector, etc. Without these electronic devices, people's lives would be totally different; human society would regress many years in standard of living. Electronic devices have already penetrated into all aspects of our lives.

When in operation, almost all electronic devices rely on a very important signal: the clock. This is simply due to the fact that electronic devices are made of very-large-scale-integration (VLSI) chips, which are primarily designed on the synchronous principle. For any chip, simple or complex, its designed functionality is achieved by millions of events that occur inside it. These events do not happen randomly but in a predetermined, orderly sequence. The clock signal is the conductor of the orchestra to produce harmony. For successful

Nanometer Frequency Synthesis Beyond the Phase-Locked Loop, First Edition. Liming Xiu.

Published 2012 by John Wiley & Sons, Inc.

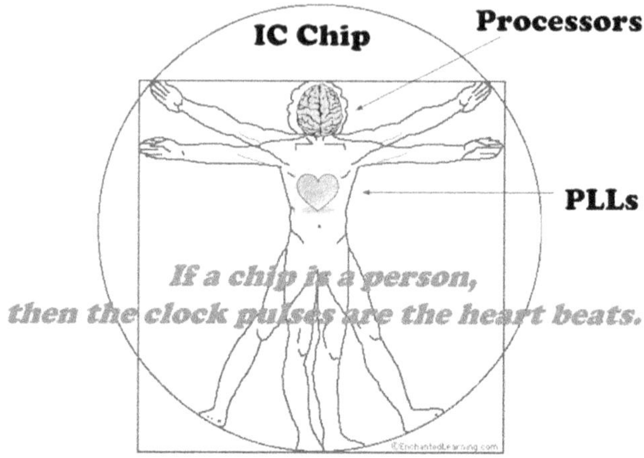

Fig. 1.1. The importance of clock pulses: they are the heartbeats.

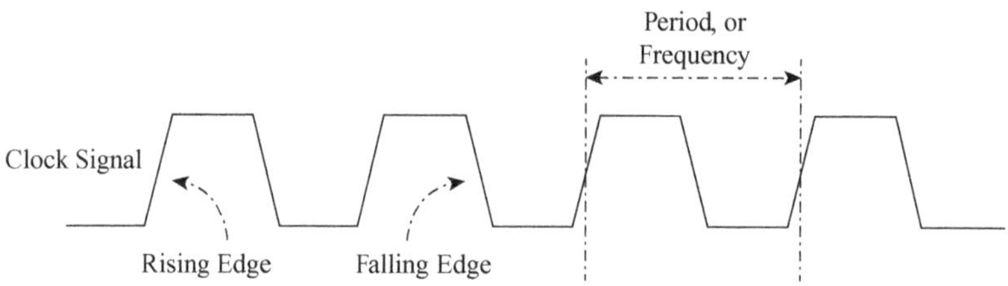

Fig. 1.2. Clock signal is an electrical pulse train.

operation in a large chip, many clock signals (as many as hundreds) could be required simultaneously. Usually, phase-locked loop (PLL) is used on-chip to generate these crucial clock signals. If a VLSI chip could be treated as a person and the on-chip processor were regarded as the brain, then the clock pulse is the heartbeat, the clock signal is the blood, and the clock distribution network (clock tree) is the vessel. This analogy is graphically demonstrated in Fig. 1.1.

In the field of VLSI circuit design, the clock signal is an electrical pulse train of square waveform as shown in Fig. 1.2. It has two distinguishable voltage levels: high and low. The basic unit in this pulse train comprises one occurrence of high level voltage and one occurrence of low level voltage. The transitions between the low-to-high and high-to-low are termed the clock edges. They are called "rising edge" and "falling edge," respectively. The length-in-time used by this basic unit is defined as the clock period; its inversion is the frequency that is often used by people to gauge the working speed of an electronic device.

One of the most important characteristics of the clock signal is that the basic unit, often called the cycle, has to be able to repeat itself indefinitely.

In other words, in this pulse train, every cycle has to be exactly the same. This is because that clock signal is the driver of the chip. The billions of operations (can also be viewed as events) inside a VLSI chip are all coordinated by clock signal. Structurally, the circuit inside the chip is designed in such way that these operations are triggered by either the rising edge or the falling edge, or both, of the clock signal. Therefore, it is essential that the occurrences of these edges in time are precisely predictable. The easiest way of achieving this goal is to make every cycle the same. A clock signal with this predictability in its waveform has enabled an important VLSI circuit design method: synchronous design. The synchronous design methodology is a milestone technology that allows the VLSI chip design industry to make great strides.

The physical medium inside the electronic circuit is electrical voltage or current. The electronic circuit is naturally suitable for handling the *magnitude* of this medium. (In all VLSI chips, information is represented through the magnitude of this medium.) By manipulating the magnitude, VLSI chips can process information and produce result for us to use. Manipulating the medium's magnitude for representing information is natural for an electronic circuit, since magnitude is directly proportional to the number of electronics flowing inside electronic devices. On the other hand, an electronic circuit is not naturally born for managing the other important variable: *time*. Instead, electronic systems use voltage transition to represent timing information. Therefore, it is not an easy task to generate the period of the basic unit (clock cycle) any way you want. It usually requires external help of a timing reference source, such as a mechanical crystal oscillator. Then, a special circuitry of PLL is used to produce other time scales based on this precise reference. This field of work is called frequency synthesis, and it is one of the most actively researched and engineered areas in VLSI circuit design.

1.1.2 The Aim of This Book

Due to the difficulty of using electronic circuits to manipulate the time scale, the capability of PLLs is limited. In many cases, it is extremely difficult and costly for the clock circuit design engineer to produce the clock frequencies that the system engineer prefers. Most of the time, the system engineer has to use whatever frequencies the PLL circuit designer is able to offer. Moreover, when a PLL is used as the clock source, it is difficult to switch from one frequency to another in a short time (a short time in comparison to the clock period). Consequently, these problems have limited our options for designing better and cheaper electronic products.

Throughout the history of frequency synthesis development, there are three distinguished approaches: direct analog synthesis, direct digital frequency synthesis (DDFS), and PLL-based indirect frequency synthesis. Among these, the PLL-based method is the most popular one for on-chip clock generation. There are several styles in the PLL-based approach:

integer-N PLL, fractional-N PLL, sigma-delta fractional-N PLL, and all digital PLL (ADPLL). All the aforesaid techniques are built around one basic consensus: constructing the clock waveform with equal lengths in time for all the cycles. In other words, the basic unit of the clock waveform is repeatable; all the units have to be exactly the same. This feature is ideal for the clock that is being used as the driver signal for chip operation because the location in time of every edge is precisely predictable. Unfortunately, this is also the single most influencing factor that makes the task of clock generation (frequency synthesis) difficult.

History shows that major science and technology advancements often start with adventurous thinking. Breakthroughs usually happen when traditional thinking is detoured. Moreover, most of the time, crucial advancement is initialized at the conceptual level. After a long period of time sticking with the belief that "all cycles shall have same length-in-time," it is worth focusing our attention back to the two fundamental issues:

1. **In the field of electronic circuit design, what does frequency mean?**
2. **In circuit design practice, how is the clock signal used?**

The process of searching the answers for these two questions has induced the formal introduction of the time-average-frequency concept (Xiu 2008a). This rigorously formed concept lays down the foundation for a new frequency synthesis technique: flying-adder direct period synthesis architecture. Together, time-average-frequency and flying-adder architecture are the two cornerstones of a new circuit component: digital-to-frequency converter (Xiu 2008b). These breakthrough innovations, as illustrated in Fig. 1.3, are the focus of this book.

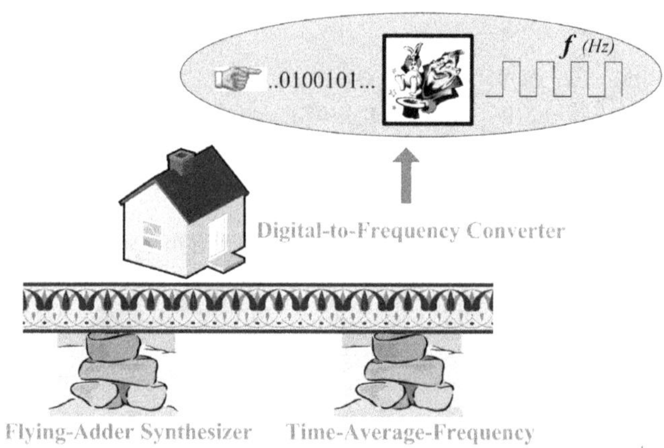

Fig. 1.3. Time-average-frequency, flying-adder synthesizer, and digital-to-frequency converter are the focus of this book.

1.2 THE CHARACTERISTICS OF CLOCK SIGNAL

The clock signal used in electronic system has two functional characteristics: frequency and phase. It also has one quality-related characteristic: jitter (phase noise). A clock period is defined as the time used by one clock cycle. The frequency, which is the mathematical inverse of the period, is used to describe the number of clock cycles (clock pulses) that exist in the time frame of 1 second. In modern synchronous design practice, all the events that happen inside a chip are triggered by either the rising edge or the falling edge, or both, of the clock pulses. Therefore, frequency determines the number of operations carried out within 1 second. It is the gauge of chip speed. For example, a CPU running at 2 GHz has 2 billion clock pulses within 1 second. Consequently, there will be 2 billion coordinated operations that occur within 1 second. Frequency is the most important characteristic of the clock signal. When more than two clock signals exist in a system and interact with each other (through the data they drive), in addition to their frequencies, the relative positions of their functional edges are of interest to system designer as well. This relative position is represented through a parameter called the clock phase. The precision associated with the position of the clock's functional edge is qualified by another parameter of jitter.

1.2.1 Jitter and Phase Noise

1.2.1.1 "Jitter" is Used to Describe the Clock Edge Uncertainty The term "jitter" is used to describe the nonidealness of the clock edges' positions in time. Ideally, all clock edges shall occur in precisely determinable positions when both the frequency and the initial position are given. Their positions should be mathematically traceable. However, in real practice, the implementation of clock generation circuit (e.g., a PLL) inevitably has some imperfections. This results in some degree of uncertainty in the position of the clock edges, as illustrated in Fig. 1.4. People use the term "jitter" to quantitatively describe the degree of this uncertainty.

1.2.1.2 Timing Error is Caused by Voltage Noise An electrical circuit is naturally suitable for representing information by using magnitude (voltage or current). Timing information is not inherently attached to the electrical circuit. In circuit practice, timing information is converted from voltage or

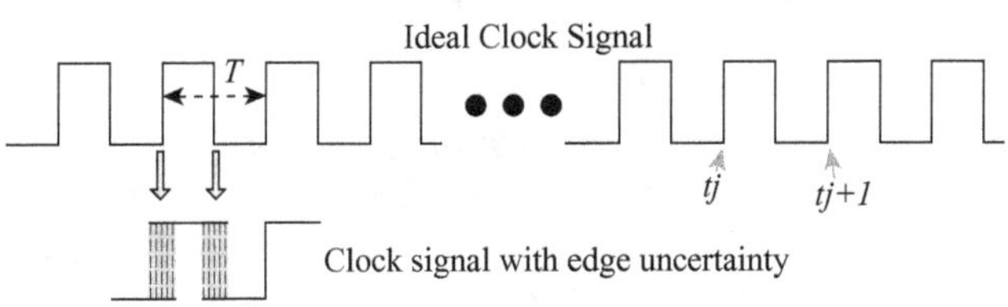

Fig. 1.4. Clock edge uncertainty is called jitter.

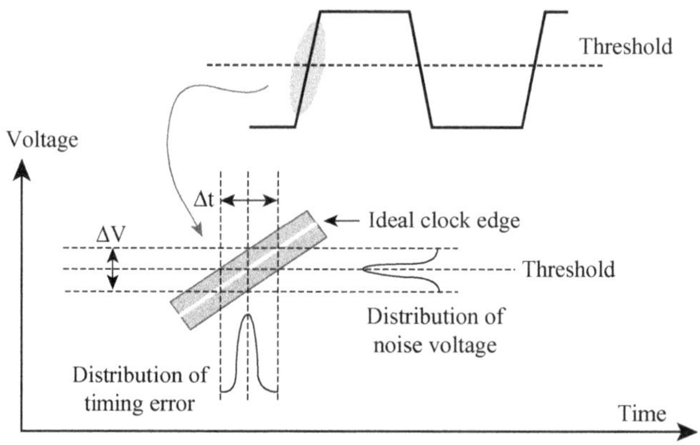

Fig. 1.5. Voltage noise is converted into timing error.

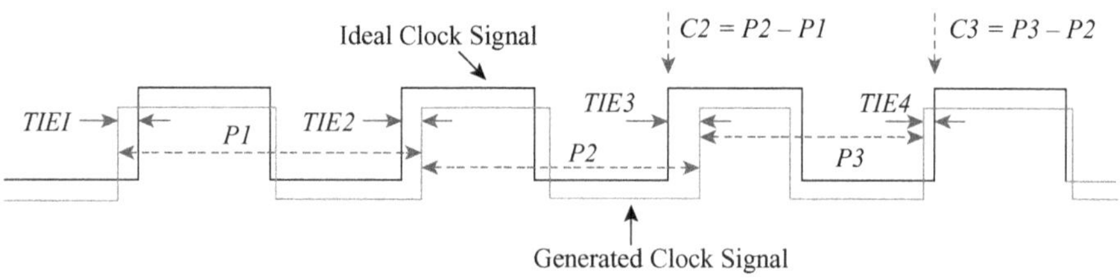

Fig. 1.6. Period jitter, cycle-to-cycle jitter, and time-interval-error.

current transient events. As shown in Fig. 1.5, the "time" in an electronic circuit is represented by the moment at which the voltage crosses a predefined threshold. In a synchronous system, jitter is the deviation of clock edges from their ideal positions. It is a form of noise, since any voltage noise that corrupts the waveform will be converted proportionately into a timing error, as also shown in Fig. 1.5. This edge fluctuation usually is a random process and must be characterized in terms of its statistics (mean value, standard deviation, confidence level, etc). There are many terminologies used in the literature to describe this clock edge uncertainty: period jitter, absolute jitter, cycle-to-cycle jitter, long-term jitter, accumulated jitter, random jitter, deterministic jitter, root mean square (*rms*) jitter, peak-to-peak jitter, periodic jitter, total jitter, etc. The fact that so many terms are used for one phenomenon is simply due to the reason that clock edge uncertainty is both an important and complex subject in academic research and engineering practice.

1.2.1.3 Look at Clock-Edge-Uncertainty in Time Domain: Period Jitter, Cycle-to-Cycle Jitter, and Time Interval Error The three most commonly used jitter terms in engineering practice are period jitter, cycle-to-cycle jitter and time interval error (TIE). As depicted in Fig. 1.6, period jitter (P1, P2, P3, etc.) is the simple measurement of the period of each clock cycle.

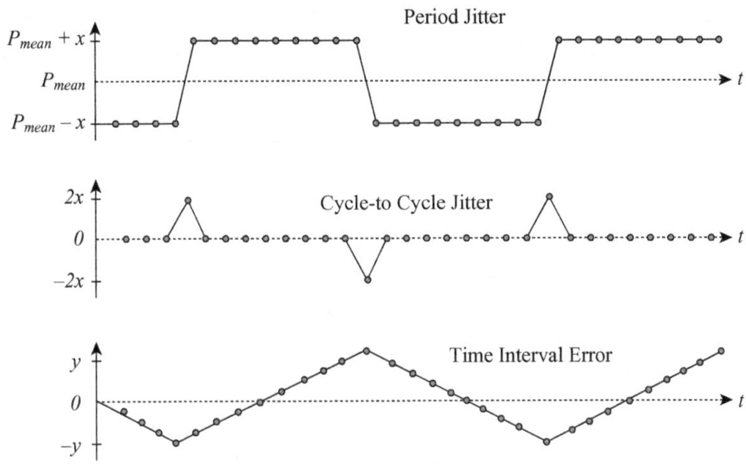

Fig. 1.7. The illustration of period jitter, cycle-to-cycle jitter, and TIE.

Cycle-to-cycle jitter measures the degree of the clock period's changes between any two adjacent cycles. By these definitions, it can be understood that no knowledge of an ideal clock signal is needed when calculating the period jitter or the cycle-to-cycle jitter. On the other hand, the TIE is defined as the measurement of how far each clock edge varies from its ideal position. Therefore, for this measurement, the ideal clock edge position must be known or estimated.

The relationships among the previously defined three jitter terms can be understood from their definitions. Figure 1.7 can help further illustrate the points where a clock signal's cycle length (period) alternates between two values: $P_{mean} \pm x$. As implied in their definitions and shown in Fig. 1.7, period jitter is the direct measurement of a clock cycle's length. It has great significance for digital operation since setup constraint is constructed under the influence of this period jitter. Meanwhile, cycle-to-cycle jitter is the first-order-difference operation to period jitter. It shows the instantaneous dynamic of the clock signal, which is very important to the PLL designer if this clock signal is used as the input of a PLL. The TIE can be regarded as the integrating operation over the period jitter (after each period is first subtracted from the ideal clock period). The TIE is significant because it shows the cumulative effect of the period jitter. It is the long-term characteristic of the clock signal. In summary, period jitter is important to digital design where only the jitter's static characteristic is of interest. Both cycle-to-cycle jitter and the TIE are important to applications where the jitter's dynamic characteristic is also critical in determining system performance, such as in clock data recovery (CDR), frequency conversion, and when used as reference.

It is worth mentioning that the term "jitter accumulation" has two completely different meanings when used in different situations. One is related to the long-term jitter, where period jitter accumulates over many clock cycles (TIE).* The other refers to the scenario that a clock signal propagates through multiple circuit stages (such as in a clock tree) and the noise generated at each stage is "added" to the clock signal. In this case, the term "accumulated jitter" is used to represent all the noises that the clock signal picks up along its propagation paths.

1.2.1.4 Distinguish the Jitter: Random or Deterministic? The period jitter, cycle-to-cycle jitter, and the TIE are used to quantitatively describe the clock edge uncertainty. However, these terms do not provide any insight to the causes of the jitter. To better describe the jitter, two additional terms are often used to distinguish the causes of the jitter: "random jitter" and "deterministic jitter." Further, the sum of random jitter and deterministic jitter is termed "total jitter." Random jitter is the timing noise that cannot be predicted. It does not have any discernable pattern. The primary source of the random jitter in electrical circuits is the thermal noise, also called Johnson noise or shot noise. It is the electronic noise generated by the thermal agitation of the electron inside the electrical conductor at equilibrium. It always happens regardless of the voltage applied on the circuits/devices. The random jitter bears the characteristic of Gaussian distribution (or normal distribution), which is shown in Fig. 1.8. As shown, this kind of stochastic process can be characterized by two values: the mean μ and the standard deviation σ. Mathematically, the root mean square (*rms*) is a statistical measure of the magnitude of a varying quantity: $x_{rms}^2 = \mu^2 + \sigma^2$. Electrical engineers often use the term "root mean square" as a synonym for standard deviation when referring to the square root of the mean squared deviation of a signal from a given baseline (AC-only *rms* of a signal). Therefore, standard deviation σ of a period jitter distribution (or cycle-to-cycle, TIE) is also called *rms* jitter. For a Gaussian distribution, one σ away from the mean (baseline) accounts

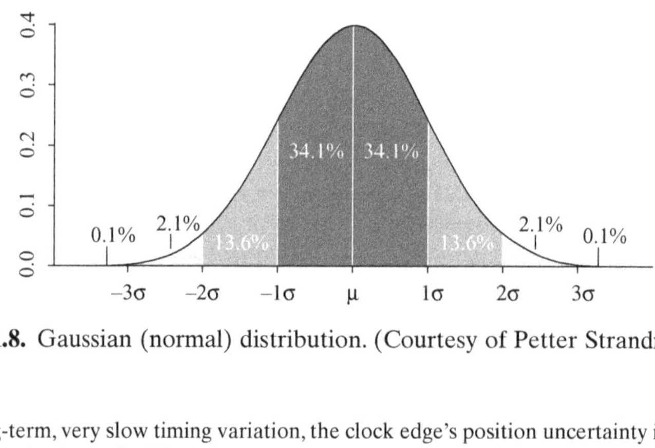

Fig. 1.8. Gaussian (normal) distribution. (Courtesy of Petter Strandmark.)

* For a long-term, very slow timing variation, the clock edge's position uncertainty is often called frequency wander instead of jitter.

for about 68% of the total; three σ away account for 99.7%. It is important to recognize that random jitter is unbounded due to the nature of the Gaussian distribution.

Deterministic jitter is the clock edge timing uncertainty that is repeatable and predictable. The root cause of deterministic jitter is usually associated with some traceable sources or events. The magnitude of the deterministic jitter is bounded. Deterministic jitter can further be categorized into periodic jitter, data-dependent jitter, and duty-cycle dependent jitter. Jitter that repeats itself in a cyclic fashion is called periodic jitter, also called sinusoidal jitter. It is typically caused by external traceable noise sources, such as a switching power supply or a local radio frequency (RF) carrier that coupled into the system.

In wired datalink communication, the jitter that correlates with the bit sequence is termed "data-dependent jitter." It is usually caused by the frequency response of the transportation media (such as cable). Different data sequences result in different electrical waveforms due to the frequency response of the cable or device. These different waveforms introduce timing differences (and hence jitter) when the threshold is crossed. Duty-cycle-dependent jitter is used to differentiate the timing difference caused by either the rising or the falling edge of the waveform. It can be introduced for two reasons: (1) the slew rates of the rising and falling edge are different and (2) the decision threshold for a waveform is either higher or lower than it should be. Data-dependent jitter and duty-cycle-dependent jitter are mostly used in CDR applications to characterize the timing information embedded in the data stream (Tektronix).

1.2.1.5 Look at the Clock-Edge-Uncertainty in Frequency Domain: Phase Noise and Spurs

In addition to being studied in the time domain, the timing irregularity of a clock signal can also be investigated from the frequency domain. Phase noise is the frequency domain representation of the rapid short-term fluctuation in the phase of an electrical wave. For a pure sinusoid wave, the signal can be described by the following equation:

$$v(t) = A * \cos(2\pi ft) \tag{1.1}$$

Phase noise is added to this signal by adding a stochastic process represented by φ(t) in the phase part as shown in Eq. 1.2. This fluctuation in phase (hence phase noise) will cause uncertainty at the exact moment at which this waveform crosses a predefined voltage threshold (jitter). The term "phase noise" is typically used by radio frequency engineers, and the term "jitter" is mainly used by digital engineers, all for the convenience of serving on what they are doing. The two terms are related; they describe the same physical phenomenon from different angles.

$$v(t) = A * \cos(2\pi ft + \varphi(t)) \tag{1.2}$$

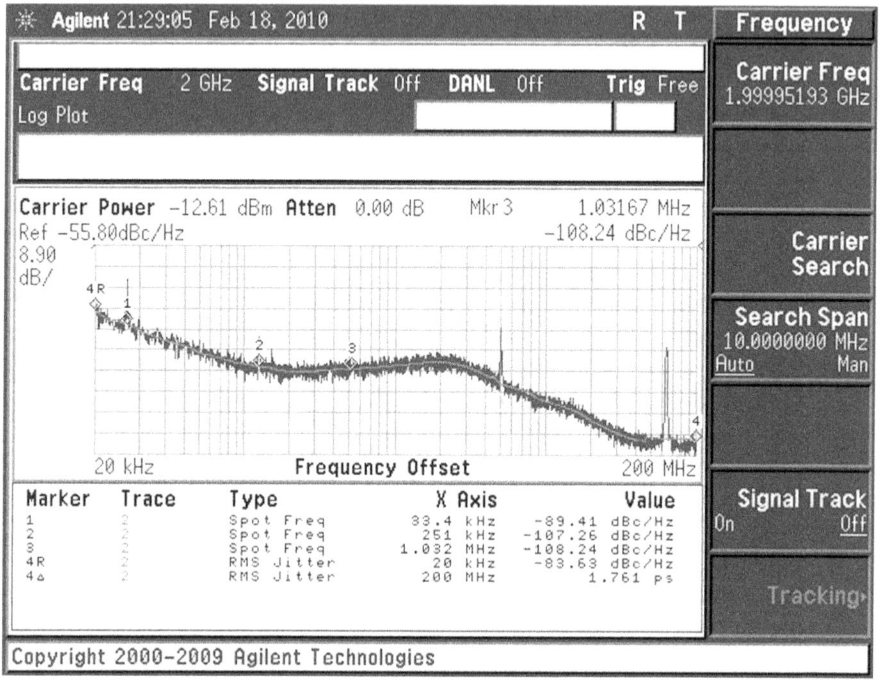

Fig. 1.9. Phase noise measurement of a 2-GHz clock signal.

Phase noise is often expressed as the ratio of sideband power in a 1-Hz bandwidth to the signal power, in units of dBc/Hz, at a given offset from the carrier frequency (Poore 2001). It is often measured by using spectrum analyzer. Figure 1.9 is an example of phase noise measurement plot of a 2-GHz clock signal. The x-axis is the frequency offset from the carrier. The y-axis represents the noise strength at that offset frequency. Phase noise can also be expressed as a value integrated over a certain range of the offset frequency. This integrated phase noise can be converted into time domain *rms* jitter. In this figure, the integrated *rms* jitter from 20 KHz to 200 MHz is 1.76 ps.

In engineering practice, a histogram is often used to graphically characterize the time jitter. Figure 1.10 is the period jitter histogram of a 2.75-GHz clock signal (refer to Fig. 1.8). The number of samples in this histogram is 1.9 million. The standard deviation σ is 2.85 ps, and the peak-to-peak range is 25.6 ps. As expected, this distribution bears approximately a Gaussian-like shape.

The shortcoming of the jitter histogram is that it does not show the temporal order in which the measurements occur. Therefore, it lacks the capability of identifying any repeating patterns that might indicate some deterministic modulation sources. A plot of jitter versus time (jitter–trend plot) can make such patterns visible. This feature can help us identify the sources of the

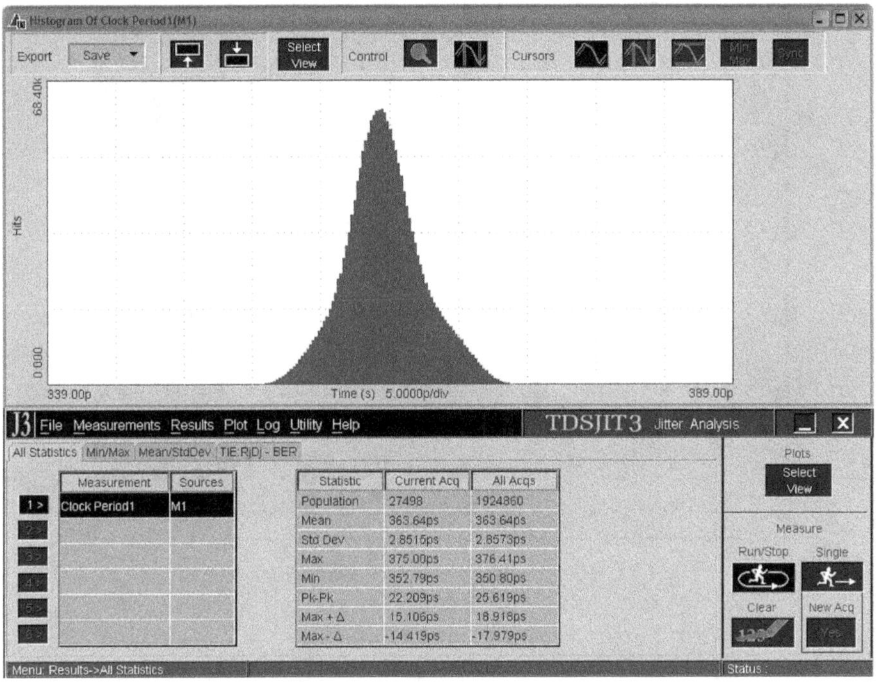

Fig. 1.10. The period jitter histogram of a 2.75-GHz clock signal.

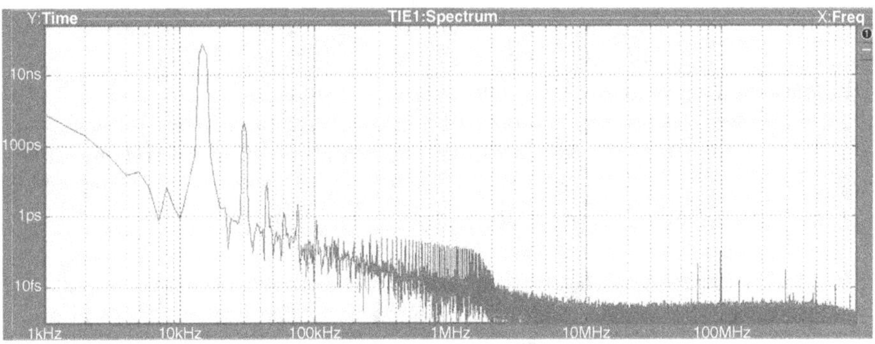

Fig. 1.11. The jitter spectrum plot.

disturbances. The extension of this jitter-vs-time measurement is to apply fast Fourier transform (FFT) to it. The result, displayed in the frequency domain, is the jitter spectrum. The benefit of jitter spectral analysis is that any periodic components (periodic jitter) embedded in the noise can potentially be distinguished. Hence, the triggering source could be identified. Figure 1.11 shows one such jitter spectrum plot.* Clearly, there is a 15-KHz fundamental

* Borrowed from Tektronix.

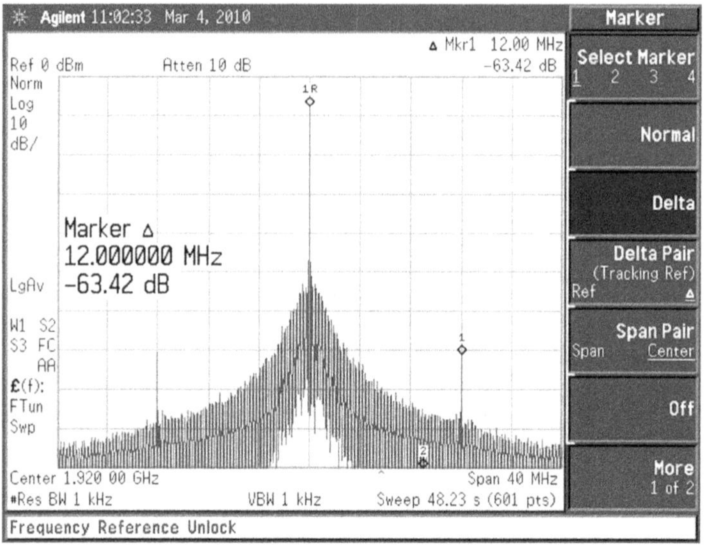

Fig. 1.12. The spectrum of a 1.92-GHz clock signal.

frequency in the noises. The second (30 KHz), and third (45 KHz) harmonics can also be seen easily. This suggests that a 15-KHz nearby signal could be coupled into the clock signal.

Another very important method used by circuit designers for studying clock signal quality is to directly perform FFT on a clock signal waveform. Figure 1.12 is an example of the FFT result performed on a 1.92-GHz clock. Clearly, the clock energy is concentrated at 1.92 GHz as designed. The spurious tone at the 12-MHz offset is an indication that there is a signal of 12 MHz modulating the 1.92-GHz clock. Indeed, in this case, the 12 MHz is the reference clock for the PLL. Clearly, it leaks to the output through the PLL.

1.2.1.6 Sources of Jitter From a solid-state physics point of view, all the voltage noises that occur inside a circuit can be traced back to thermal noise and *1/f* flicker noise. From a system perspective, there are two types of systems that bear unique jitter characteristics. The first type is the autonomous system, which can oscillate on its own. The jitter associated with those systems accumulates. There is no inherent force that counteracts the wander tendency of its oscillating frequency. It has the characteristics of frequency modulation (FM jitter). In a typical PLL system, the input oscillator and the VCO (voltage control oscillator) are autonomous components. The other type is the driven system, which can only be activated by outside signals. Its edge uncertainty has a reference point. It syncs with the driving signal, and hence the jitter does not accumulate. This kind of behavior bears the phase modulation characteristic (PM jitter). The dividers and phase detector inside a PLL belong to this category.

When an electronic system is investigated as a whole, components that can contribute to total jitter though jitter accumulation are as follows:

- all transistors used in the circuit
- all passive components (resistor, capacitor, and inductor) used in the circuit
- random thermal and mechanical noise from crystal
- parasitic components from signal interconnections (within the integrated circuit [IC])
- trace, cable, and connector used in the printed circuit board (PCB) level.

1.2.1.7 Summary Table 1.1 lists all the methods for studying clock quality. They are different ways of looking at the same thing: clock edge uncertainty. Digital designers prefer to use the term "jitter" while RF designers typically use the term "phase noise." They are related and can be converted to/from each other. When clock edge uncertainty is caused by stochastic processes, its distribution in the time domain histogram is Gaussian-like. In the frequency domain, it raises the noise floor. When clock edge uncertainty is sourced from periodic events, spurs (spurious tones) appear in its frequency spectrum. In the time domain, its histogram will deviate from Gaussian distribution because of those periodic events.

1.2.2 Clock Phase

When a clock signal is used to drive an analog-to-digital converter (ADC), another clock characteristic called *clock phase* is important. An example is shown in Fig. 1.13. In this system, an analog signal and a clock signal are transmitted from transmitter to receiver through different cables. Thus, they experience different delays. Moreover, the analog signal is originated from a digital-to-analog converter (DAC). There is an area of overshoot and ringing within each data boundary. Clearly, on the receiving side, the exact moment at which the ADC takes the sample has great impact on the value converted. It is desirable that some tuning capability is available inside the receiving side's clock circuitry so that the position of the clock edge that will trigger the ADC can be adjusted. Within such a system, the exact sampling moment is called the clock phase, as illustrated in Fig. 1.14. In this scenario, phase is proportional to time. Different phases correspond to different time delays from a reference point. In many such systems, there could be 4, 8, 16, or 32 phases available within one clock cycle to help achieve the optimal result.

Clock phase is also important in digital communication when data are moved between blocks, modules, and chips. In such applications, information is exchanged between different domains, and each domain has its own clock. The relative position of the clock edges, which is represented using the clock phase of one of the involved clocks, plays a crucial role in the success of the data transfer. Examples include double data rate (DDR) memory interface,

TABLE 1.1. The Various Approaches of Studying Clock Quality

Domain	Methods	Parameters	Quantifier	Used by	Purposes
Time Domain	Histogram	Period jitter Cycle-cycle-jitter Time interval error	*rms* mean value Standard deviation Peak-to-peak value	Digital designer	Digital design
	Jitter-vs-time (Jitter trend)	N/A	Rate of change Pattern of change Magnitude of change	PLL designers RF system designer	Identify noise source
Frequency Domain	Clock spectrum	Noise floor Spur location Spur magnitude	Spur location SFDR	PLL designers RF system designer	Identify noise source
	Phase noise plot	Noise floor Spur location Spur magnitude	Spur location SFDR	PLL designers RF system designer	Identify noise source, PLL loop study
	Jitter spectrum	Noise floor Spur location Spur magnitude	Spur location SFDR	PLL designers RF system designer	Identify noise source

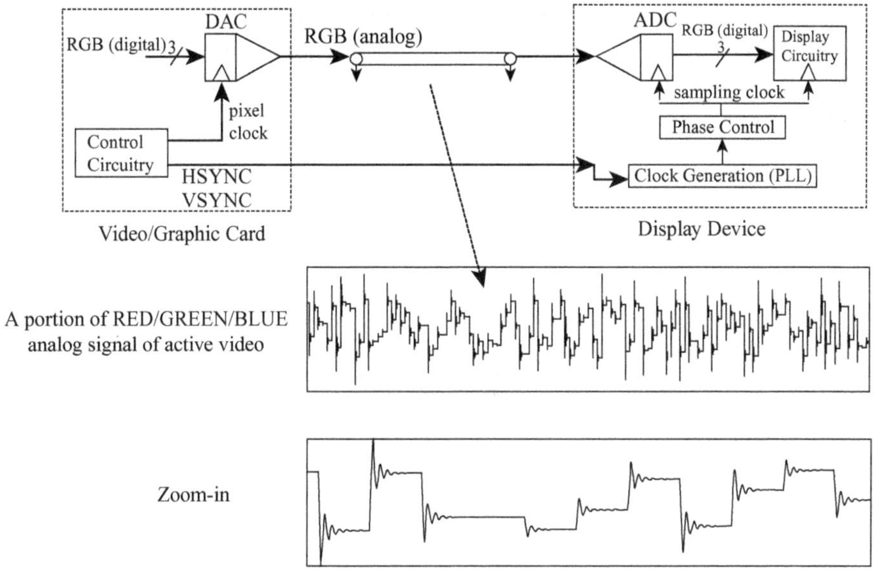

Fig. 1.13. The analog signal and the clock signal in a system with an ADC.

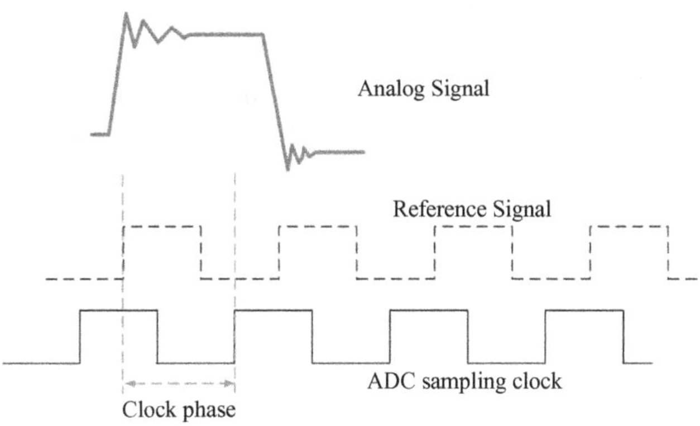

Fig. 1.14. The clock phase.

datalink IP (such as HDMI, USB, PCI), etc. In these applications, the phase-adjustment capability is used to move the clock edge (of the receiver) to the center of the incoming data (from transmitter) for maximum timing margin.

1.2.3 Clock Skew

In today's digital circuit implementation, the base cells are not transistors but components. All these components can be classified into two classes: logic (combinational) cell and sequential cell. Logic cells are used for performing logic operations (or computations in a broader view). A logic cell is a type of circuit where its output depends only on its inputs. Its function is to perform Boolean algebra on signals presented at its input ports. Examples of combinational logic cells include: inverter, buffer, nand, or, xor, etc. Conversely, a sequential cell is a type of circuit whose output depends not only on its inputs but also on its present state. In other words, a sequential cell has memory; it is used for storing information. The clock signal is used for controlling its write and read operations. Examples of sequential cell are: latch, flip-flop, static random-access memory (SRAM) and dynamic random-access memory (DRAM), etc. In large chips there could be millions of these cells, both logic and sequential, coexisting in a die. One of the challenges associated with the synchronous design method is that the clock signals have to be distributed to all sequential cells in the chip. For a large clock domain with hundreds of thousands of such cells, the construction of this distribution network (clock tree, which will be explained later) is not a trial task. This is due to the following reasons:

- within a chip, the clock signals are typically loaded with the greatest fan-out
- clock signals travel the longest distances
- clock signals operate at highest speeds of the chip.

The primary target of the clock tree is the minimization of clock skew. The secondary objective is to minimize/balance the clock tree delays. Skew (global skew) is defined as the maximum time difference among all the clock paths from the root (clock source) to all the leaves (clock sinks). The clock-path-induced time difference between sequentially adjacent sequential cells (having data communication between them) is called local skew. Usually, from the chip design perspective, system-wide global skew is used to constrain the design. Clock delay is the propagation delay induced by the clock tree. The concepts of clock tree, clock skew, and clock tree delay are graphically illustrated in Fig. 1.15. Figure 1.16 shows a clock tree in three-dimensional (3D) fashion (it clearly demonstrates the point of why clock distribution network is called clock tree). The X-Y plan is the chip's physical dimension. The clock source, which is the system-PLL in this case, is located at one corner of the chip. All the clock sinks are highlighted in red. The z-axis represents the time required by the clock signal to reach each clock sink (due to the RC delay induced by the clock paths in the clock tree). The maximum delay difference among all the sinks is the clock skew (global skew).

Within a chip, the clock skew can be caused by any of the following reasons: (1) the differences in metal lines' lengths from clock source to clock sinks; (2) the differences in the delays of the active buffers used in the clock tree; (3) the differences in passive interconnect parameters such as metal line resistivity, dielectric constant and thickness, via and contact resistance, line and

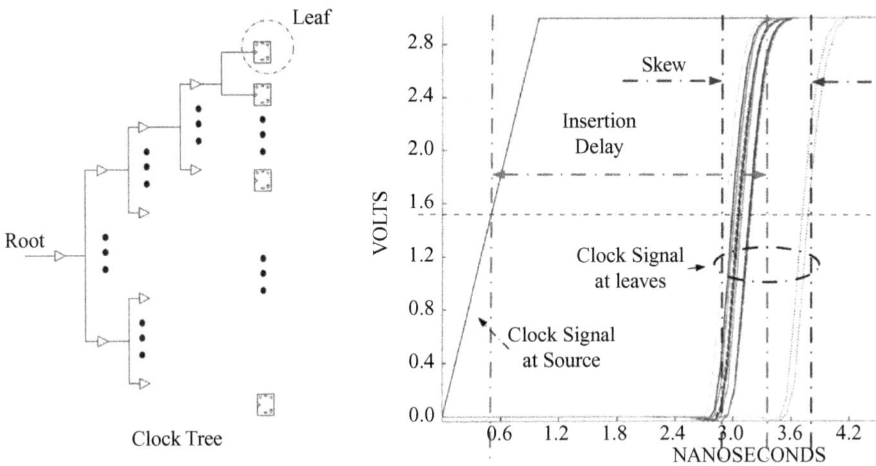

Fig. 1.15. The clock tree, the clock skew, and the clock tree delay.

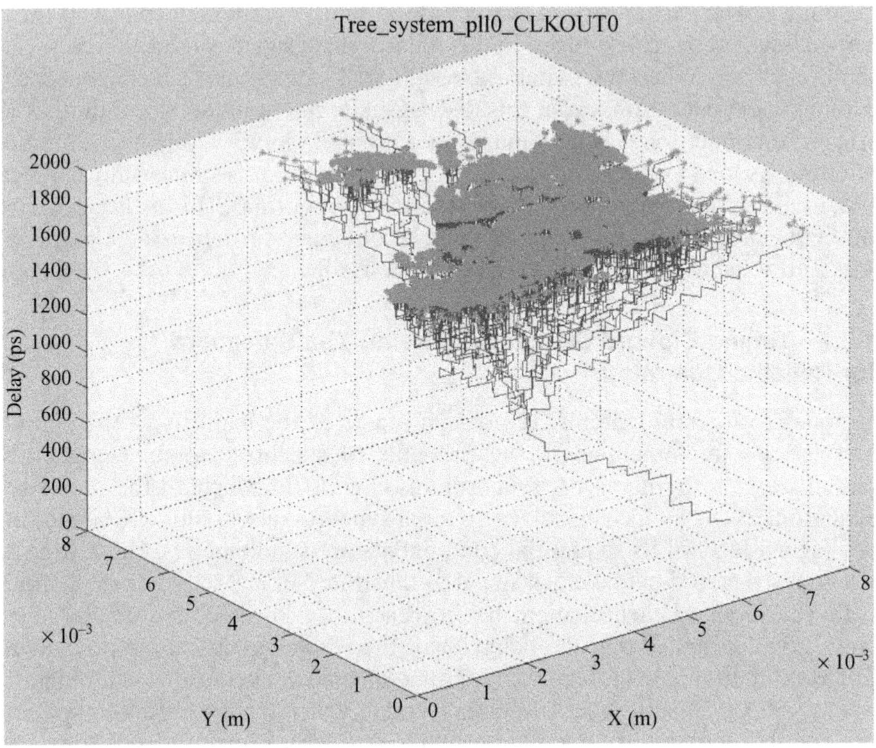

Fig. 1.16. A clock tree shown in 3D.

fringing capacitance, line dimension, etc.; and (4) the differences in active device parameters such as transistor threshold voltage and channel mobility. The task of minimizing clock skew (and clock tree delay in some cases) has been an ever-increasing challenge due to the continuous shrinking of transistor geometry. The higher clock rate and the larger die size of modern designs have made this problem very difficult. Moreover, as previously mentioned, the delay uncertainty caused by process and environment variations further complicate the issue.

1.3 CLOCK SIGNAL DRIVING DIGITAL SYSTEM

1.3.1 Clock Signal as a Trigger

In an electronic system, the clock signal is created to control pace and record time. Electrically, it is used to drive two types of circuits: (1) as a trigger to fire logic circuits and (2) as a switch to take a sample (ADC), or to construct a waveform (DAC). As illustrated in Fig. 1.17, a whole digital block's operation can be divided into groups of local operations. Within each group, the logic operation is performed by combinational logic cells. The groups' boundaries are established by sequential cells. Between the groups, there are information exchanges. The exchanges are accomplished by the sequential cells, and they only happen at the clock edges. In this regard, clock signal can be viewed as an ignition switch. When the switch is closed (clock edge occurs), it triggers each group's logic operation. From this discussion, it can be understood that, as a trigger, the main focus is how many operations it ignites within a given time window. As long as the requested number of operations is successfully carried out, the precise moment of each ignition is not important. This is the scenario of digital circuit operation. This fact will be discussed further in later chapters when establishing the base for time-average-frequency.

1.3.2 Timing-Closure Design Constraint: The Safeguard for Reliable Operation

In today's integrated circuit (IC) design practices, the majority of the digital systems are implemented on the principle of synchronization. Hence, the circuit is given the name of "synchronous circuit," in which all the parts are synchronized by a clock signal. In an ideal synchronous circuit, all changes in the logical level of its sequential components are simultaneous. These transitions are triggered by the clock signal as illustrated in Fig. 1.17. For sequential cells, the input to each element has to reach its final value before the next clock edge occurs so that the behavior of the whole circuit can be predicted. For combinational cells, a certain amount of time is needed for each logical operation. This results in a maximum speed at which this synchronous system can run. The method of static timing analysis (STA) is often used to determine the maximum operating speed.

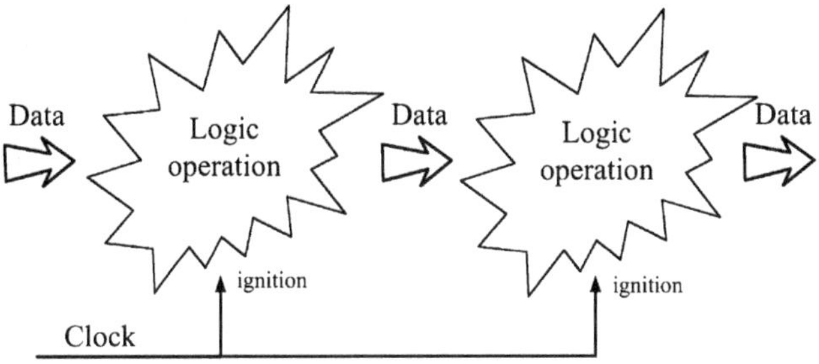

Fig. 1.17. Clock signal as trigger.

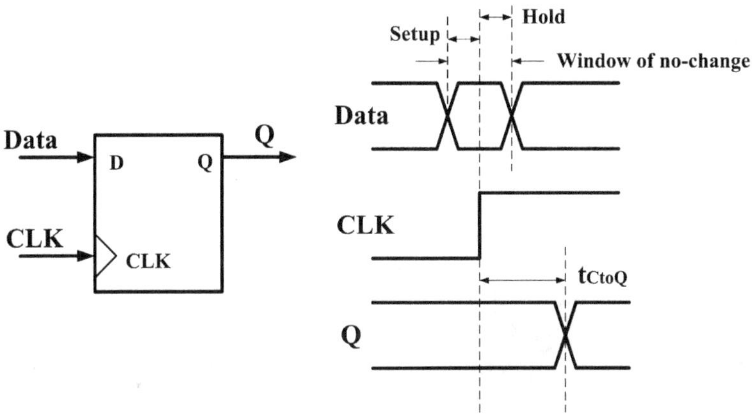

Fig. 1.18. The setup and hold constraints: a local view.

The main advantage of synchronization is that it simplifies digital design. All the operations inside a synchronous system must be completed within a fixed interval of time between the two clock edges of a clock cycle. As long as this condition is met, the circuit is guaranteed to be reliable. In circuit design practice, the safeguard for ensuring the satisfaction of this condition is the *setup* and *hold* check, which is the backbone of STA. As shown in Fig. 1.18, the data presented at a sequential cell's input are not allowed to change within the time window during which the clock signal is changing state (clock edge).

The spirit of synchronous system is that information is manipulated and transported cycle by cycle. At each stage (within each clock cycle), this information processing (performed by combinational cells) can neither be too fast not too slow. Compared to clock speed, if it is too slow, the generation of new information cannot be finished. Compared to the sequential cells' switching speed, if this processing is too fast, new information will pass through. This scenario is graphically illustrated in Fig. 1.19, where the houses are used to represent the sequential cells. Most of the time, the houses' doors are closed. The door's open-then-close action corresponds to the clock edge. The physical

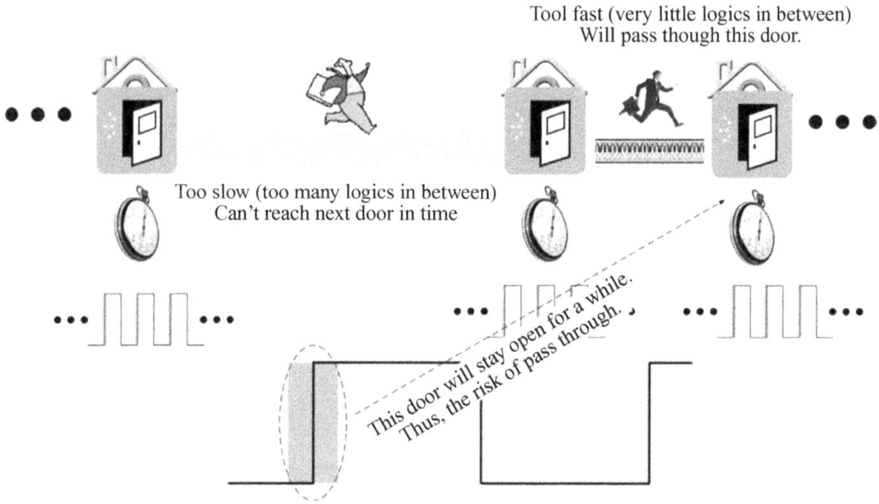

Fig. 1.19. The setup and hold constraints in a circuit environment.

distance between any two houses is proportional to the complexity of the logic operation in between the corresponding two sequential cells. There are two important comparisons in this analogy. The distance (logic complexity) is compared against the clock period (setup constraint), and the distance is compared against the speed of the open-then-close action (hold constraint). Both cases illustrated in the figure have to be avoided for correct operation. This figure is especially useful for understanding the difficult-to-understand hold check. For example, when the middle house's door opens, the stored information is immediately released. It is desired that this information (which can be processed and transformed into new information by the logic cells in between) be captured by the next door at ***next*** clock transition, not the current clock transition. However, since there is a finite time window during which the doors remain open, there is a risk that this information will reach the next door before it closes if the information travels too fast. It is the designer's responsibility to slow down this path (add delay cells in this path) to prevent this from happening. This task is called meeting the hold constraint in SoC timing closure. Clearly, from this picture, hold check has nothing to do with clock speed (the clock period) but is closely tied to the door's switching speed (the sequential cells' open-then-close speed).

On the other hand, the setup check is clock period dependent since it uses two consecutive clock edges (current and previous edges). It compares the circuit speed against the clock speed. One of the key purposes of setup check is to avoid a problem called metastability, which can occur in a sequential cell's operation. If the data are changing at the same instant when the clock is making a transition (within the no-change window in Fig. 1.18), the behavior of the sequential cell's output could be unpredictable. It might take very long time for it to settle down to its final (intended) value. In the worst case, it may oscillate and take infinite time to settle. This causes a logic error in circuit operation.

Due to the importance and the difficult-to-understand nature of these concepts, and for the purpose of introducing the time-average-frequency concept, it is worth summarizing them in the following statements:

- The setup constraint is used for comparing the circuit speed with the clock speed.
- The hold constraint is used for comparing the circuit speed with the sequential cells' switching speed.

For any synchronous circuit, regardless of its complexity at a functional level, the setup and hold constraints are the only safeguard needed for the circuit's reliable operation at the electrical level. Meeting these constraints will ensure its correct operation at the designed speed.

1.3.3 Clock Jitter and Design Constraint

In the previous section, the setup and hold concepts were introduced. Their impact on circuit operation was explained. However, their relationship with clock jitter was not discussed. The following statements describe the interaction: (1) clock jitter deducts the same amount (itself) from the timing budget of setup constraint; (2) clock jitter has no impact on hold check.

Since setup constraint uses two consecutive clock edges, any clock jitter (edge uncertainty) will make the current clock cycle's length-in-time longer or shorter (sometimes longer and sometimes shorter). To be safe, we have to use the shorter scenario to constrain the circuit. In other words, we have to speed up the logic between the sequential cells. If we want to keep the circuit untouched, we have to slow down the clock. On the other hand, the hold constraint uses only one clock edge (the current edge). The comparison between the circuit speed and the sequential cells' switching speed happens at the same clock edge. Hence, it could not sense clock jitter because there is no reference.

1.3.4 Clock Skew and Design Constraint

The concept of clock skew is explained in Section 1.2.3. Its relationship with setup and hold check is stated below:

- Clock skew affects setup check. It can impact circuit operation in either a positive or negative way.
- Clock skew affects hold check. It can impact circuit operation in either a positive or negative way.

The key difference between clock jitter and clock skew is that clock jitter originates at clock source and clock skew is caused by a clock distribution

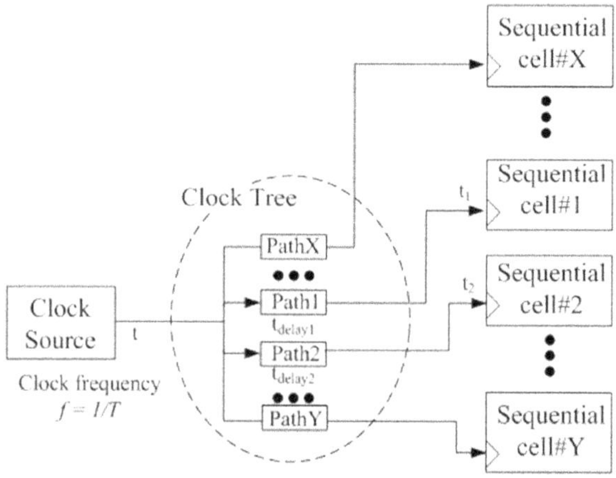

Fig. 1.20. Clock skew and design constraints.

network (clock tree). Since jitter is initiated at the source, all the sequential cells (clock sinks) attached to this source sense the same impact. Skew is caused by the physical distribution network; each individual clock sink feels a different impact owing to its unique path. (Refer to Fig. 1.20 where there are a group of sequential cells attached to a clock source.) The clock signal from the source is distributed to all the sequential cells through the clock tree; each cell has its own unique physical distribution path and thus unique timing delay associated with it. We use cell #1 and cell #2 to illustrate the interaction between clock skew and design constraint.

For this investigation, there are two clock edges and two cells involved: the current clock edge and the previous clock edge, the launching cell (the cell that launches data), and the receiving cell (the cell that receives data). The following is the list of symbols that we will use for discussion (refer to Fig. 1.20).

- t_c: the moment that *current* clock edge emerges from the clock source
- t_p: the moment that *previous* clock edge emerges from the clock source
- t_{1c}: the moment that current clock edge reaches cell #1, the launching cell
- t_{1p}: the moment that previous clock edge reaches cell #1
- t_{2c}: the moment that current clock edge reaches cell #2, the receiving cell
- t_{2p}: the moment that previous clock edge reaches cell #2
- t_{skew}: $t_{skew} = t_{delay2} - t_{delay1}$

By definition, we have

$$t_c - t_p = T \tag{1.3}$$

$$t_{1c} = t_c + t_{delay1},\ t_{1p} = t_p + t_{delay1} \tag{1.4}$$

$$t_{2c} = t_c + t_{delay2},\ t_{2p} = t_p + t_{delay2} \tag{1.5}$$

For a setup check, data are launched from cell #1 at the **previous** edge. They are received at cell #2 at the **current** edge. Therefore, the impact of skew on the timing budget (allocated for logic operation in between the two adjacent sequential cells), t_{s_delta}, is calculated in Eq. 1.6.

$$t_{s\_delta} = t_{2c} - t_{1p} = t_c + t_{delay2} - t_p - t_{delay1} = T + t_{skew} \tag{1.6}$$

For the hold constraint, instead of the previous edge, data are launched from cell #1 at the **current** edge. They are received at cell #2, also at the **current** edge. Thus, the skew's impact on timing budget t_{h_delta} can be expressed in Eq. 1.7:

$$t_{h\_delta} = t_{2c} - t_{1c} = t_c + t_{delay2} - t_c - t_{delay1} = t_{skew} \tag{1.7}$$

From Eqs. 1.6 and 1.7, it is clear that clock skew t_{skew} has an impact on both the setup and hold checks. Depending on the sign of t_{skew}, it can play a positive or negative role in circuit operation. For example, when t_{skew} is positive (t_{delay2} is larger than t_{delay1}), the current clock edge will arrive at cell #2 later than scheduled. This gives more time for the logic operation to be performed between cell #1 and cell #2. It eases the setup check. On the other hand, since the current clock edge arrives later than scheduled, cell #2 will consequently close its door later than normal. This fact increases the risk of data pass through for the data launched from cell #1 at the current edge. In other words, it makes it more difficult to satisfy the hold constraint. In the case where t_{skew} is negative, a similar analysis can be carried out.

In the above analysis, the clock source is assumed to be ideal since $t_c - t_p = T$. If clock jitter is included, Eq. 1.3 would be modified to $t_c - t_p = T + t_{jitter}$, where t_{jitter} is the amount of clock jitter. And Eq. 1.6 needs to be revised as $t_{s\_delta} = T + t_{jitter} + t_{skew}$. From here, it is clear that jitter has impact on the setup check as stated in the previous section. From Eq. 1.7, however, the hold check is not related to clock period T. This explains why the hold check is not clock speed dependent.

Since the concepts of jitter, skew, setup, and hold are important and their relationship to clock frequency is difficult to be understood, Table 1.2 is created for reference. This understanding is crucial for the time-average-frequency concept that will be introduced in later chapters.

TABLE 1.2. Jitter, Skew and Setup, Hold Check

	Cause	Impact on setup check (current and previous edge)	Impact on hold check (current edge)
Jitter	Clock source (PLL/DLL)	Yes	No
Skew	Physical distribution path	Yes	Yes

1.4 CLOCK SIGNAL DRIVING SAMPLING SYSTEM

In modern electronic designs, sampled data systems can be found in many places. This is due to the fact that the world in which humans directly interact is analog in nature. In the meantime, the computation (the task of information processing) performed by computer hardware is carried out in binary fashion. Consequently, sampled data systems, which convert information from the analog domain to binary digital domain, are ubiquitous in modern designs. A typical information processing flow is illustrated in Fig. 1.21. The ADC and DAC stand for analog-to-digital converter and digital-to-analog converter, respectively. They are key components in this flow.

1.4.1 Clock Signal as a Switch

In digital systems where clock is used as a trigger (Section 1.3.1, Fig. 1.17), only the clock signal has anything to do with the absolute wall time; the data signal has no sense of wall time. All data-related actions are controlled by the clock and its reference to the absolute wall time is accomplished through this clock. The clock is the trigger of the system. In contrast, when an ADC and DAC are used in the system, the clock is used as a switch. In this case, there are two time-sensitive signals involved: the signal of clock and the signal of interest as illustrated in Fig. 1.22. In this application, the exact moment of the switch-close is important since both the signal of interest and the signal of clock are referenced to the absolute time (wall time). The clock's threshold-crossing moment affects the level of the other signal that will be captured. This issue is nonexistent in the previous clock-as-trigger case because the digital signal only has two levels (low and high). As long as the clock's threshold-crossing-moment is outside the setup-hold prohibited window, the output will be the same regardless of where the clock edge is.

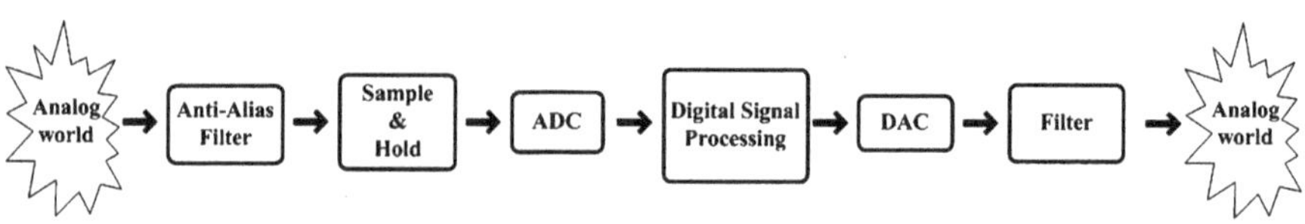

Fig. 1.21. A typical system based on sampled data.

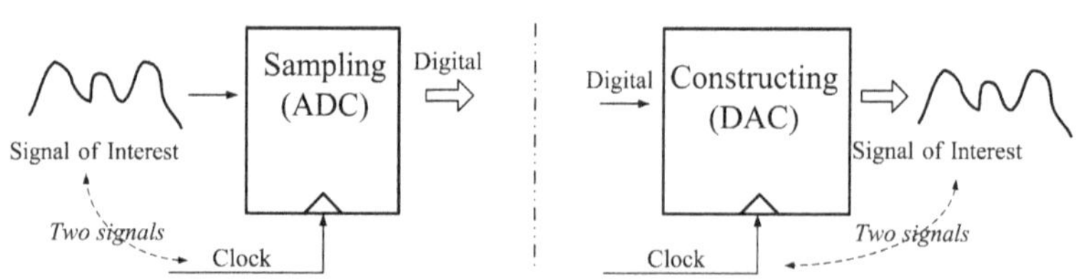

Fig. 1.22. Clock as a switch: both signals in this system are wall time sensitive.

In this clock-as-switch application, the issue of the clock-affecting signal cannot be analyzed easily in the time domain. Short-term behavior alone is unable to provide clear picture. The study must be further carried out in long-term fashion. Hence, this subject is often investigated in the frequency domain. The clock spectral purity is of high concern.

1.4.2 Clock Signal and Analog-to-Digital Converter

The ADC is an important component for a signal processing system. There are two key concepts involved in the actual ADC conversion process: discrete time sampling and finite amplitude resolution (quantization). In implementation, there are many varieties in ADC architecture. However, the ADC's performance can be summarized by a relatively small number of parameters: resolution (number of bits per sample), signal-to-noise ratio (SNR), spurious-free dynamic range (SFDR), and power dissipation. The noise spectrum that affects the ADC performance contains contributions from such mechanisms as quantization noise, thermal noise, comparator ambiguity, and aperture jitter (aperture uncertainty). Among these, the aperture jitter, which is defined as a sample-to-sample variation of the instant at which the sampling operation occurs (switch-close), has great impact on SNR, SFDR, and ENOB (effective number of bits).

Figure 1.23 shows the diagram of a typical sample and hold circuitry of an ADC. As shown, a clock signal controls the sampling switch. The variation on the instant of the switching can affect the analog voltage taken (left illustration of Fig. 1.23), which could make the converted digital code deviate from its expected value. For example, assume that the input signal is a sinusoidal wave

$$V(t) = Asin(2\pi ft) \tag{1.8}$$

Its first derivative is

$$dV/dt = 2\pi Afcos(2\pi ft) \tag{1.9}$$

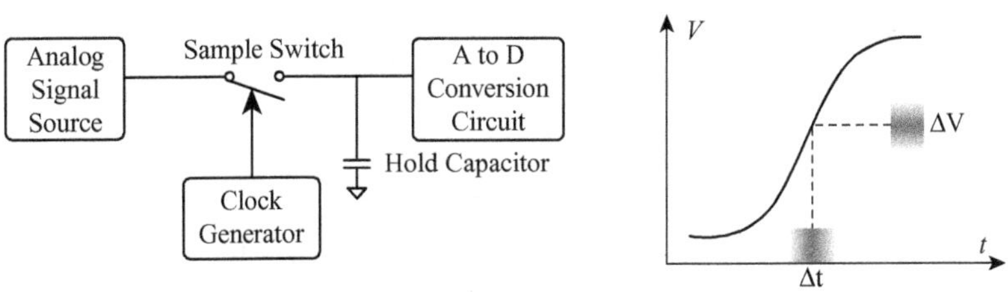

Fig. 1.23. Sample and hold circuitry in an ADC.

Therefore, the maximum time-error-introduced magnitude error occurs when $cos(2\pi ft) = 1$ and $dv/dt = 2\pi Af$. Conceptually, if dt is the aperture jitter t_a, dV is the error in the sampled voltage, which is termed V_e. Then, we have

$$V_e = 2\pi A f t_a \tag{1.10}$$

This simple model indicates that the sampling voltage error increases linearly both with the input signal's frequency and the size of the jitter. The SNR can also be calculated from this simple model:

$$SNR = 20\log\left(\frac{A}{V_e}\right) = -20\log(2\pi f t_a) \tag{1.11}$$

Figure 1.24 shows the SNR degradation due to the aperture jitter, calculated from this simple model. The left axis is the resolution limited from quantization noise.

Besides degrading SNR through clock jitter, the spurious frequency contents in the sampling clock (clock spurs) can cause spurious tones in the ADC output. This phenomenon is illustrated in Fig. 1.25. The impact can be calculated as follows (refer to Neu 2009 for more detail): Assume that the input signal is a sinusoidal wave with frequency f_i (ω_i) and the clock frequency is f_c (ω_c). Also assume that there is a spurious component of frequency f_s (ω_s) presented in the clock.

$$s(t) = A_i \sin(\omega_i t) \tag{1.12}$$

$$c(t) = A_c \sin(\omega_c t) + B_s \sin(\omega_s t) \tag{1.13}$$

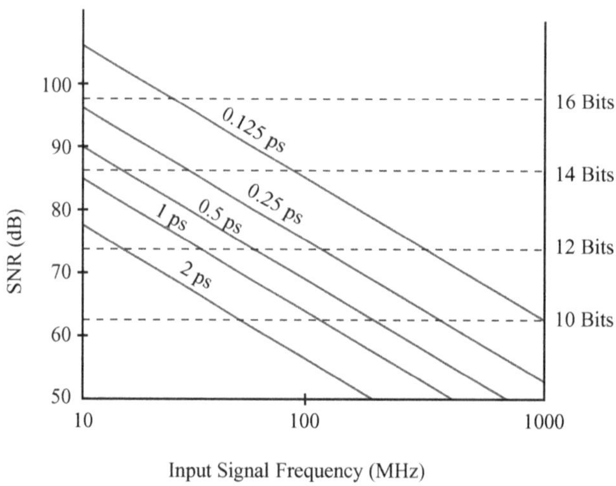

Fig. 1.24. The SNR degradation due to the aperture jitter.

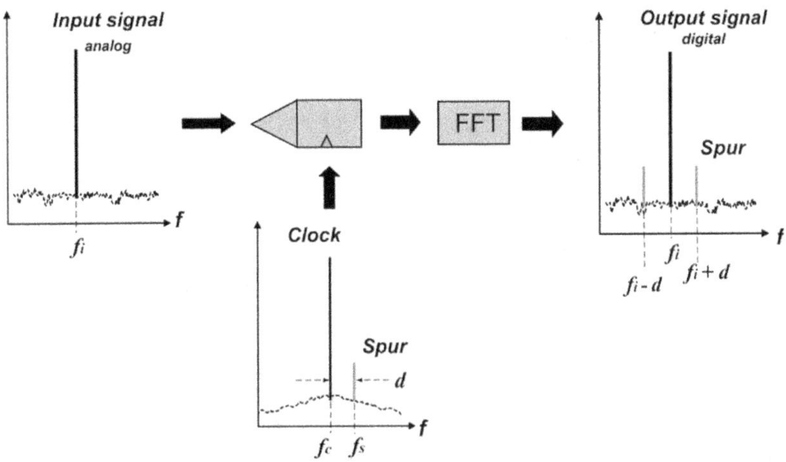

Fig. 1.25. Clock spurs introduce ADC spurs.

In Fig. 1.23, we assume that the sampling switch closes at the moment that the clock signal crosses zero. If the original (spur-free) zero-crossing moment is t, the actual zero-crossing moment (with the spur presented in the clock) will be $t + \Delta t$, which satisfies this equation:

$$c(t) = A_c \sin(\omega_c(t+\Delta t)) + B_s \sin(\omega_s(t+\Delta t)) = 0 \tag{1.14}$$

Usually, the magnitude of the spurious component is much smaller than the magnitude of the clock's main tone: $A_c \gg B_s$. Further, the disturbance caused by the spur is also small: $\Delta t \approx 0$. Under these conditions, Δt can be solved as

$$\Delta t = -\frac{B_s \sin(\omega_s t)}{A_c \omega_c} \tag{1.15}$$

The input signal $s(t)$, instead of being sampled at t, will be sampled at moment of $t + \Delta t$. Hence,

$$\begin{aligned} s(t+\Delta t) &= A_i \sin(\omega_i(t+\Delta t)) \\ &= A_i \sin(\omega_i t)\cos(\omega_i \Delta t) + A_i \cos(\omega_i t)\sin(\omega_i \Delta t) \\ &\approx A_i \sin(\omega_i t) + A_i \cos(\omega_i t)\omega_i \Delta t \end{aligned} \tag{1.16}$$

In Eq. 1.16, the first term is the ideal sample with no spurs effect. The second term is due to the spur presented in the clock. If Δt of Eq. 1.15 is substituted into this term, we have the error signal $S_{spur}(t)$ as

$$
\begin{aligned}
S_{spur}(t) &= A_i \cos(\omega_i t)\,\omega_i \frac{B_s \sin(\omega_s t)}{A_c \omega_c} \\
&= A_i \left(\frac{B_s \omega_i}{2A_c \omega_c} \right) \{\sin[(-\omega_s + \omega_i)t] + cos[(-\omega_s - \omega_i)t]\}
\end{aligned}
\tag{1.17}
$$

Compared with Eq. 1.12, the scaling factor of error signal is $B_s\omega_i/2A_c\omega_c = B_s f_i/2A_c f_c$. Thus, its magnitude increases linearly both with the input frequency f_i and the magnitude-of-clock-spur B_s. If expressed in decibels, the magnitude can be shown as

$$
Mag(S_{spur}) = B_s - A_c + 20\log[f_i/(2f_c)] \tag{1.18}
$$

The spur locations are at $-\omega_s + \omega_i$ and $-\omega_s - \omega_i$, or $f_{s1} = -f_s + f_i$ and $f_{s2} = -f_s - f_i$. We can move the clock spur f_s by multiples of clock f_c. In other words, if there is a clock spur at $-f_s$, we can also find spurs at $-f_s + f_c$. Therefore, Eq. 1.19 can be derived where d is used to represent the distance between the clock's main tone and its spur: $d = f_s - f_c$.

$$
\begin{aligned}
f_{S1} &= -f_s - f_i + f_c = -(f_i + f_s - f_c) = -(f_i + d) = f_i + d \\
f_{S2} &= -f_s + f_i + f_c = (f_i - f_s + f_c) = (f_i - d) = f_i - d
\end{aligned}
\tag{1.19}
$$

When the clock spur is far away from the clock's main tone, the FFT plot of the ADC output can be confused if care is not taken. In these cases, the generated ADC spurs can be pushed outside the plot boundaries, either to the negative side or beyond the $f_c/2$. The spurs will be aliased back and produce asymmetric plots, as demonstrated in Fig. 1.26.

1.4.3 Clock Signal and Digital-to-Analog Converter

As shown in Fig. 1.21, the ability to convert a digital signal back to analog is also very important. The digital-to-analog conversion process is essentially the inverse of the analog-to-digital process. There are various types of DACs: current-scaling DAC, voltage-scaling DAC, charge-scaling DAC, and serial DAC. A generic DAC block diagram, which is the representative of all the types, is depicted in Fig. 1.27. As shown, the clock signal plays a crucial role in the DAC as well: the digital word is synchronously clocked and the analog output is sampled and held by the clock.

Just as in the case of the ADC, one of the fundamental problems in the DAC is the timing accuracy of the conversion. For an N-bit discrete-time signal to be converted into a continuous-time signal, usually $2^N - 1$ equally designed elements (current sources) are required. These current sources are switched on and off depending on the input data. Ideally, the switched element shall all turn on/off at the same moment as defined by the clock edge. But in reality,

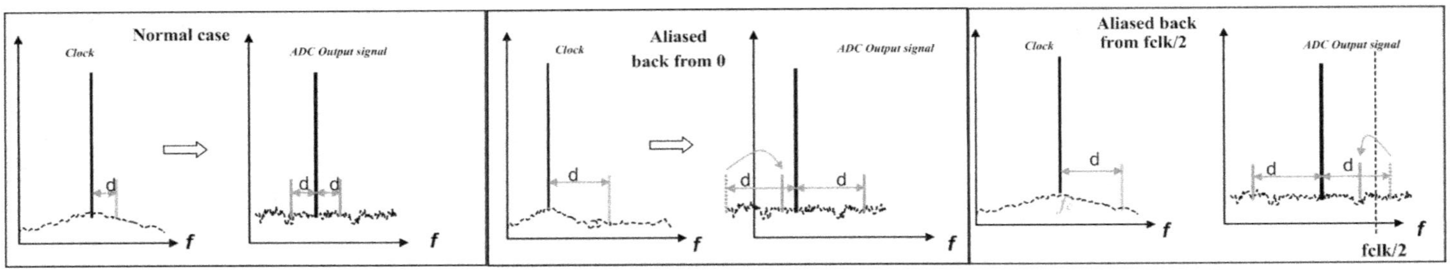

Fig. 1.26. The spurs are aliased back in the FFT plot.

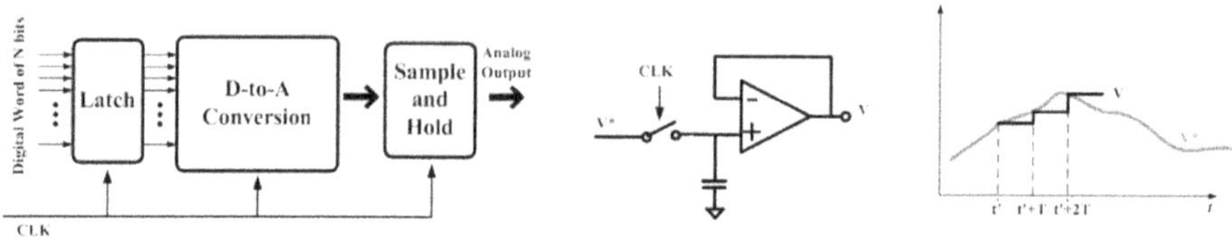

(a) a DAC system (b) sample-and-hold (c) output voltage

Fig. 1.27. The clock signal and the digital-to-analog converter.

timing errors exist. This impacts the performance of the digital-to-analog conversion process. These timing problems can be classified into global and local categories. Global timing error, such as clock jitter, is associated with the clock generator, which has same impact on all the elements. Local timing error is related to individual clocked units, such as the physical mismatches of switches, the different RC time constants of interconnections, etc. To some extent, it is similar to the clock skew in the clock distribution network for digital systems.

In some cases, such as in direct digital frequency synthesis (DDFS), the DAC is used to produce the single tone sinusoidal waveform. The spurious tone contained in the reference clock (which drives the DAC) could appear in the output signal's spectrum.

Sigma-delta modulation is a technique of encoding a high-resolution signal into a low-resolution signal by using pulse density modulation. This technique has been used widely in data conversion circuitry. Within these sigma-delta-based ADC and DAC, the sigma-delta modulators are driven by a higher speed clock. The quality of this clock signal has a significant impact on the quality of the corresponding data converter.

1.5 EXTRACTING CLOCK SIGNAL FROM DATA: CLOCK DATA RECOVERY

All VLSI chips are designed for processing information. This task can be further divided into two categories: computation and communication. In computation (CPU, DSP, and microcontroller), the clock is used to control the pace of the operation. In communication, information is exchanged between blocks, modules, or chips. The clock is used for controlling the rate of information flow. In wired communication (optical communication, backplane routing, chip-to-chip interconnects, etc.), many industry standards have been developed over the years for different applications, such as SATA, SONET, PCI Express, IEEE 1394b, USB 3.0, HDMI, DVI, DisplayPort, etc. The backbone behind those standards is the serializer/deserializer technology. A serializer/deserializer (SerDes) is a pair of functional blocks that convert data from serial to parallel and vice versa, as shown in Fig. 1.28.

As can be appreciated from this figure, the benefits of serialization are few wires, smaller board space, longer communication distance, and lower power

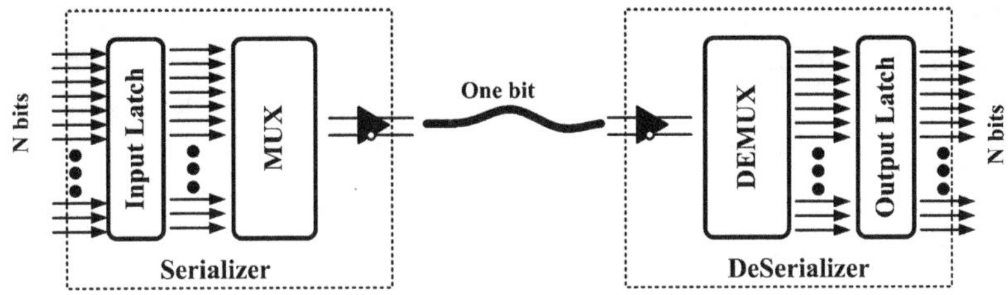

Fig. 1.28. Serializer and deserializer.

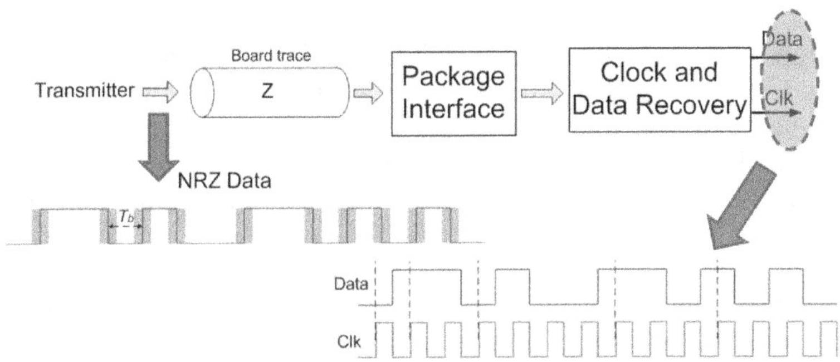

Fig. 1.29. The clock signal in clock data recovery.

consumption. The information transmitted over a SerDes is a string of "0" and "1." The clock plays crucial part in transmitting and receiving these bits. There are several SerDes architectures: parallel clock SerDes, embedded clock SerDes, 8b/10b SerDes, and bit interleaved SerDes. In a parallel clock SerDes, a clock signal is transmitted along with the data but in a different channel. In an embedded clock SerDes, the clock signal is explicitly embedded in the data stream. In the other two methods, the clock information is not explicitly presented but embedded in the "0" → "1" and "1" → "0" signal transitions. The 8b/10b SerDes has been adopted widely in many popular communication standards, such as PCI Express, 1394b, USB 3.0, HDMI, DVI, and DisplayPort. The most popular interface signaling technology used in SerDes is LVDS (low voltage differential signaling). However, for high-speed signaling, CML (current mode logic) and LVPECL (low-voltage positive emitter-coupled logic) are often used.

When serial data streams are sent without an accompanying clock signal, the receiver must first generate a clock from an approximate frequency reference and then frequency-align and phase-align it to the transitions embedded in the data stream with a PLL. This process is commonly known as clock data recovery (CDR). It is a critical block in 8b/10b SerDes. In CDR applications, there are three important issues related to the clock: frequency generation, clock-data alignment, and jitter transfer. As illustrated in Fig. 1.29, the recovered clock has to bear the frequency that matches the incoming data rate. The incoming data are driven by a clock that is invisible to the CDR. The task of the CDR is to find its frequency through received data. Additionally, the phase of this clock has to lie in the center of the data time window for a maximum safety margin. Furthermore, in the process of clock generation, the timing jitter embedded in the incoming data has to be rejected as much as possible.

1.6 CLOCK USAGE IN SYSTEM-ON-CHIP

The task of on-chip clock generation (frequency synthesis) is to generate required frequencies for supporting chip operation. In today's system-on-a-chip (SoC) environment, more and more functions are integrated into one single chip. To support this large number of functions, hundreds of frequencies could be required for successful operation. To make the situation even more difficult, all the frequencies are preferred to be generated from one single reference source (usually a crystal) for cost consideration. Besides the quality requirement (low jitter, ample frequencies), it is also demanded that the clock circuitry should use as few resources as possible (area, power). This is especially important for consumer electronic market where price is the most effective tool that can be used to win in competition. From the functional perspective, as illustrated in Fig. 1.30, clock circuitry can be responsible for the following:

- driving digital processing units (CPU, DSP, microcontroller, etc.)
- driving on-chip ADC and DAC

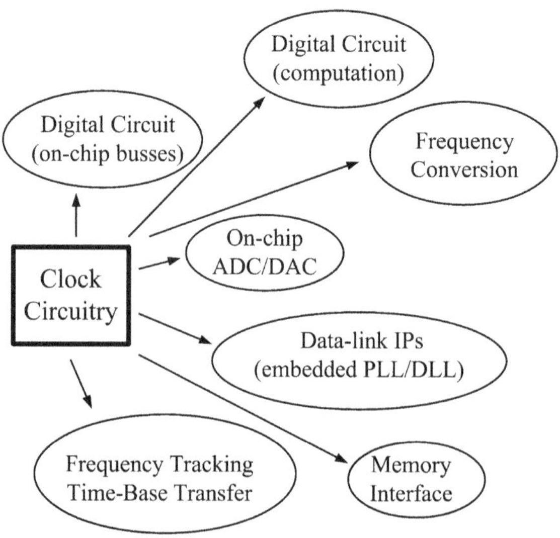

Fig. 1.30. The clock challenges in the system-on-chip environment.

- providing frequency reference for on-chip IPs (USB, DDR, LVDS, HDMI, etc.)
- local oscillator (LO) for frequency down-conversion or up-conversion
- frequency tracking

Overall, digital circuits account for the majority of SoC clock loading. The most important concerns in this task are jitter and skew. On the other hand, the tasks of driving ADC/DAC, providing references to IP addresses, and performing frequency conversions require spectral purity in the clock signal. When clock circuitry is used for frequency tracking (also called time-based transfer or timing recovery), the desirable frequency is not predetermined, but only decided in real time from tracking certain target.

1.7 TWO FIELDS: CLOCK GENERATION AND CLOCK DISTRIBUTION

In the construction of a synchronous system, the clock is the signal that requires the highest priority. In clock implementation, there are two different fields: clock generation and clock distribution. This is illustrated in Fig. 1.31. Clock generation refers to the task of generating the necessary frequencies for supporting the various on-chip functions. This is also commonly called frequency synthesis. The key circuit component used in this field is the PLL. The important issues are high frequency, low jitter/noise, fine frequency resolution, and fast switching. Clock distribution is the work of distributing the generated clock signal to all the clock sinks attached to this clock source, which could be spread all over the chip. The key challenges in this task are the minimization of the clock skew, controlling the slew rate, and the balancing and minimization of clock tree insertion delays. These two fields are major focuses in both research and engineering. As design complexity continually increases, these fields are ever-changing. They are among the most actively researched areas and will remain so for the foreseeable future.

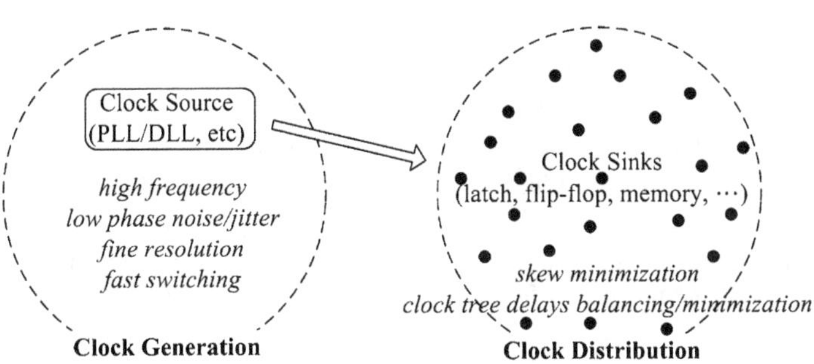

Fig. 1.31. The two clock-related fields: clock generation and clock distribution.

BIBLIOGRAPHY

Phase Noise and Jitter

Abidi, A. A. 2006. “Phase Noise and Jitter in CMOS Ring Oscillators,” *Solid-State Circuits IEEE J.*, vol. 41, no. 8, pp. 1803–1816.

Blakkan, K. and M. Soma. 2009. “A Time Domain Method to Measure Oscillator Phase Noise,” *VLSI Test Symposium, 2009. VTS ’09. 27th IEEE*, 3–7 May, pp. 297–302.

Chin, J. and A. Cantoni. 1998. “Phase Jitter = Timing Jitter?” *Commun. Lett. IEEE*, vol. 2, no. 2, pp. 54–56.

Demir, A. 2002. “Phase Noise and Timing Jitter in Oscillators with Colored-Noise Sources,” *Circuits Syst. I Regular Papers IEEE Trans.*, vol. 49, no. 12, pp. 1782–1791.

Demir, A. 2006. “Computing Timing Jitter from Phase Noise Spectra for Oscillators and Phase-Locked Loops with White and 1/f Noise,” *Circuits Syst. I Regular Papers IEEE Trans.*, vol. 53, no. 9, pp. 1869–1884.

Hajimiri, A. and T. H. Lee. 1998. “General Theory of Phase Noise in Electrical Oscillators,” *Solid-State Circuits IEEE J.*, vol. 33, no. 2, pp. 179–194.

Kim, Y. W. and J. D. Yu. 2008. “Phase Noise Model of Single Loop Frequency Synthesizer,” *Broadcast. IEEE Trans.*, vol. 54, no. 1, pp. 112–119.

Kundert, K. S. 1999. “Introduction to RF Simulation and Its Application,” *IEEE J. Solid-State Circuits*, vol. 34, pp. 1298–1319.

Kundert, K. S. “Predicting the Phase Noise and Jitter of PLL-Based Frequency Synthesizers,” http://www.designers-guide.com.

Lecroy. “Clock Recovery Methods for Jitter Analysis,” Technical brief.

Lee, T. H. and A. Hajimiri. 2000. “Oscillator Phase Noise: A Tutorial,” *Solid-State Circuits IEEE J.*, vol. 35, no. 3, pp. 326–336.

Liang, D. and R. Harjani. 2000. “Comparison and Analysis of Phase Noise in Ring Oscillators,” *Circuits and Systems, 2000. Proceedings. ISCAS 2000 Geneva. The 2000 IEEE International Symposium on*, vol. 5, 28–31 May, pp. 77–80.

Liang, D. and R. Harjani. 2002. “Design of Low-Phase-Noise CMOS Ring Oscillators,” *Circuits Syst. II Analog Digit. Signal Processing IEEE Trans.*, vol. 49, no. 5, pp. 328–338.

Navid, R., T. H. Lee, and R. W. Dutton. 2005. “Minimum Achievable Phase Noise of RC Oscillators,” *Solid-State Circuits IEEE J.*, vol. 40, no. 3, pp. 630–637.

Mak, T. M. 2008. “Jitters in High Performance Microprocessors,” *Test Conference, 2008. ITC 2008. IEEE International*, 28–30 Oct., pp. 1–6.

Poore, R. 2001. “Overview on Phase Noise and Jitter,” Agilent Technologies, http://cp.literature.agilent.com/litweb/pdf/5990-3108EN.pdf.

Razavi, B. 1996. “A Study of Phase Noise in CMOS Oscillators,” *Solid-State Circuits IEEE J.*, vol. 31, no. 3, pp. 331–343.

Shimanouchi, M. 2001. “An Approach to Consistent Jitter Modeling for Various Jitter Aspects and Measurement Methods,” *Test Conference, 2001, Proceedings, International*, pp. 848–857.

Tektronix. “Understanding and Characterizing Timing Jitter,” application note.

Clock Distribution and Clock Skew

Friedman, E. G. 2001. "Clock Distribution Networks in Synchronous Digital Integrated Circuits," *Proc. IEEE*, vol. 89, no. 5, pp. 665–692.

Harris, D., M. Horowitz, and D. Liu. 1999. "Timing Analysis including Clock Skew," *IEEE Trans. CAD*, vol. 18, no. 11, pp. 1608–1618.

Jiang, X. and S. Horiguchi. 2001. "Statistical Skew Modeling for General Clock Distribution Networks in Presence of Process Variations," *IEEE Trans. VLSI Syst.*, vol. 9, no. 5, pp. 704–717.

Ramanathan, P., A. J. Dupont, and K. G. Shin. 1994. "Clock Distribution in General VLSI Circuits," *Circuits Syst. I Fundam. Theory Appl. IEEE Trans.*, vol. 41, no. 5, pp. 395–404.

Zanella, S., A. Nardi, A. Neviani, M. Quarantelli, S. Saxena, and C. Guardiani. 2000. "Analysis of the Impact of Process Variations on Clock Skew," *Semicond. Manuf. IEEE Trans.*, vol. 13, no. 4, pp. 401–407.

Clock Jitter on Data Converter

Analog Devices. "Fundamentals of Sampled Data Systems," Application Note, AN-282.

Angrisani, L. and M. D'Arco. 2009. "Modeling Timing Jitter Effects in Digital-to-Analog Converters," *Instrum. Meas. IEEE Trans.*, vol. 58, no. 2, pp. 330–336.

Brannon, B. "Sampled System and the Effects of Clock Phase Noise and Jitter," Analog Device, AN-756.

Brannon, B. and A. Barlow. "Aperture Uncertainty and ADC System Performance," Analog Device, AN-501.

Da Dait, N., M. Harteneck, C. Sandner, and A. Wiesbauer. 2001. "Numerical Modeling of PLL Jitter and the Impact of Its Non-white Spectrum on the SNR of Sampled Signals," *Mixed-Signal Design, 2001. SSMSD. 2001 Southwest Symposium on*, 25–27 Feb., pp. 38–44.

Da Dalt, N., M. Harteneck, C. Sandner, and A. Wiesbauer. 2002. "On the Jitter Requirements of the Sampling Clock for Analog-to-Digital Converters," *Circuits Syst. I Fundam. Theory Appl. IEEE Trans.*, vol. 49, no. 9, pp. 1354–1360.

Doris, K., A. van Roermund, and D. Leenaerts. 2002. "A General Analysis on the Timing Jitter in D/A Converters," *Circuits and Systems, 2002. ISCAS 2002. IEEE International Symposium on*, vol. 1, 26–29, May, pp. I-117–I-120.

Hai, T., L. Toth, and J. M. Khoury. 1999. "Analysis of Timing Jitter in Band Pass Sigma-Delta Modulators," *Circuits Syst. II Analog Digit. Signal Processing IEEE Trans.*, vol. 46, no. 8, pp. 991–1001.

Jenq, Y.-C. 1997. "Direct Digital Synthesizer with Jittered Clock," *Instrum. Meas. IEEE Trans.*, vol. 46, no. 3, pp. 653–655.

Neu, T. 2009. "Impact of Sampling-Clock Spurs on ADC Performance," *Analog Appl. J.*, 3rd, 2009, pp. 5–12. Texas Instruments.

Shinagawa, M., Y. Akazawa, and T. Wakimoto. 1990. "Jitter Analysis of High-Speed Sampling Systems," *IEEE JSSC*, vol. 25, no. 1, pp. 220–224.

Clock and SerDes, CDR

Cho, L. C., C. Lee, C. C. Hung, and S. I. Liu. 2009. "A 33.6-to-33.8 Gb/s Burst-Mode CDR in 90 nm CMOS Technology," *JSSC*, vol. 44, no. 3, pp. 775–783.

Horowitz, M., C. K. K. Yang, and S. Sidiropoulous. 1998. "High-Speed Electrical Signaling: Overview and Limitation," *IEEE Micro.*, vol. 18, no. 1, pp. 12–14.

Kim, J., J. Yang, S. Byun, H. Jun, J. Park, C. S. G. Conroy, and B. Kim. 2005. "A Four-Channel 3.125 Gb/s/ch CMOS Serial-Link Transceiver with a Mixed-Mode Adaptive Equalizer," *JSSC*, vol. 40, no. 2, pp. 462–471.

Kim, J. K., J. Kim, G. Kim, and D. K. Jeong. 2009. "A Fully Integrated 0.13 um CMOS 40-Gb/s Serial Link Transceiver," *JSSC*, vol. 44, no. 5, pp. 1510–1521.

Lee, J., K. S. Kundert, and B. Razavi. 2004. "Analysis and Modeling of Bang-Bang Clock and Data Recovery Circuit," *JSSC*, vol. 39, no. 9, pp. 1571–1580.

Lewis, D. 2004. "SerDes Architecture," National Semiconductor Corporation.

Loke, A. L. S., R. K. Barnes, T. T. Wee, M. M. Oshima, C. E. Moore, R. R. Kennedy, and M. J. Gilsdorf. 2006. "A Versatile 90 nm Charge-Pump PLL for SerDes Transmitter Clocking," *JSSC*, vol. 41, no. 8, pp. 1894–1907.

"*LVDS Owner's Manual Design Guide*," 2001. National Semiconductor Corporation.

Razavi, B. 2002. "Challenges in the Design of High-Speed Clock and Data Clock Data Recovery Circuits," *IEEE Communication Magazine*, Aug..

Time-Average-Frequency and Digital-to-Frequency Converter

Xiu, L. 2008a. "The Concept of Time-Average-Frequency and Mathematical Analysis of Flying-Adder Frequency Synthesis Architecture," *IEEE Circuit And System Magazine*, 3rd quarter, pp. 27–51, Sep..

Xiu, L. 2008b. "Some Open Issues Associated with the New Type of Component: Digital-to-Frequency Converter," *IEEE Circuit And System Magazine*, 3rd quarter, pp.90–84, Sep..

CHAPTER 2

CLOCK GENERATION: EXISTING FREQUENCY SYNTHESIS TECHNIQUES

As discussed in Chapter 1, clock signal is of vital importance to all electronic systems. When a clock signal is used in those systems, the concerns around this crucial signal are:

- available frequencies
- frequency resolution
- flexibility (how fast can frequency be switched from one to another)
- clock edge uncertainty (jitter/phase noise)
- frequency purity (spurious tones)
- the cost of building the clock-generation circuit

Among these concerns, the most important one is frequency accuracy. Hence, clock generation is also termed "frequency synthesis." The task of frequency synthesis can be roughly stated as generating other frequencies from a source frequency (a fixed time base) or from a group of frequencies. In this chapter, we will briefly review the distinguished techniques developed over the past several decades.

All the techniques can roughly be classified into three main groups: direct analog frequency synthesis, direct digital frequency synthesis, and indirect method (phase-locked loop [PLL] based). The word "direct" refers to the fact that the output clock waveform is directly constructed; there is no feedback

Nanometer Frequency Synthesis Beyond the Phase-Locked Loop, First Edition. Liming Xiu.

Published 2012 by John Wiley & Sons, Inc.

mechanism used in the corresponding methods. A PLL-based system produces its clock from an autonomous oscillator (such as a voltage-controlled oscillator), whose output is compared against an input reference. In this approach, the output is driven toward the input (most of the time with a multiplying factor) through feedback. Hence, it is called "indirect." Since this is the most popular method for on-chip clock generation, a slightly more detailed description is given in Section 2.3.

In the last part of this chapter, an overview of existing frequency synthesis techniques is presented. From there, two not-very-well-solved problems are recognized. These problems are the driving force behind our search for a new direction in frequency synthesis. This search leads to the time-average-frequency concept, which will be introduced in Chapter 3.

2.1 DIRECT ANALOG FREQUENCY SYNTHESIS

In direct analog frequency synthesis, frequencies are created by operations of mixing, filtering, multiplying, and dividing. The operation of frequency mixing can be seen in Fig. 2.1 and Eqs. 2.1–2.3.

$$v_1 = A_1 \sin(2\pi f_1 t) \tag{2.1}$$

$$v_2 = A_2 \sin(2\pi f_2 t) \tag{2.2}$$

$$v_o = v_1 v_2 = \frac{1}{2} A_1 A_2 [\cos(2\pi(f_1 - f_2)t) - (\cos 2\pi(f_1 + f_2)t)] \tag{2.3}$$

From Eq. 2.3, it can be seen that two new frequencies, $f_1 - f_2$ and $f_1 + f_2$, have been generated from the two original frequencies, f_1 and f_2. Usually, f_1 and f_2 are much higher frequencies than the difference $f_1 - f_2$. Extracting this frequency difference, which is often the principal purpose of using a frequency mixer, can be done by filtering out the higher frequencies. There are many ways of multiplying signals in real implementation. Using an electronic device called a *diode* is one of the simplest methods. Because of its nonlinear characteristic, a diode can be used to perform frequency multiplication.

Theoretically, these operations of mixing, filtering, and dividing can be repeated at arbitrary times to achieve fine frequency resolution. Figure 2.2 is

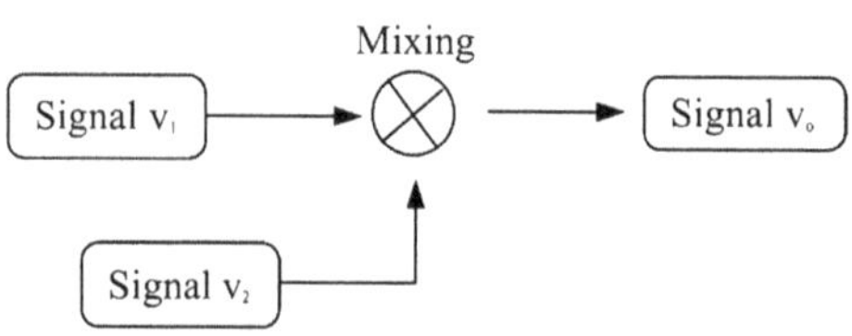

Fig. 2.1. A frequency mixer.

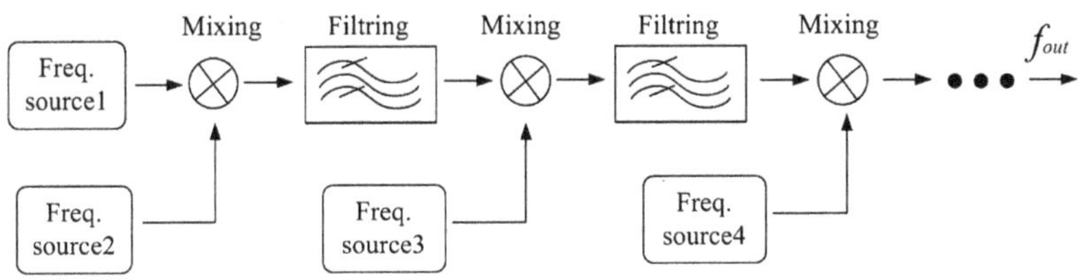

Fig. 2.2. Direct analog frequency synthesis.

a block diagram of conventional direct analog frequency synthesizer with several stages.

The advantages of direct analog synthesis are very fast switching speed and arbitrarily fine frequency resolution (at least in theory). Moreover, by using high-quality frequency sources, high spectral purity can be achieved in the final output. The major drawback of this method is that it requires a large amount of hardware. The cost, size, and weight of the mixing/filtering/dividing circuitry could grow rapidly when a fine frequency step is required. Furthermore, since many frequencies coexist simultaneously in direct analog frequency synthesis, extensive shielding is a must. Otherwise, significant spurious tones could be present in the output. Due to this high cost, direct analog synthesis is hardly chosen for on-chip clock generation. It is applicable only to applications where high price can be tolerated, such as in high-end instruments and radar systems.

2.2 DIRECT DIGITAL FREQUENCY SYNTHESIS

Direct digital frequency synthesis (DDFS) is a method of producing a time-varying signal in digital form. There are two types of direct digital synthesis: DDFS and pulse rate digital frequency synthesis. Both of them relay on a reference clock signal. In DDFS, the time-varying waveform is first generated in a digital domain. Then, digital-to-analog conversion is performed to produce the actual signal, which could be any arbitrary waveform such as square wave, triangular wave, or sinusoidal wave. In pulse rate digital frequency synthesis, the output is a square-wave clock pulse whose high and low states are composed of certain numbers of reference pulses. It is a direct construction of a pulse waveform; no digital-to-analog conversion is required.

Figure 2.3 is the generic block diagram of DDFS. The major blocks included in this system are accumulator, angle-to-amplitude converter, digital-to-analog converter (DAC), and reconstruction filter (RCF). In operation, a frequency tuning word M is first fed into the system. Then, this tuning word is accumulated in an N-bit accumulator at every cycle of the reference clock f_s. The output of the accumulator, which could possibly be truncated into a P-bit, will be converted to signal amplitude by the angle-to-amplitude converter. Its output, in the size of a D-bit, is sent to the DAC to produce the corresponding

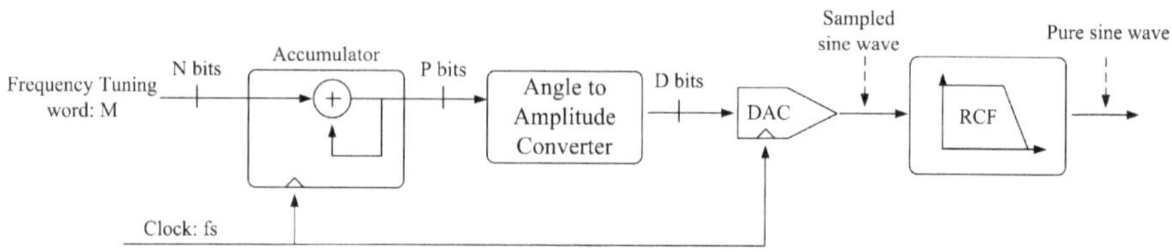

Fig. 2.3. The block diagram of DDFS.

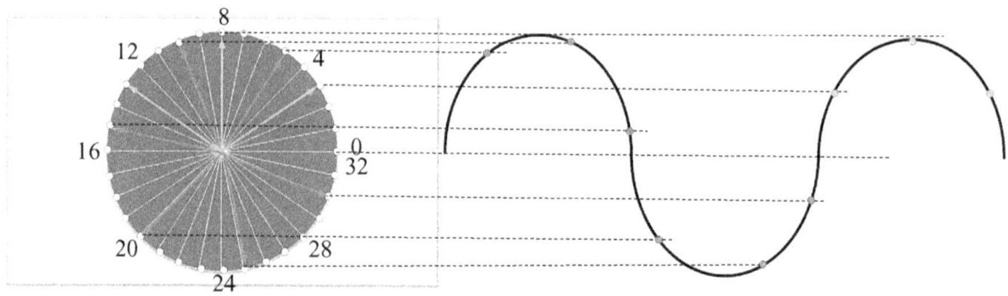

Fig. 2.4. The concept of phase wheel and angle-to-amplitude conversion.

analog waveform. The RCF filter is used to filter out the sampled images at higher frequencies.

One of the key concepts in DDFS is the phase wheel, which is illustrated in Fig. 2.4. The phase of one cycle of any repetitive signal can be represented by a wheel, which covers the range of 2π. In Fig. 2.4, we assume that the phase resolution is 1/32, or $\pi/16$. This corresponds to $N = 5$ in Fig. 2.3. Assume that the frequency tuning word $M = 5$ and that the accumulator's initial value is zero. Then, at every clock cycle, the accumulator will output a value from the following sequence: 0, 5, 10, 15, 20, 25, 30, 3 (roll over after 32), 8, 13, and so on. These values can be used as indexes for the angle-to-amplitude converter, which is usually implemented as a lookup table. If complete sine-wave-shape data are stored in the table, the corresponding amplitude value at each clock is shown in the plot on the right. The red color represents the dots in the first cycle; the green is for the second cycle. As can be seen, the larger the frequency tuning word M is, the faster the cycle will be completed, and thus, the faster the output frequency will be.

This amplitude data, when fed into the DAC, will produce a sine-wave output. The data format at each stage is shown further in Fig. 2.5. In certain systems, the accumulator's output could be truncated from N-bit to P-bit. As a result, some information is lost when data reach the input of angle-to-amplitude converter. This can result in additional spurious components at the DDFS output.

The phase resolution is determined by the accumulator's size: 2^N. The DDFS output frequency depends on three parameters: reference clock frequency f_s, accumulator size, and frequency tuning word M. The output frequency f_o can be easily worked out as

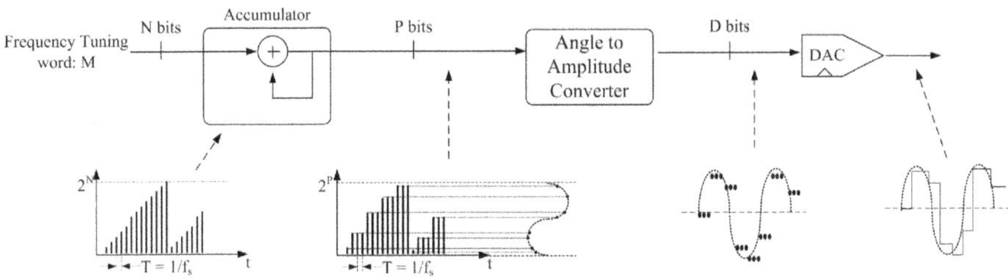

Fig. 2.5. The data format at each stage of DDFS.

$$f_o = (M * f_s)/2^N \quad (2.4)$$

The output of DDFS is a single tone waveform, or a sine wave at a specific frequency. Its frequency is digitally tunable, as is its phase. Since the control parameters are numerically determined, there is no problem of temperature- and aging-induced drift. DDFS also has fast frequency switching capability. However, DDFS requires a high-frequency reference clock. Its output frequency must be less than one-half of the reference frequency by Nyquist criterion. Furthermore, there are artifacts in the DDFS output spectrum such as phase truncation spurs, DAC nonlinearity, and DAC switching noise. Due to its analog complexity and its high cost, DDFS is not commonly used for on-chip clock generation. It is often found in instruments.

2.3 INDIRECT METHOD (PHASE-LOCKED LOOP BASED)

2.3.1 Brief History

In the early 1930s, the superheterodyne radio receiver was very popular in the field of radio electronics. Edward Howard Armstrong was one of the leading contributors in this field, and the superheterodyne radio receiver is one of his many inventions. In 1932, a team of British scientists was experimenting with a new method that would surpass the superheterodyne. This new type of radio receiver, called the synchrodyne, consists of a local oscillator, a mixer, and an audio amplifier. When the input signal and the local oscillator are mixed at the same phase and frequency, the output is an exact audio representation of the modulated carrier. The initial tests were encouraging. But it was later found that, after a period of time of operation, the synchronous reception became difficult due to the slight drift in the frequency of the local oscillator. To counteract this frequency drift, the frequency of the local oscillator was compared with a fixed input by a phase detector so that a correction voltage could be generated and fed back to the local oscillator. This kept the oscillator on its original frequency. This feedback circuit marks the beginning of the PLL evolution. It is believed that the British scientists developed this feedback system based on a paper written in 1932 by French scientist H. de Bellescise.

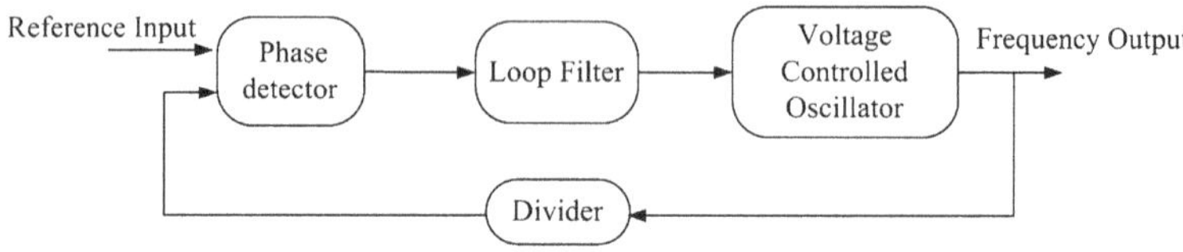

Fig. 2.6. The structure of the phase-locked loop.

Although the synchronous radio receiver was superior to the superheterodyne version, the cost of the PLL circuit outweighed its advantages. Because of this prohibitive cost, the widespread use of this principle did not occur until the successful development in other fields of application. In the 1940s, the first widespread use of the PLL was in the synchronization of the horizontal and vertical sweep oscillators to the transmitted sync pulses in television receivers. Those circuits were called synchrolock and/or synchroguide. Since then, the electronic PLL principle has been extended to many other applications. In the modern electronics industry, most electronic devices would not be possible without PLL technology.

2.3.2 The Basic Structure of the Phase-Locked Loop (PLL)

From the circuit perspective, the PLL is an amazing system that blends digital and analog techniques beautifully in one package. The basic structure is shown in Fig. 2.6. It is a system whose aim is to generate a signal that has a fixed phase and frequency relationship to an input reference signal. From the control point of view, it is a negative feedback system that responds to both the frequency and the phase of the input signals. It can potentially reach a steady state, or establish equilibrium around certain balanced point. The name *indirect* comes from the fact that the output frequency is related to the input frequency only through the feedback loop. In many PLL systems, the reference input is a crystal oscillator that has high frequency stability and good frequency purity. A PLL combines the frequency flexibility of the voltage control oscillator (VCO) with the frequency stability of the crystal oscillator to produce the desired output frequency.

The phase detector is used to compare the two signals of the reference input and the VCO feedback. Based on these, the detector produces a signal whose magnitude is proportional to their frequencies and/or phase differences. This difference signal acts as a correction mechanism and will be applied, after the filter, to the VCO and drives its oscillation frequency toward the input reference. There are two basic types of phase detectors: type I and type II. A type I detector responds to the voltage levels of the two compared signals. A type II detector only responds to the signal transitions (low-to-high or high-to-low edges). A type I detector includes a frequency mixer and digital XOR gate. A tri-state PFD (phase frequency detector) is a typical type II detector. The symbols of these detectors are illustrated in Fig. 2.7.

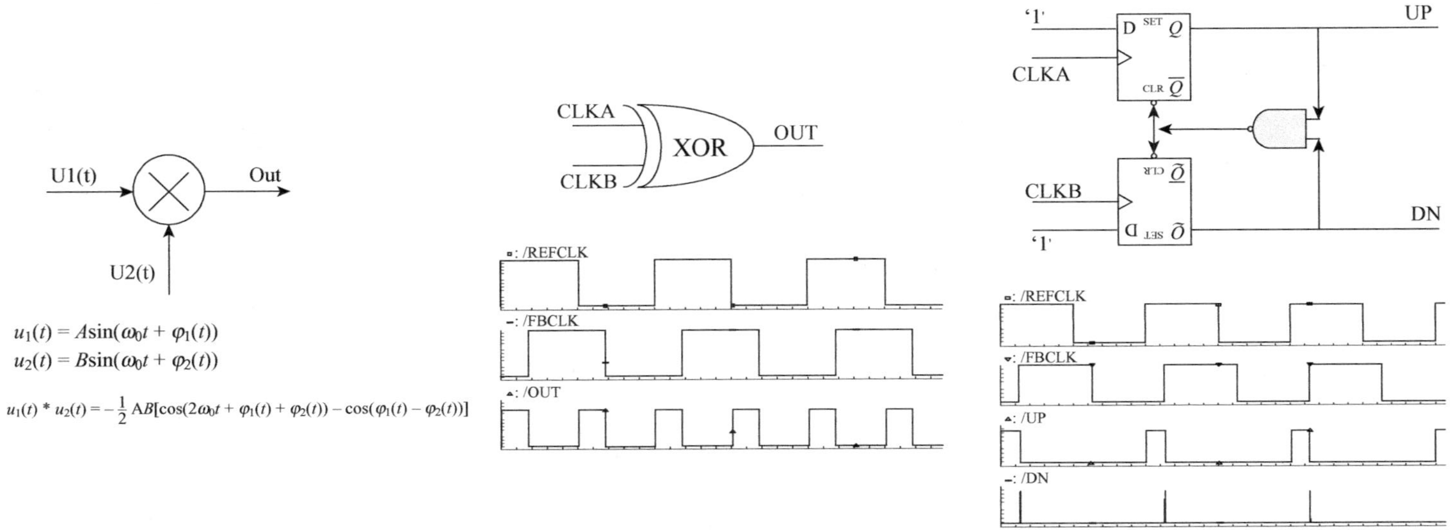

Fig. 2.7. Phase detector: frequency mixer (left), XOR gate (middle), and tri-state PFD (right).

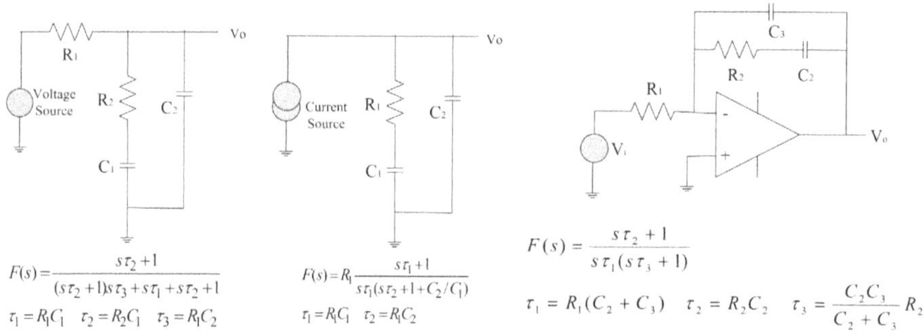

Fig. 2.8. Second order loop filter: passive (left and middle); active (right).

The loop filter in a PLL has low-pass characteristics. Roughly speaking, it averages out the phase detector output and extracts the DC component. It is used to control the dynamic behavior of the feedback loop. The loop filter also provides some kind of short-term memory to ensure a rapid recapture of the state if the system is thrown out of lock by a noise transient. The loop filter plays a crucial role in determining the PLL's order and type. It can be implemented either as passive filter or active filter. Some second-order examples are given in Fig. 2.8.

The VCO is the heart of the PLL. Its oscillation frequency depends on the magnitude of the control parameter (usually a voltage) applied on its control terminal. Mathematically, its transfer function is expressed in Eq. 2.5, where Δf is the VCO frequency change, ΔV_c is the control voltage change, and K_{vco} is the VCO gain in Hz/V. The last equation $\theta(s)$ is expressed in Laplace representation.

$$\begin{aligned}\Delta f &= \frac{d\theta}{2\pi dt} = K_{vco}\Delta V_c \\ \theta(t) &= \int 2\pi K_{vco}\Delta V_c dt \\ \theta(s) &= \frac{2\pi K_{vco}\Delta V_c}{s}\end{aligned} \tag{2.5}$$

In circuit implementation, VCO can be realized by an LC resonant oscillator, metal–oxide–semiconductor field-effect transistor (MOSFET) varactor, and ring oscillator, as shown in Fig. 2.9. Numerous variations of these three basic structures have been used in real applications. Each has its advantages and disadvantages.

A divider is often used inside the PLL to assist the VCO in achieving certain frequencies. Its transfer function can be simply expressed, as in Eq. 2.6, if the divide ratio is N.

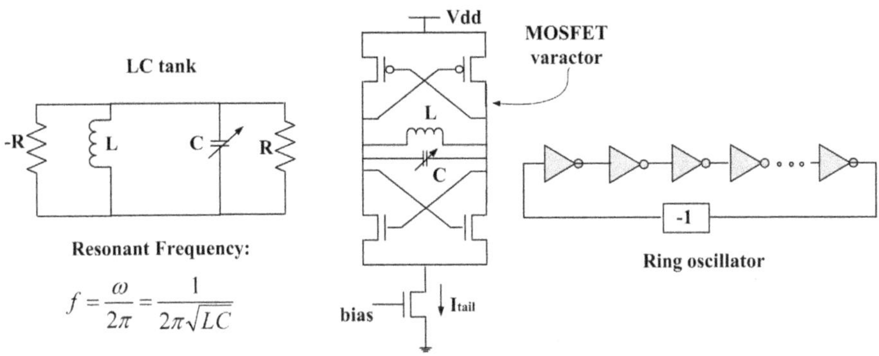

Fig. 2.9. The basic structures of the voltage controlled oscillator.

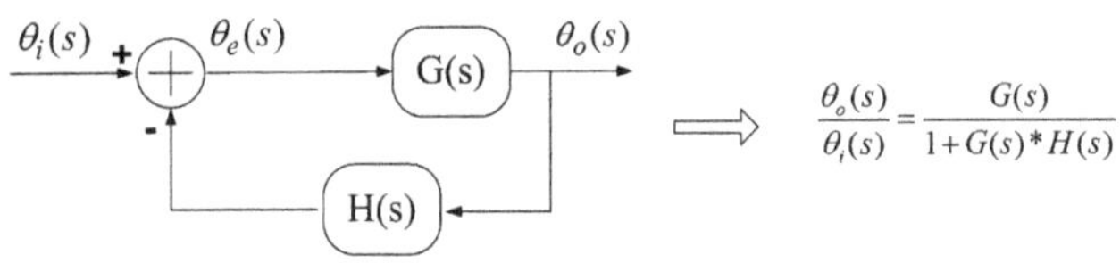

Fig. 2.10. The PLL transfer function in Laplace representation.

$$H(s) = \frac{\theta_{out}}{\theta_{in}} = 1/N \tag{2.6}$$

When in the neighborhood of equilibrium (lock), the PLL's behavior can be linearly modeled in Laplace representation. If the PLL's forward gain is called $G(s)$ and the feedback gain is $H(s)$, the output–input transfer function can be derived as in Fig. 2.10.

2.3.3 An Example of Third-Order Type-II Charge Pump PLL

In modern monolithic PLL implementation, the charge pump PLL is very popular. One such system is depicted in Fig. 2.11. The charge pump (CP), which sits between the phase frequency detector (PFD) and the loop filter, is used to create an integrator that helps automatically shift the DC level of the PFD output. The VCO is another integrator (Eq. 2.5). Usually, a second-order loop filter (the middle structure in Fig. 2.8) is employed after the CP. Thus, the loop forward gain $G(s)$ has three poles (see Eq. 2.7 below); two of them are located in the origin. Due to these three poles, the system is classified as third order. Further, because of the two poles at the origin, the system is termed type II. The transfer function of each component is labeled in the figure. The op-amp is configured as a unity gain buffer. It is used to provide the large current required by the VCO. In the interested frequency range, its gain is 1.

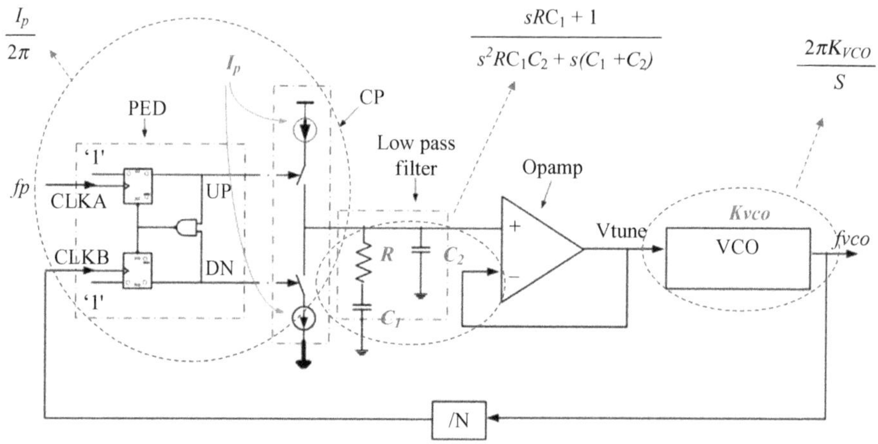

Fig. 2.11. A charge-pump PLL example.

From Fig. 2.11, the forward gain can be calculated as in Eq. 2.7:

$$G(s) = \frac{I_p}{2\pi} \times \frac{sRC_1 + 1}{s^2RC_1C_2 + s(C_1 + C_2)} \times \frac{2\pi K_{vco}}{s} = \frac{I_pK_{vco}(sRC_1 + 1)}{s^3RC_1C_2 + s^2(C_1 + C_2)} \tag{2.7}$$

According to Fig. 2.10, with feedback gain $H(s) = 1/N$, the PLL transfer function can be derived as:

$$TF(s) = N\frac{K_{vco}I_p(sRC_1 + 1)}{Ns^3RC_1C_2 + Ns^2(C_1 + C_2) + K_{vco}I_pRC_1s + K_{vco}I_p} \tag{2.8}$$

Equation 2.8 shows that this PLL is a third-order system. If C_2 is ignored (usually $C_2 \ll C_1$), Eq. 2.8 can be reduced to second order as derived in Eq. 2.9. In Eq. 2.10, ω_n and ξ are the *nature frequency* and *damping factor* often used in control system.

$$\begin{aligned} TF(s) &= NK_{vco}I_p\frac{sRC_1 + 1}{Ns^2C_1 + K_{vco}I_pRC_1s + K_{vco}I_p} \\ &= N\frac{(K_{vco}I_pR/N)s + K_{vco}I_p/NC_1}{s^2 + (K_{vco}I_pR/N)s + K_{vco}I_p/NC_1} = N\frac{2\xi\omega_n s + \omega_n^2}{s^2 + 2\xi\omega_n s + \omega_n^2} \end{aligned} \tag{2.9}$$

$$\omega_n = \sqrt{\frac{K_{vco}I_p}{NC_1}} \quad \xi = \frac{R}{2}\sqrt{\frac{K_{vco}I_pC_1}{N}} \tag{2.10}$$

A design example uses 12 MHz as the input reference. The required VCO frequency is 1080 MHz ($N = 90$). Assume that VCO gain $K_{vco} = 1$ GHz/V. The design target of the PLL bandwidth is 700 KHz. The free design parameters

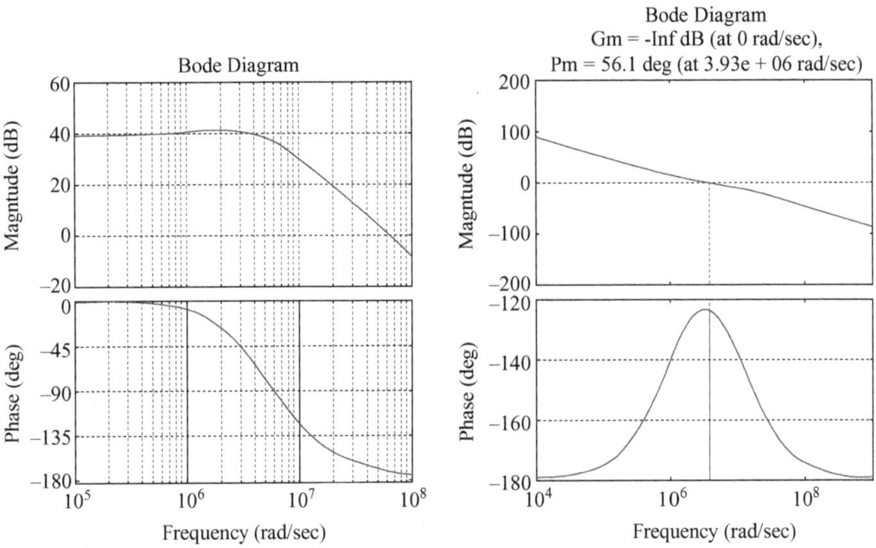

Fig. 2.12. The PLL's closed loop (left) and open loop (right) responses.

are C_1, C_2, R, and CP current I_p. Using some of the PLL's design tools, we can choose $C_1 = 50$ pf, $C_2 = 5$ pf, $R = 20$ K Ohm, and $I_p = 20$ uA to make the loop filter's zero and pole at 175 K Hz and 1.26 MHz, respectively. The desired loop bandwidth of 700 KHz is just in between the zero and the pole, which can make the loop stable with a decent margin. We can also use Matlab's built-in functions *bode* and *margin* to double-check the result. Figure 2.12 is the closed loop and open-loop response. The open-loop response is used to check loop stability. As can be seen from the plot, the phase margin is about 56°. The closed-loop response is used for studying the loop's response to a step change in its input. It is also used to observe jitter peaking. As can be seen, the gain's 3-db drop point (from the gain at DC) is about 665 KHz (~4e6 rad/sec), which is close to our original bandwidth target of 700 KHz.

2.3.4 Major PLL Architectures

Within the indirect frequency synthesis category, there are several popular PLL architectures that have been widely used in industry. Structurally, the most straightforward one is integer-N PLL. In this structure, as shown in the left drawing of Fig. 2.13, the divider used in the loop takes only integers. Therefore, the output frequency can only be an integer multiple of the input frequency. To reach finer frequency resolutions, the input frequency has to be lowered. This, however, requires a reduction in PLL bandwidth, which has several drawbacks. In this resolution versus bandwidth battle, fractional-N PLL architecture emerges (the right drawing in Fig. 2.13). In this structure, the divider takes two or more integers to assist the VCO in generating a frequency. The switching among the integers is accomplished by the dithering block. The

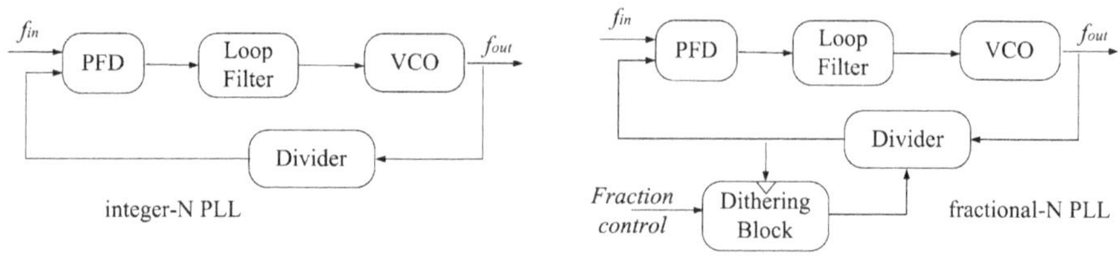

Fig. 2.13. The structure of integer-N PLL and fractional-N PLL.

frequency produced at the VCO output depends on the time average of these values. This average is determined by the weights assigned to all the integers. Thus, a fractional multiple of the input frequency can be achieved. From another perspective, for a given frequency resolution, a higher input frequency can be used, and consequently a larger PLL bandwidth becomes possible.

There are several developments within fractional-N PLL architecture, as shown in Fig. 2.14. The classical one uses a simple accumulator. When a carry is generated from the accumulator (a first-order delta-sigma modulator), the divider ratio is switched from N to N+1, or vice versa (Fig. 2.14a). This classical fractional-N PLL has the problem of poor spectrum purity at its output. The constant switching between N and N+1 alters the PFD output at a certain rate. It consequently modulates the VCO, which generates spurious tones. To cure this problem, the phase-interpolation-fractional-N PLL is developed. A DAC, whose input is derived from the accumulator, is used to produce a signal that counteracts the disturbance caused by the divider ratio switching (Fig. 2.14b). This will diminish the fractional spurs at the output. Another elegant solution is to employ a delta-sigma modulator to control the pattern of the divider ratio switching (Fig. 2.14c). While achieving the desired average value, the delta-sigma modulator moves the quantization noise to higher frequency band, which could potentially be filtered by the low-pass PLL. As such, this scheme provides another possibility of improving the VCO's frequency spectrum purity.

In all the previously discussed fractional-N PLL architectures, there is a phase jump of the size of one VCO period that occurs at the PFD input whenever the divider ratio changes. In recent developments, the VCO of multiple outputs (phases) has been used in fractional-N PLL. By using this multiphase VCO, the phase jump at the PFD input can be reduced to less than one VCO period. This can further reduce the magnitude of the disturbance, resulting in a cleaner spectrum.

In the family of PLL-based approaches, the latest member is all digital PLL (ADPLL), as shown in Fig. 2.15. In this architecture, the analog VCO is replaced by a digital controlled oscillator (DCO), whose input control is in the digital domain. The conventional PFD is replaced by a time-to-digital converter (TDC), which converts the time difference between the edges of the two PFDs inputs into a digital value. Moreover, the analog loop filter is

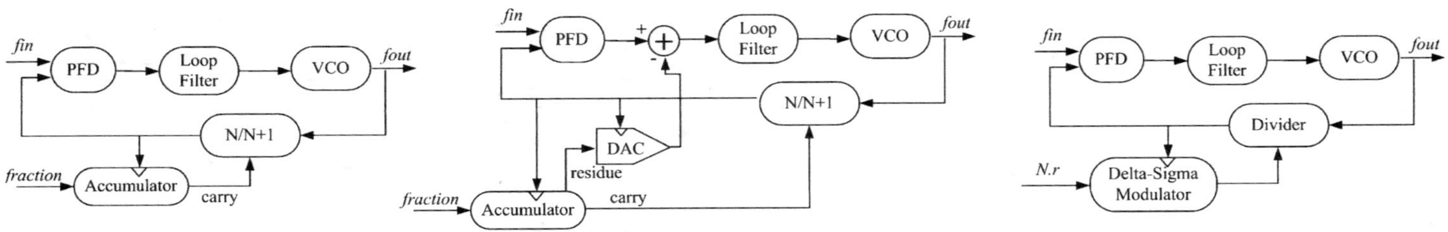

(a) Classical fractional-N PLL

(b) Phase interpolation fractional PLL

(c) Delta-Sigma fractional PLL

Fig. 2.14. The various types of fractional-N PLL.

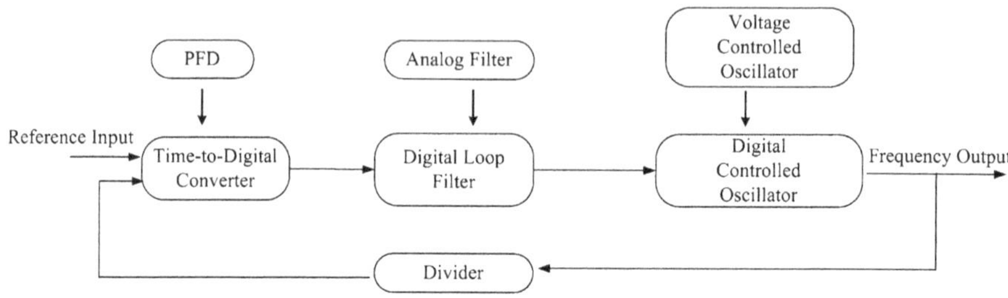

Fig. 2.15. The concept of all digital PLL.

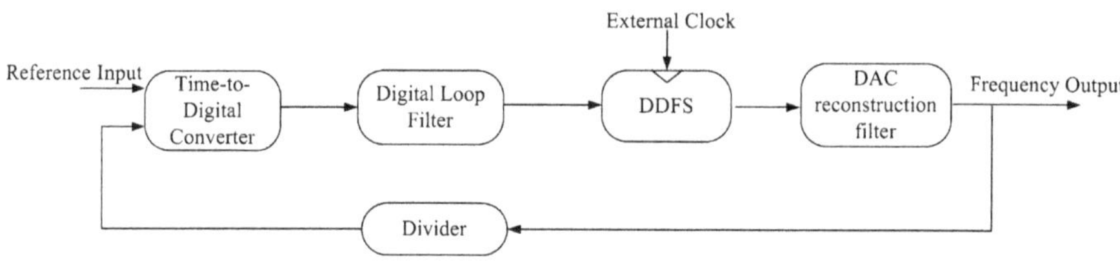

Fig. 2.16. The DDFS-based PLL.

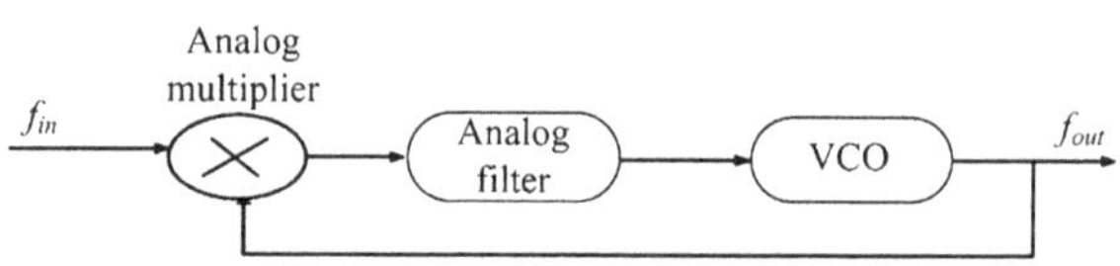

Fig. 2.17. The analog PLL: everything is analog including the phase detector.

replaced by a digital filter. As can be appreciated from this discussion, all the loop variables in ADPLLs are in the digital domain, which provides great flexibility in building the PLL loop.

DDFS has been used together with PLL to form a hybrid structure as illustrated in Fig. 2.16. The output from the DDFS (after the DAC as shown) is fed back to the TDC for frequency comparison. As a result, the output frequency is the desired multiple of the input frequency when lock is reached.

In their very early years, PLLs were built in pure analog fashion, as shown in Fig. 2.17. In this structure, all components are analog including the phase detector (which is implemented through an analog multiplier). This is the so-called all analog PLL (APLL), or linear PLL (LPLL). The PLL structures listed in Figs. 2.13 and 2.14 are traditionally called digital PLL due to the use of the digital phase detector (the middle and right structures in Fig. 2.7) and, optionally, the digital divider. The structure in Fig. 2.15 is termed an all digital PLL because all its loop variables are digital values.

Figure 2.18 shows a timing circuit that is functionally close to PLL: the delay-locked loop (DLL). The DLL does not have an oscillator in its loop.

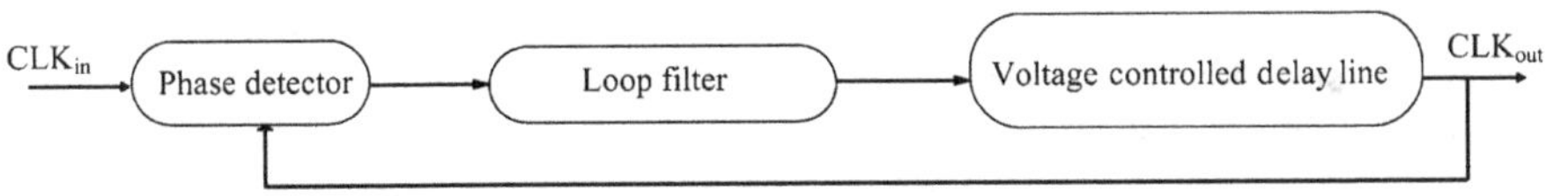

Fig. 2.18. The structure of the delay-locked loop (DLL).

Instead, a voltage control delay line (VCDL) is adopted. As a result, it cannot generate frequency on its own. Its main purpose is to provide delay capability on its input signal (phase movement). It is useful in many applications.

2.4 THE SHARED GOAL: ALL CYCLES HAVE SAME LENGTH-IN-TIME

Although significantly different in style, all the frequency synthesis methods introduced in this chapter share the same goal: to make all the cycles in this clock pulse train have exactly the same length-in-time.

Having the same size clock cycle in the entire train is ideal so it can be used as the driver of VLSI circuits. It can help synchronous circuits achieve their maximum efficiency. However, this belief of "all cycles have same length-in-time" is also the key reason that makes frequency synthesis difficult. Regardless of the great techniques described in this chapter, there are still two problems that have not been solved to our full satisfaction:

- It is difficult to generate a frequency that is arbitrarily desired.
- It is difficult to switch between frequencies at a fast pace.

Achieving these two goals—arbitrary frequency generation and fast switching—at the same time is an even more difficult task. This difficulty leads us to investigate this crucial question: **All cycles have same length-in-time: Is this the only way to make the clock signal?**

In the field of frequency synthesis, this question has never been asked before. Throughout the rest of this book, it will be our main focus.

BIBLIOGRAPHY

Direct Analog Frequency Synthesis

Gilbert, B. 1982. "A Monolithic Microsystem for Analog Synthesis of Trigonometric Functions and Their Inverses," *JSSC*, vol. 17, no. 6, pp. 1179–1191.

Rokita, A. 1998. "Direct Analog Synthesis Modules for an X-Band Frequency Source," Microwaves and Radar, 1998. MIKON '98., 12th International Conference on, 20–22 May, vol. 1, pp. 63–68.

Stork, M., D. Mayer, and J. Hrusak. 2009. "A Novel Type of Mixer Used for Direct Frequency Synthesis." Digital Signal Processing, 2009 16th International Conference on, 5–7 July, pp. 1–6.

Direct Digital Frequency Synthesis

A Technical Tutorial on Digital Signal Synthesis, 1999. Analog Devices.

Cordesses, L. 2004. "Direct Digital Synthesis: A Tool for Periodic Wave Generation (Part 1)," *Signal Processing Magazine, IEEE*, vol. 21, no. 4, pp. 50–54.

Goldberg, B. G. 1999. *Digital Frequency Synthesis Demystified*, LLH Technology Publishing.

Kroupa, V. F. 1998. *Direct Digital Frequency Synthesis*, IEEE Press.

Indirect Frequency Synthesis (Phase-Locked Loop Based)

Best, R. 2007. *Phase Locked Loops 6/e: Design, Simulation, and Applications*, 6th edition, McGraw-Hill Professional.

Egan, W. F. 1999. *Frequency Synthesis by Phase Lock*, 2nd edition, Wiley-Interscience.

Egan, W. F. 2007. *Phase-Lock Basic*, 2nd edition, Wiley-IEEE Press.

Gardner, F. M. 2005. *Phaselock Techniques*, 3rd edition, Wiley-Interscience.

Goldman, S. J. 2007. *Phase-Locked Loops Engineering Handbook for Integrated Circuits*, Artech House Publishers.

Razavi, B. 2008. *Phase-Locking in High-Performance Systems: From Devices to Architectures*, Wiley-IEEE Press.

CHAPTER 3

TIME-AVERAGE-FREQUENCY

3.1 THE SCALE OF LEVEL AND THE SCALE OF TIME

Viewed from a higher level, all electronic circuits are used for processing information. Within a circuit, information takes the form of a signal. To describe a signal, two scales of measurement are required, as illustrated in the left drawing of Fig. 3.1: the scale of level and the scale of time. Level is used to measure the magnitude (or strength) of the signal. Time is used to record the moment at which that particular magnitude occurs. The medium used for carrying information is electrical voltage or current. An electronic circuit is naturally suitable for handling the *magnitude* of this medium since magnitude is directly proportional to the number of electrons flowing inside the electronic device. Inside VLSI chips, as a result, information is represented through the magnitude of this medium. In other words, an electronic circuit bears an inherent mechanism to differentiate magnitudes. It can use the scale of level readily. By manipulating the magnitude, the VLSI chip can process information and produce results for us to use.

The other scale—time—is very interesting. It is a major subject of religion, philosophy, and science. Among great thinkers, there are two distinct viewpoints on time. One view is that time is part of the fundamental structure of the universe, a dimension in which events occur in sequence. The opposing view is that time does not refer to any kind of container that events and objects move through. Instead, time is part of a fundamental intellectual structure

Nanometer Frequency Synthesis Beyond the Phase-Locked Loop, First Edition. Liming Xiu.

Published 2012 by John Wiley & Sons, Inc.

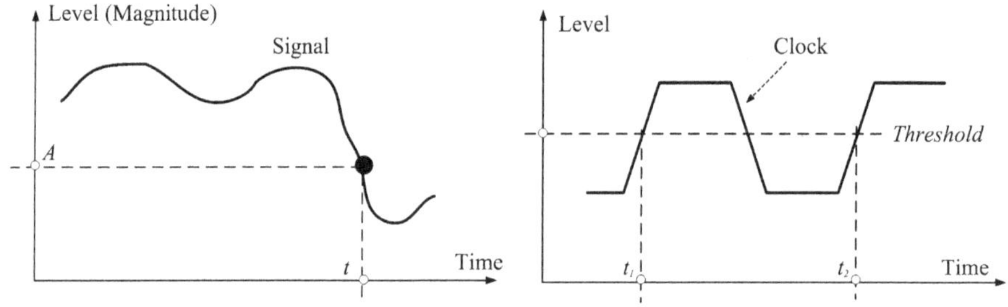

Fig. 3.1. The scale of level and the scale of time (left); the creation of clock signal (right).

(made of *space*, *number*, and *time*) within which humans sequence and compare events. In this second view, time is neither an event nor a thing, and thus is not itself measurable (Jones 2000; Landes 1983; Taylor and Thompson 2008).

In our daily life, time is used to organize events, and this function is achieved through the use of a piece of hardware called a clock or watch. The *second* is a unit of time. It is the base unit of time in the International System of Units (Taylor and Thompson 2008). It can be measured by using a clock or watch. With the technology of the modern atomic clock, it became feasible to define the second itself based on some fundamental property of nature. Since 1967, the second has been defined as the duration of 9,192,631,770 periods of the radiation corresponding to the transition between the two hyperfine levels of the ground state of the cesium 133 atom (http://www.nist.gov/pml/div688/grp50/primary-frequency-standards.cfm).

From this discussion, it is understandable that the electronic circuit is not fundamentally equipped with any mechanism for managing this important scale of *time*. Just like the clock and watch we use in our daily lives, we need something in the circuit world to sense the flow of time. Since level is readily available and can be accurately measured in an electronic circuit, we use level's transition to indirectly imply time. To achieve this, a special signal called "clock" is created and its level threshold crossing is used to record the time. As shown in the right drawing of Fig. 3.1, the *moment* t_1, t_2 (and the like) are used for organizing the events that occur inside the circuit. Within the electronic circuit, how accurately can we record the time solely depends on this clock signal's quality. Frequency synthesis, as the technique of creating the clock signal, is a fundamentally difficult task since it deals with its primary target, time, only indirectly.

3.2 WHAT IS FREQUENCY?

Conventionally, frequency is regarded as the number of occurrences of a repeating event in per unit time (such as 1 second). Period is the duration of one occurrence of such an event. Period is the reciprocal of frequency.

Calculating the frequency of a repeating event can be accomplished by counting the number of times that event occurs within a specific time duration, then dividing the count by the length of this duration.

3.2.1 How Is Frequency Implemented In Circuit Design?

In circuit design, frequency is implemented through the clock signal created in Fig. 3.1. Within the time frame of 1 second, the number of clock pulses occurring in the pulse train is the frequency. The length-in-time of each pulse is the period. From the discussion in Sections 1.3 and 1.4, it is understood that the clock signal controls all the activities inside the chip. Therefore, the frequency defined on the clock signal is the time reference for the entire circuit.

3.2.2 How Is Frequency Used in Electronic System?

From Sections 1.3 and 1.4, the clock signal is used for two functions in an electronic circuit: as a trigger to ignite a logic operation and as a switch to control a sampling system. Therefore, frequency, which is tied to the number of clock pulses found in the time frame of 1 second, serves two purposes:

- Frequency is the indication of the number of operations performed within the time frame of 1 second, such as the number of CPU instructions executed in 1 second, the number of data exchanges that occur between blocks/modules/chips in 1 second, etc.
- Frequency is the number of samples created (continual information becoming discrete samples) between the analog and digital domains within the time frame of 1 second.
- Frequency is the number of discrete samples used within the time frame of 1 second to construct a continual information flow.

3.2.3 "Instantaneous Frequency" and "Instantaneous Period"

As described in the previous section, in an electronic circuit, the concept of frequency is created to describe how fast the clock waveform repeats itself. It is defined as the number of repeatable patterns that exist within a given time window. By definition, it is a concept of long term since a time window of significant duration must be involved. The term "instantaneous frequency" thus is invalid by this definition. But "instantaneous period" is meaningful since it refers to the length-in-time of one particular cycle. Although it is a source of confusion, in many cases, instantaneous frequency has often been used as the reciprocal of instantaneous period for convenience.

3.3 REINVESTIGATING THE FREQUENCY CONCEPT: THE BIRTH OF TIME-AVERAGE-FREQUENCY

The clock signal created in Fig. 3.1 is the driver of all processing circuits on chip. It establishes the time base. It is used to identify when events happened in the past and to schedule when events shall happen in future. The timing information is implied by the exact moment that this clock signal crosses a predefined voltage (or current) threshold. This exact moment of clock edge (such as the t_1 or t_2 in Fig. 3.1) is the triggering point that controls other processing circuits' operation. Among its many characteristics, the most important one for the clock is its waveform repeatability. All the crossing-over moments are required to happen exactly at the same time (relative to each cycle). Any deviation from the ideal position will result in timing error (jitter) and have negative impact on system operations.

When a clock signal is used in an electronic system as the driver of processing units, a system architect prefers the following two features: its frequency shall be able to be arbitrarily set, and the switching from one frequency to another shall be achievable in a short time. These two features could enable the system designer to create a better and more efficient system. In reality, however, these goals have not been fulfilled fully, regardless of all the great circuit techniques developed so far. Direct analog synthesis can switch its output frequency at a fast pace. But the frequency step is coarse, and the associated cost is very high. Direct digital synthesis can achieve fine resolution, but cost has prevented it from being a major player in on-chip frequency synthesis. The indirect phase-locked loop-based (PLL-based) technique inherently has the issue of speed impotence since it uses feedback to direct its output toward input. All things considered, it is believed that the key reason that prevents us from solving these two problems from the root is the belief that "all cycles shall have same length-in-time." This belief makes clock circuit implementation a very difficult task.

> ***"All cycles shall have same length-in-time"; is this the only way in making a clock signal?***

In 2008, the concept of time-average-frequency (TAF) was proposed (Xiu 2008a). It removes the constraint of "all cycles shall have same length-in-time." This move is based on the understanding that frequency is a long-term concept. Its spirit is to differentiate actions by counting the number of events, or operations, that happen within a given time window. As long as a given number of operations are successfully carried out, the entity cannot tell, or doesn't care about, how the operations are performed. Looking at the frequency concept from this angle, "all cycles shall have same length-in-time" is just one way of implementing the time-counting mechanism (one way of constructing the clock signal in Fig. 3.1). Time-average-frequency, which utilizes two or more different types of cycles, is another style that can achieve the same ultimate goal of counting. The definition of frequency and time-average-frequency are listed side by side in below for comparison.

Conventional clock signal	Time-average-frequency-based clock signal
1. The waveform of the clock signal is constructed by a defined pattern that repeats itself indefinitely.	1. The waveform of the clock signal is composed of an infinite number of cycles.
2. The defined pattern has two, and only two, distinguishable voltage levels: one is regarded as high and the other is low.	2. Each cycle has two, and only two, distinguishable voltage levels: one is regarded as high and the other is low.
3. The defined pattern has one rising edge, which is the intermediate state when the signal transits from its low voltage level to high voltage level, and one falling edge that is the intermediate state when the signal transits from a high to low level.	3. Each cycle has one rising edge, which is the intermediate state when the signal transits from a low voltage level to a high voltage level, and one falling edge that is the intermediate state when the signal transits from a high to a low level.
4. Within a given time frame, such as 1 second, the number of times that the defined pattern repeats is defined as the frequency of this clock signal, f.	4. Within a given time frame, such as 1 second, the number of cycles that exist in this clock waveform is defined as the time-average-frequency of this clock signal, f_{TAF}.
5. The time required to complete one such defined pattern is termed as the period of this clock signal, T. By definition, $T \equiv 1/f$.	5. Between any two adjacent rising (falling) edges, the time occupied is defined as the instant period, T, of that cycle, which lies between these two edges.
	6. The size of the instant period, T, of each cycle must be controllable, or must be known by its creator (the clock synthesizer) when in construction.

The trade-off of the extra mental burden associated with time-average-frequency is the implementation flexibility. Mathematically, with the conventional "same length-in-time" approach, there is only one implementation possibility for a given frequency. For example, a frequency of 49.9 MHz can only be realized by using a cycle of 20.04 ns. Employing the time-average-frequency concept, we can use cycle A of 20 ns and cycle B of 20.1 ns to make the 49.9 MHz by the pattern of ABAAB (20*3/5 + 20.1*2/5 = 20.04 ns). We can also use cycle A of 20 ns and cycle B of 20.08 ns by the pattern of AB. In all the cases, the number of clock cycles found within the time window of 1 second is 49,900,000, which is exactly the meaning of frequency 49.9 MHz. Mathematically, there are countless possibilities of constructing this 49.9 MHz

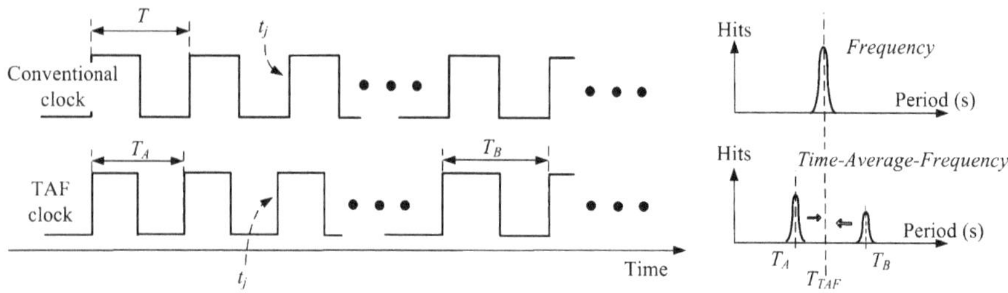

Fig. 3.2. The time-average-frequency time domain characteristic.

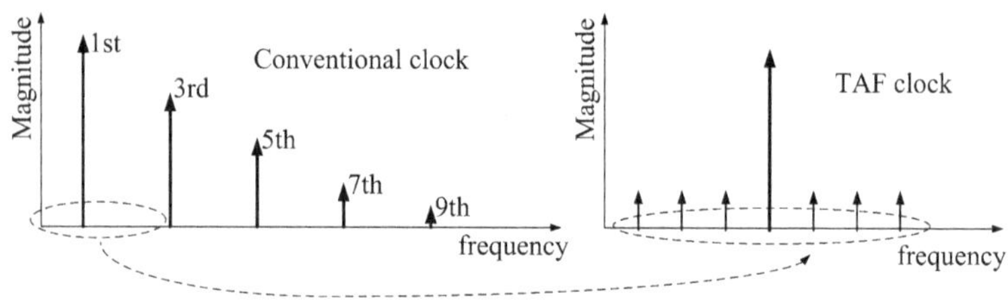

Fig. 3.3. The time-average-frequency frequency domain characteristic.

frequency. This greatly improves the flexibility in circuit implementation. It is worth noting that, at the practical level, cycles A and B shall be carefully chosen so that no significant side effect results when this type of clock is used to drive circuit operation (this will be discussed more in later chapters). Also worth mentioning is the fact that fractional-N PLL does not follow the time-average-frequency concept, although averaging is also used in there. In fractional-N PLL, averaging is used in a closed loop structure; the final output from VCO still targets the "same length-in-time."

Figure 3.2 illustrates the time-average-frequency time domain characteristics through an example. In this case, the time-average-frequency is made of two types of cycles: T_A and T_B. The weight assigned to T_A and T_B are $1\text{-}r$ and r, respectively. According to its definition, the time-average-frequency (period) is calculated as $T_{TAF} = (1 - r)*T_A + r*T_B$.

Time-average-frequency also has its unique frequency domain characteristics. For a conventional clock signal (50% duty cycle square-wave pulse of equal length-in-time), its energy is distributed among first, third, fifth, . . . harmonics as illustrated in Fig. 3.3. For time-average-frequency, besides these harmonics, a certain amount of its energy leaks to some other frequencies (spurious frequency tones). This is due to the fact that the different cycles appear in periodic patterns and they intermodulate each other.

In previous discussions, the time-average-frequency is constructed by using two types of cycles. Generally, more types of cycles can be employed to achieve the desirable average frequency. The general form is expressed below:

$$T_{TAF} = a_1 * T_1 + a_2 * T_2 + a_3 * T_3 + \ldots + a_n * T_n \\ \text{and } a_1 + a_2 + a_3 + \ldots + a_n = 1 \tag{3.1}$$

This general form could be very useful in the analysis of certain applications, such as for frequency matching in digital communication and for spread spectrum clock generation. Unlike many other engineering cases where approximations have to be made here or there, TAF-related analysis can be carried out precisely, without any approximation. This is made possible by the precision built within the time-average-frequency definition and the flying-adder's open loop structure. It is further assisted by two important mathematical tools: *number theory* and *Fourier analysis*. Together, they enable rigorous mathematical treatment of time-average-frequency-based (TAF-based) applications (which will be discussed later).

The time-average-frequency approach is a fundamental breakthrough in frequency synthesis. Its uniqueness lies in the fact that it originates at the concept level. It changes people's historical perspective on clock frequency. Operationally, unlike all previous techniques, it uses mathematically rigorous averaging and open-loop counting to truly solve the arbitrary frequency generation problem: $T_{TAF} = (1 - r)*T_A + r*T_B$. Based on the carefully selected two cycles, T_A and T_B, the weight r can be precisely set to achieve any arbitrarily desired frequency.

3.4 TIME-AVERAGE-FREQUENCY IN CIRCUIT IMPLEMENTATION

To implement the time-average-frequency concept in a clock signal, the circuit architecture style must be in direct (open-loop) fashion as depicted in Fig. 3.4. The capability of directly manipulating the cycle period (length-in-time) is essential. Only in this manner can we precisely control the distribution of the cycles. Moreover, an open loop enables us to adjust the waveform configuration immediately after control command is received, resulting in a fast response. This is impossible with a feedback-based indirect approach.

Before time-average-frequency can be used in real circuit design, a very important issue that needs to be addressed is its impact on timing-closure design constraints. Section 1.3.2 presents the detailed discussion on the

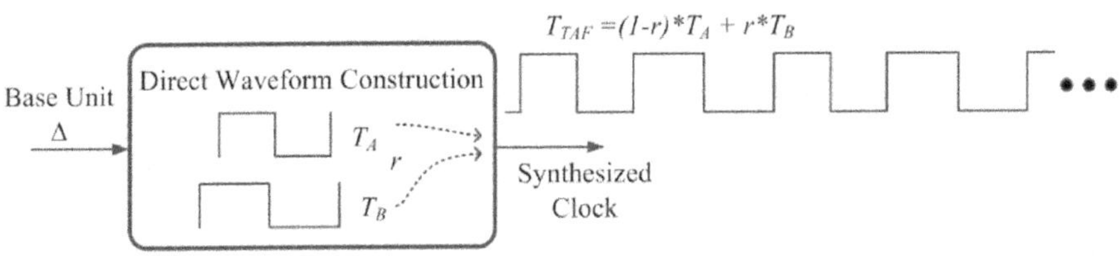

Fig. 3.4. Open loop structure enables precise control and fast response.

relationship between setup/hold check and clock. Based on that discussion, the following can be said:

- The TAF-based clock has no impact on hold check. This is due to the fact that the hold check only uses one edge; it is not period related.
- For a setup check, the constraint should be created based on the shortest cycle in the time-average-frequency clock (the T_A in Fig. 3.2). As such, when the clock is used as a trigger, the circuit will behave exactly the same as when it is driven by a conventional clock of the same frequency.

Flying-adder architecture, which will be presented in Chapter 4 is the first implementation technique that realizes the time-average-frequency in real practice. Early experimental circuits emerged more than a decade ago (Mair and Xiu 2000; Xiu and You 2002). Since then, numerous circuit-level improvements have been developed; flying-adder and flying-adder-like techniques have been used as enablers for innovations in commercial applications for over a decade, especially in video applications where frequency requirements are very complex. Along this path of flying-adder usage, however, the average frequency concept is only used subconsciously. During flying-adder technology development, like many other engineering cases in history, theory lags behind practice. Only after a long period of knowledge accumulation from real practice, the time-average-frequency concept was created in 2008. Although seemingly insignificant, this concept breakthrough is a giant step philosophically. It begins a new way of thinking and points to a new direction that could result in both more powerful and more efficient implementation. In Fig. 3.5, the relationship between flying-adder circuit technology and time-average-frequency theory is illustrated.

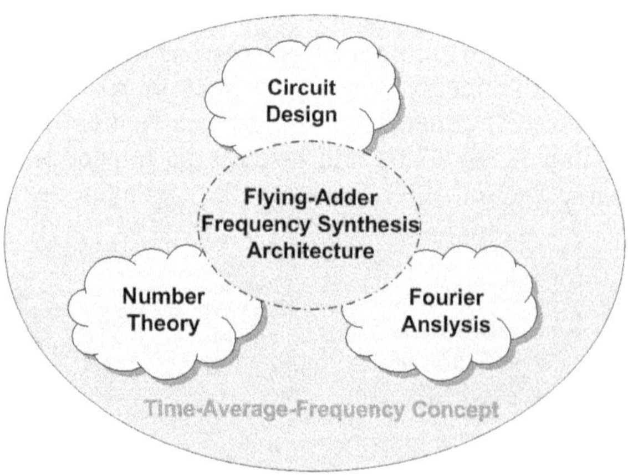

Fig. 3.5. The key components in time-average-frequency-based flying-adder technology.

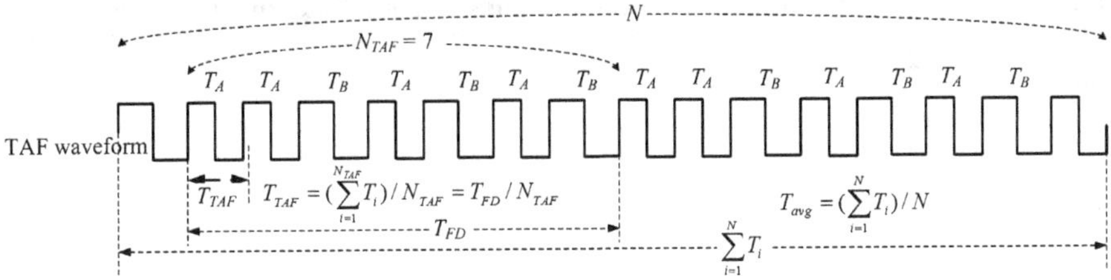

Fig. 3.6. The time-average-frequency-based clock pulse train.

3.5 AVERAGE FREQUENCY, TIME-AVERAGE-FREQUENCY, AND FUNDAMENTAL FREQUENCY

The time-average-frequency concept is rigorously formulated in Xiu (2008a). In this new clocking approach, there are two important concepts: the time average frequency and the fundamental frequency. Figure 3.6 is an illustration of a TAF-based clock pulse train. This pulse train is made up of two types of cycles: T_A and T_B. For a given number of cycles N, according to the discussion in Section 3.2, the conventional *average frequency* (period) f_{avg} is given by

$$1/f_{avg} = T_{avg} = \left(\sum_{i=1}^{N} T_i\right) \Big/ N, \; where \quad T_i = T_A \quad or \quad T_B \tag{3.2}$$

If N_{TAF} is the minimum number of cycles that makes the clock waveform repeat, the *time average frequency* (period) f_{TAF} can be calculated as

$$1/f_{TAF} = T_{TAF} = \left(\sum_{i=1}^{N_{TAF}} T_i\right) \Big/ N_{TAF}, \; where \quad T_i = T_A \quad or \quad T_B \tag{3.3}$$

The *fundamental frequency* (period) f_{FD} is defined as

$$1/f_{FD} = T_{FD} = \sum_{i=1}^{N_{TAF}} T_i, \; where \quad T_i = T_A \quad or \quad T_B \tag{3.4}$$

From the above discussion, it can be seen that the average frequency can be applied on any number of cycles. However, time-average-frequency is strictly defined on N_{TAF}, which makes up the minimum repeatable pattern. Table 3.1 lists a few examples that further illustrate this point.

For the most often used time-average-frequency implementation of two types of cycles, the T_{TAF} can be expressed through the weight r assigned to the cycles. Since it will be used intensively in later chapters, we list it here as Eq. 3.5.

$$T_{TAF} = (1-r) \cdot T_A + r \cdot T_B \tag{3.5}$$

TABLE 3.1. Cycle Pattern, Time-Average-Frequency, and Fundamental Frequency

Repeatable cycle pattern	N_{TAF}	T_{TAF}	T_{FD}	r
ABABABABAB . . .	2	$(T_A + T_B)/2$	$T_A + T_B$	0.5
AAABAAABAAAB . . .	4	$(3T_A + T_B)/4$	$(3T_A + T_B)$	0.25
ABBABBABBABB . . .	3	$(T_A + 2T_B)/3$	$(T_A + 2T_B)$	0.6666667
AAAAAAAAABAAAAA AAAAB . . .	10	$(9T_A + T_B)/10$	$(9T_A + T_B)$	0.1
AAAAAAABBBBBBAA AAAAABBBBBB . . .	13	$(7T_A + 6T_B)/13$	$(7T_A + 6T_B)$	0.4615385

For any number used in a computer system, it can be proven that it must be rational. Therefore, r can be expressed as $r = p/q$, where q and p are integers and their greatest common divisor (GCD) is 1. The integer q actually is the N_{TAF}. Thus, the fundamental frequency (period) T_{FD} can be expressed as

$$T_{FD} = N_{TAF} * T_{TAF} = q * T_{TAF} \tag{3.6}$$

The clock energy distribution profile (frequency spectrum) is completely determined by r, T_A, and T_B. In Sotiriadis (2010a), it is rigorously proven that the majority of energy is located at frequency $f_{TAF} = 1/T_{TAF}$ (for most practical cases). The spurious tones caused by the T_A and T_B intermodulation is spaced at $f_{FD} = f_{TAF}/q$. Further, the strength associated with these frequency stems can be calculated precisely (Sotiriadis 2010c; Xiu et al. 2010). The detailed discussion on this subject will be presented in Chapter 5 (after the circuit is presented in Chapter 4).

3.6 THE NEED OF A THEORY

The discussion in Section 3.5 is helpful in understanding the basic issues related to the time-average-frequency clocking approach. However, to explore its full potential, a theory is needed to guide us in understanding important issues such as those listed below.

1. What is the time-average-frequency clock's time domain behavior?
2. What is the time-average-frequency clock's frequency domain behavior?
3. How can we manipulate the time-average-frequency in the time domain to our advantage? In other words, can we invent some mechanisms to our benefit when a TAF-based clock is used to drive a digital circuit?
4. How can we manipulate the time-average-frequency in frequency domain to our advantage? This issue has great impact when a TAF-based clock is used to drive sampling-oriented application.

5. Can we use the time-average-frequency concept and theory to improve the direct period synthesis circuit?
6. How can we use the time-average-frequency concept and theory to guide us in finding new applications?

Only by understanding these issues at a high theoretical level can we use time-average-frequency with full confidence and apply this new clocking approach for future innovations.

3.7 THE SUMMARY: WHY DO WE NEED TIME-AVERAGE-FREQUENCY?

Chapter 3 has been devoted to the concept of time-average-frequency. The two measuring scales for describing electrical signals are investigated first. The frequency concept is then reviewed. This discussion on frequency leads to the introduction of time-average-frequency. The practical issues related to time-average-frequency are examined. After that, the average frequency, time-average-frequency, and the fundamental frequency are distinguished.

Why do we need time-average-frequency? We want to solve these two long-lasting problems: (1) arbitrary frequency generation and (2) instantaneous frequency switching.

Time-average-frequency theory establishes the foundation for us so that we may possibly solve these problems from the root. In next chapter, we will present the workhorse for time-average-frequency: the flying-adder direct period synthesis architecture.

BIBLIOGRAPHY

Jones, T. 2000. *Splitting the Second: The Story of Atomic Time*, Institute of Physics Pub.

Landes, D. S. 1983. *Revolution in Time*, Cambridge, Massachusetts: Harvard University Press.

Mair, H. and L. Xiu. 2000. "An Architecture of High-Performance Frequency and Phase Synthesis," *IEEE Journal of Solid-State Circuits*, vol. 35, pp. 835–846.

"NIST-F1 Cesium Fountain Atomic Clock: The Primary Time and Frequency Standard for the United States," http://www.nist.gov/pml/div688/grp50/primary-frequency-standards.cfm.

Sotiriadis, P. 2010a. "Theory of Flying-Adder Frequency Synthesizers, Part I: Modeling, Signals' Periods and Output Average Frequency," *IEEE Transactions on Circuits and Systems I*, vol. 57, pp. 1935–1948.

Sotiriadis, P. 2010b. "Theory of Flying-Adder Frequency Synthesizers, Part II: Time and Frequency Domain Properties of the Output Signal," *IEEE Transactions on Circuits and Systems I*, vol. 57, pp. 1949–1963.

Sotiriadis, P. 2010c. “Exact Spectrum and Time-Domain Output of Flying-Adder Frequency Synthesizers,” *IEEE Transactions on Ultrasonics, Ferroelectrics, and Frequency Control*, vol. 57, pp. 1926–1935.

Taylor, B. N. and A. Thompson. 2008. “The International System of Units (SI),” NIST Special Publication 330. pp. 53 ff., http://physics.nist.gov/Pubs/SP330/sp330.pdf.

Xiu, L. 2008a. “The Concept of Time-Average-Frequency and Mathematical Analysis of Flying-Adder Frequency Synthesis Architecture,” *IEEE Circuit And System Magazine*, 3rd quarter, pp.27–51.

Xiu, L. 2008b. “Some Open Issues Associated with the New Type of Component: Digital-to-frequency Converter,” *IEEE Circuit And System Magazine*, 3rd quarter, pp.90–84.

Xiu, L. and Z. You. 2002. “A Flying-Adder Architecture of Frequency and Phase Synthesis with Scalability,” *IEEE Transactions on Very Large Scale Integration Systems*, vol. 10, no. 5, pp. 637–649.

Xiu, L., C. W. Huang, and P. Gui. 2010. “The Analysis of Harmonic Energy Distribution Portfolio for Digital-to-frequency Converters,” *IEEE Transactions on Instrumentation Measurement*, vol. 59, no. 10, pp. 2770–2778.

CHAPTER 4

FLYING-ADDER DIRECT PERIOD SYNTHESIS ARCHITECTURE

4.1 THE WORKING PRINCIPLE

4.1.1 The First Structure

As discussed in Chapter 3 (Section 3.4), the time-average-frequency circuit implementation has to be in direct and open-loop style (Fig. 3.4). Flying-adder architecture, as the workhorse for implementing time-average-frequency in a circuit, uses a plurality of inputs to directly synthesize/construct/assemble pulse-like waveforms with desirable lengths-in-time. These input signals are illustrated in Fig. 4.1. As shown, there are K inputs with the same frequency of $f = 1/T$. These K signals are evenly spaced in one cycle, T. The time difference between any two temporally adjacent signals is, therefore, $\Delta = T/K = 1/(f \cdot K)$. The base unit is Δ, and it will be used in constructing the clock pulses. As an analogue, it is the link that is used to build the chain; it is the brick that is used to build the walls.

Figure 4.2 is the basic circuit structure that shows the flying-adder working principle. As illustrated, the key components in this structure are a $K \rightarrow 1$ multiplex (MUX), a K-bit accumulator with registered output, and a D-type flip-flop (DFF) that is configured as a toggle flip-flop. The K inputs described above are fed into this MUX whose decoding address comes from the output of the accumulator. The output of the MUX is used to trigger the DFF. The DFF's output is the synthesized pulse-like waveform. It is the output clock

Nanometer Frequency Synthesis Beyond the Phase-Locked Loop, First Edition. Liming Xiu.

Published 2012 by John Wiley & Sons, Inc.

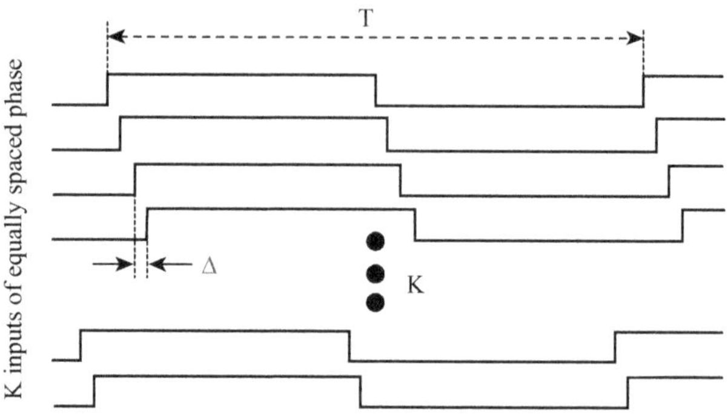

Fig. 4.1. The multiple input signals used for a flying-adder synthesizer.

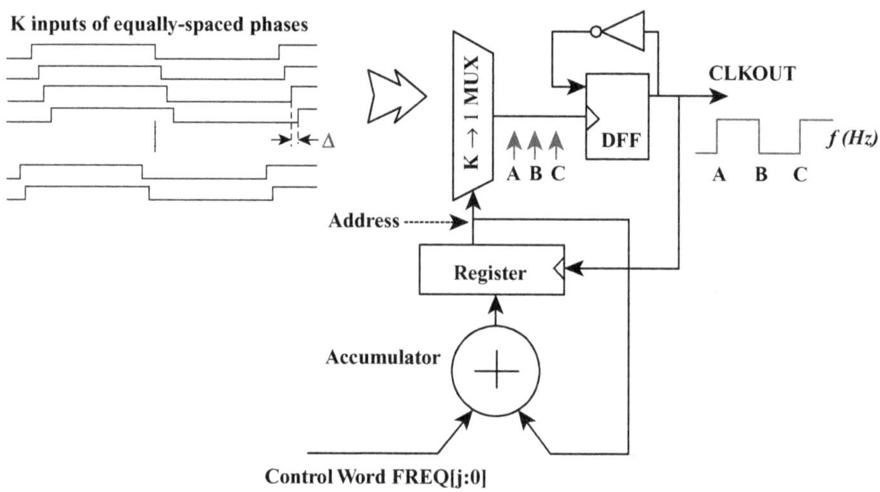

Fig. 4.2. The basic circuit structure that shows the flying-adder principle.

with the desired frequency (period). Moreover, it is also used to drive the register that forms a local feedback. The operation of this circuit is to select a particular input at an appropriate time (through a predetermined schedule) and let it pass through the MUX and trigger the DFF. The predetermined schedule is controlled by the frequency control word FREQ[j:0], where $j = \log_2(K) - 1$. Figure 4.3 is a numerical example where $K = 32$ ($j = 4$) and FREQ[4:0] = 10 = 01010b.

Assume that, initially, the address of the MUX is 0. Thus, the first selected signal out of the K inputs is the 0th. Its rising (or falling) edge will pass through the MUX and trigger the DFF. Also assume that the register is activated both at the rising and falling edge of its driving clock. Therefore, the next address for the MUX will be 0 + 10 = 10, which indicates that the next selected signal will be the 10th input. The following two will be the 20th and the 30th. The

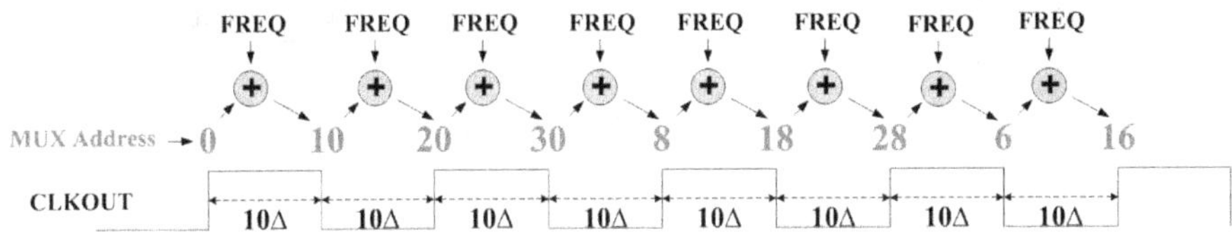

Fig. 4.3. Flying-adder example of $K = 32$ and FREQ[4:0] = 10 = 01010b.

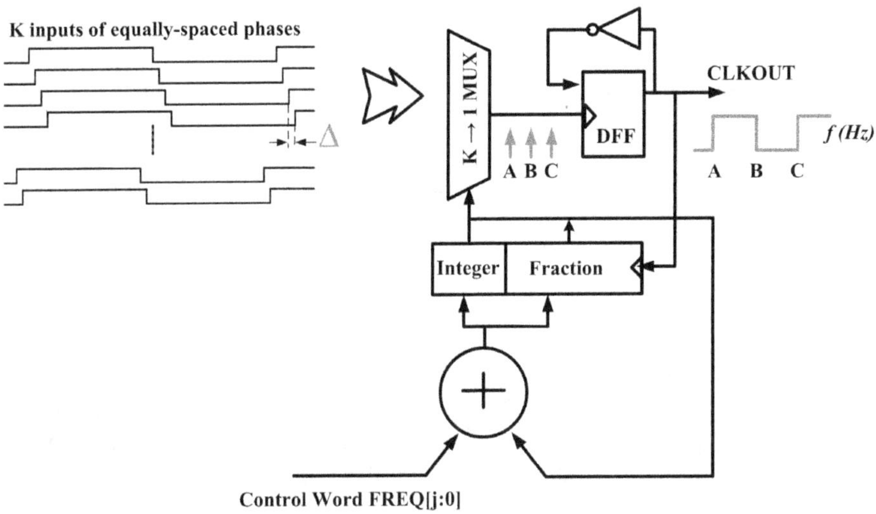

Fig. 4.4. The flying-adder structure with fraction.

next one after 30th is the 8th since it rolls over the modulus $K = 32$. As shown, the time elapse between any two adjacent edges of the output clock *CLKOUT* is 10Δ. The period of the *CLKOUT* is 20Δ. As can be imagined, a different period/frequency can be generated if FREQ is changed to another value. We have assumed that the initial MUX address value is 0, but this is not important. It can be any value since only the value increment is meaningful in operation. From this example, it can be appreciated that the Δ is the basic unit. It is the virtual building block for the waveform.

4.1.2 One Step Forward

The circuit drawn in Fig. 4.4 is one step forward from the one in Fig. 4.2. The key difference lies in the accumulator and its associated register. In this circuit, the accumulator and the register have both integer and fraction parts. However, only the result from the integer part is used to update the MUX address. The fractional result is just used for accumulating "error." In the numerical example of Fig. 4.5, the frequency control word is 10-bits wide. FREQ[9:5] is the integer,

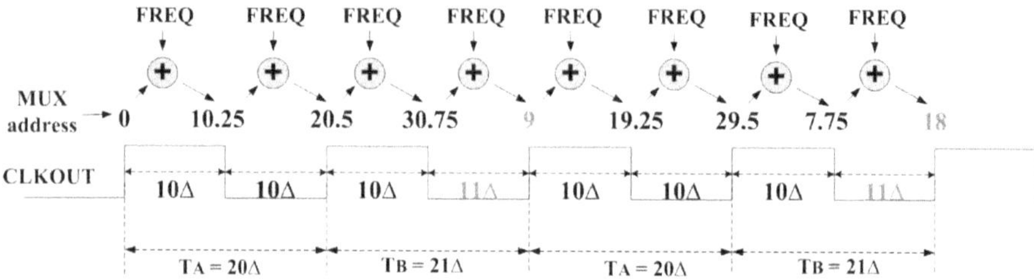

Fig. 4.5. A flying-adder example of K = 32 and FREQ[9:0] = 10.25 = 01010.01000b.

and FREQ[4:0] represents the fraction. In this example, FREQ[9:0] = 10.25 = 01010.01000b. Assume the MUX address's initial value is 0. Its update sequence then is 0, 10.25, 20.5, 30.75, 9, 19.25, ... Among these addresses, only the integer part (5-bits wide) is forwarded to the MUX's decoding circuit. Therefore, the output waveform is the same as that in Fig. 4.3 for the first four edges. As shown, after four times of accumulation, the fraction 0.25 shows its impact (overflow) and the address becomes 9, instead of 8 as in the previous case. Consequently, the synthesized clock edge moves one extra Δ into the future (it is also called cycle prolong).

In this example, there are two different types of cycles: $T_A = 20\Delta$ and $T_B = 21\Delta$. They occur in the pattern of ABABAB . . . . The time-average-frequency (TAF) frequency (refer to Chapter 3) is $T_{TAF} = (20\Delta + 21\Delta)/2 = 20.5\Delta$. The fundamental frequency is $T_{FD} = 20\Delta + 21\Delta = 41\Delta$. It can be understood that, if a different fraction is used, the prolonging of the cycle would happen in a different pattern. It results in a different output frequency (period). This one small step forward in circuit structure creates great potential for generating many more frequencies.

4.2 THE MAJOR CHALLENGES IN THE FLYING-ADDER CIRCUIT

Inspecting the structure in Fig. 4.2, it is obvious that there are at least two major design challenges. This first one is the glitch problem associated with the MUX. The second one concerns the speed of the accumulator. Additionally, although outside the flying-adder circuit itself, the generation of the K inputs presents challenges as well (which will be discussed in detail in Sections 4.2.3 and 4.18).

4.2.1 The Glitch Problem

Figure 4.6 shows a 32 → 1 MUX, which is used to illustrate the glitch problem. The symbol at left is the MUX with a 5-bit address. Assume that at moment t, its address changes from "00000" to "11111," which indicates a switching from IN_0 to IN_{31}. However, in a real circuit environment, there is no

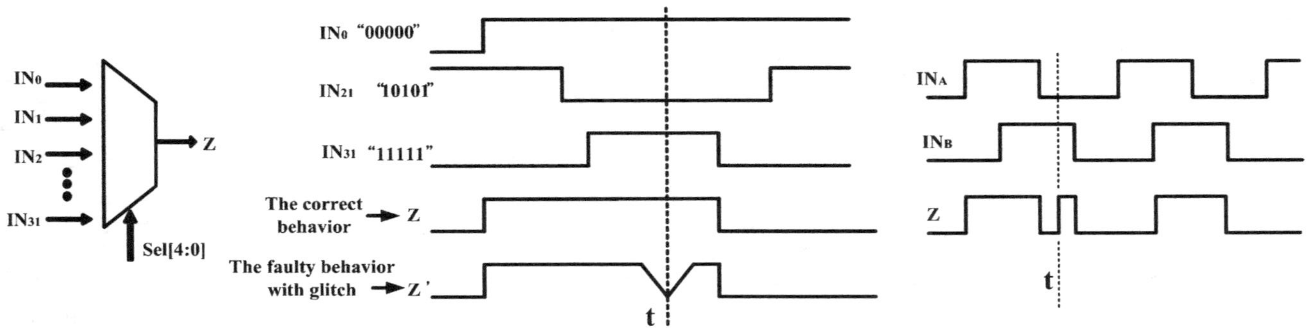

Fig. 4.6. The glitch problem in MUX: the random glitch (middle) and the inherent glitch (right).

guarantee that all the five bits will switch at the same time since each of the five signals in this 5-bit bus is generated and fed to the MUX individually. Therefore, there is the possibility that some intermediate values could result. For example, in a short period, the value presented to the MUX decoding circuit could be "10101" (IN_{21}). The waveforms corresponding to these addresses around the neighborhood of t are depicted in the middle drawing of Fig. 4.6. Ideally, the MUX's output waveform shall bear the characteristic of Z. But the real waveform could be Z′ due to the intermediate value of "10101." As such, there is a glitch produced that is unacceptable for this signal to function as a clock. Since this glitching scenario depends on uncontrollable factors (such as the random intermediate value), it is called a random glitch in this presentation.

Another type of glitch problem could be caused by certain inherent mechanisms. As shown in the right drawing of Fig. 4.6, when the MUX switches its input from IN_A to IN_B, there is a glitch-like waveform produced at its output due to the waveform characteristics of IN_A and IN_B. From a circuit operation point of view, this is a correct behavior. From the functional perspective, however, this should be avoided. This is termed "inherent glitch," which is dangerous as well for Z to function as clock.

4.2.2 The Speed of Accumulator

It is clear from Figs. 4.2 and 4.3 that the flying-adder's output depends on the MUX's address update, which in turn relies on the accumulator. The higher the output frequency is, the faster the address update needs to be. In other words, for high output frequency, a high-speed accumulator is needed. This becomes even more challenging for the structure in Fig. 4.4, where there could be a large number of fractional bits included.

4.2.3 The Generation of the *K* Inputs

The quality of flying-adder circuit's output depends heavily on the quality of the K inputs. In practice, the K inputs are usually generated from a multistage voltage control oscillator (VCO), which is locked to a reference through a phase-locked loop (PLL). This structure is shown in the left drawing of Fig. 4.7. The K inputs can also be generated from a delay-locked loop (DLL) as depicted in drawing on the right of Fig. 4.7. However, compared to a PLL-based approach, the DLL-based method has several drawbacks. First of all, a high-frequency input is needed since a DLL usually does not have frequency multiplication capability. Secondly, the amount of steady state phase error is directly transferred to the error on Δ. For example, assume that the steady state error is amounted at e_{ss}, and there are K stages in the DLL. The following equation, then, must be true, where T_{in} is the input signal's period:

$$e_{ss} + K \cdot \Delta = T_{in} \tag{4.1}$$

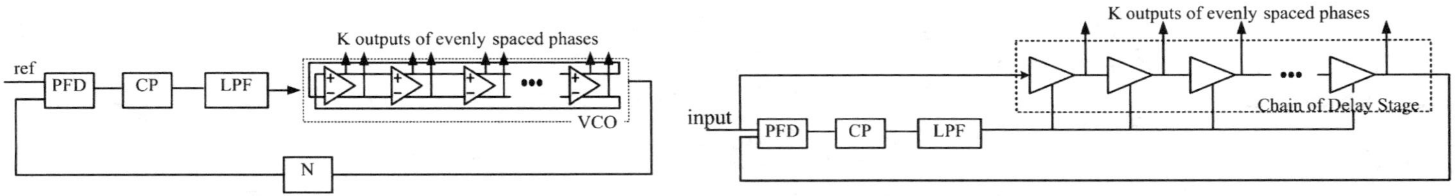

Fig. 4.7. Flying-adder K-inputs generated from a PLL (left) and a DLL (right).

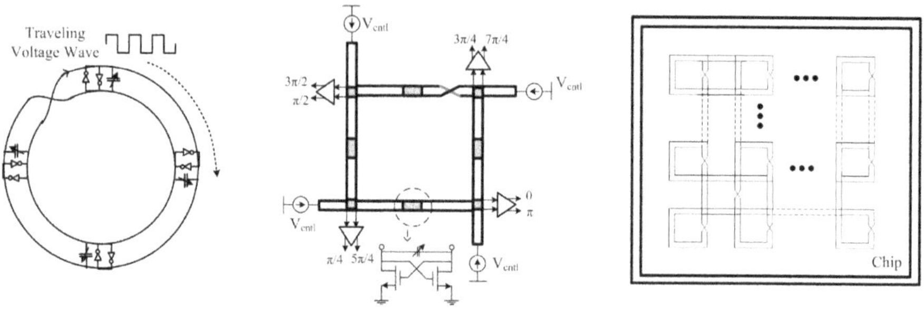

Fig. 4.8. Rotary traveling wave oscillator (RTWO) principle (left), one implementation schematic of RTWO (middle), and multiple RTWO cells to form a rotary oscillator array (right).

This gives us $\Delta = (T_{in} - e_{ss})/K$. As such, the absolute error introduced into Δ is e_{ss}/K. This results in jitter, frequency error, or both in the synthesized output. This is not the case for a PLL, since the VCO itself is an oscillator (a DLL is not a complete ring in itself).

The CMOS-ring-based oscillators in Fig. 4.7 have inferior phase noise performance compared to an LC-tank-based oscillator. To provide cleaner inputs to the flying-adder circuit, one of the options is to use a rotary traveling wave oscillator (RTWO) for generating the K inputs (Chien and Lu 2007; Takinami et al. 2011; Wood et al. 2001). As shown in the left drawing of Fig. 4.8, RTWO is made of a cross-coupled transmission line that bears LC oscillator characteristics. A noise event can trigger a signal between the differential conductors. If no loss is assumed, the resulting signal wave could travel along this ring indefinitely. It provides a full clock cycle for every other rotation. In real implementation, multiple ant-parallel inverter pairs are added to the line to overcome the losses. The oscillation frequency is determined by the transmission line's physical dimension and bias current. In principle, any arbitrary number of phases can be obtained from this oscillation. The drawing in the middle is a circuit diagram of one RTWO implementation. In this example, eight phases are extracted from one RTWO cell. These eight phases can be used as input for a flying-adder synthesizer. Multiple RTWO cells can be arranged to form a rotary oscillator array (ROA) as illustrated in the right-side drawing of Fig. 4.8. The ROA can help ease the clock distribution problem for a large chip. Conveniently, one flying-adder synthesizer can be attached to each RTWO cell in the ROA. This local flying-adder synthesizer can be used to support local circuits.

Another method of LC-based VCO for generating multiple-phase outputs is shown in Fig. 4.9 (Leung and Loung 2004). An LC-VCO can be designed to oscillate at a very high frequency. A chain of four differential latches is configured as a "divide-by-4" divider. As a result, in the perspective of circuit topology, a precise evenly spaced eight outputs can be generated since everything is fully symmetric. In principle, the LC-VCO can be chosen to

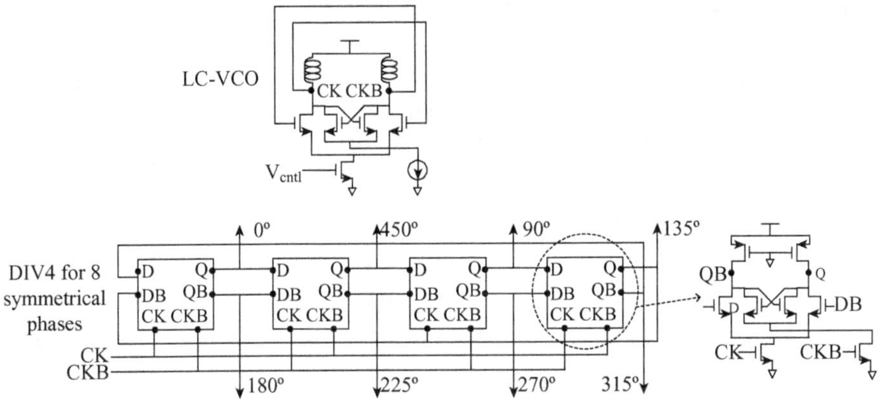

Fig. 4.9. Using an LC-VCO and a differential divider to generate K inputs for a flying-adder.

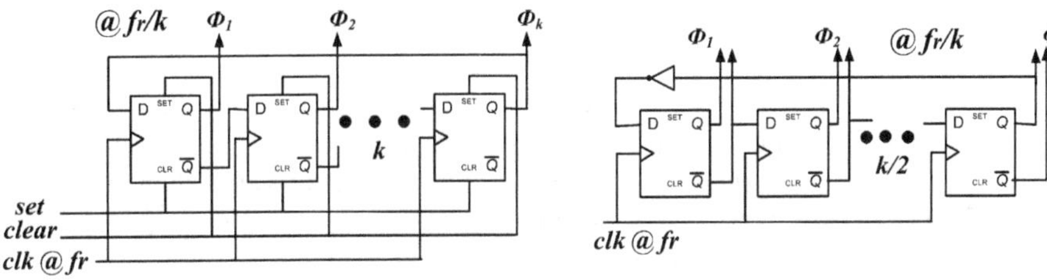

Fig. 4.10. Standard cell-based approach for generating K inputs for a flying-adder.

oscillate at a very high frequency for a small area. A "divide-by-2" divider can optionally be used in between the VCO and the chain of differential latches so that the eight outputs can be in the desired frequency range (the desired Δ value).

Sometimes, the *K* inputs can also be generated without using a PLL. As shown in Fig. 4.10, a chain of flip-flops is capable of producing a group of signals with the same frequency and evenly spaced phases. In the left-hand drawing, if the initial values of the flip-flops are set as 00001111 ($K = 8$), this chain of flop-flips will oscillate at a frequency of $f_r/8$, where f_r is the driving clock's frequency. The outputs from these flip-flops form the eight inputs required for the flying-adder. In the right-side drawing, *K* outputs are available from this chain of *K*/2 flip-flops since both Q and QB are used. The key drawback with these structures is that the *K* outputs' frequency is *K* times lower than the f_r. Also, in the structure on the right, the Δs are not perfectly matched since both Q and QB are used. Nevertheless, the structures in Fig. 4.10 can be very useful for low-frequency application.

4.3 THE CIRCUIT OF PROOF OF CONCEPT

Section 4.1 introduced the circuit of principle. Section 4.2 discussed the problems that have to be solved in real implementation. In this section, a detailed description of the circuit of proof of concept will be presented step by step.

4.3.1 Using Two Paths to Solve the Glitch Problem

The glitch problem is solved by using two paths. This idea is illustrated in Fig. 4.11, where there are two identical blocks: PATH_A and PATH_B. In each of these blocks (compared to Fig. 4.2 or Fig. 4.4), the extra component is the NAND gate. This NAND gate is used to control the $K \to 1$ MUX output. Referring to the CLK1 and CLK2 waveforms on the right, during the period of CLK1 = 1 and CLK2 = 0, the output from MUX_A is blocked, while the output from MUX_B is enabled. In the other period of CLK1 = 0 and CLK2 = 1, the situation is reversed. The motivation of this interlocking mechanism is to prevent the unstable MUX output (caused by the glitch problem discussed in Section 4.2.1) from reaching the final output. Only when the MUX finishes the decoding and its output becomes stable will the interlock release the signal to the circuit that follows. Similar to the structures of Figs. 4.2 and 4.4, there is a DFF configured as toggle flip-flop in each of the blocks. When the particular path is enabled by the NAND gate, the next rising edge from the MUX output will trigger the DFF. Outside the two blocks, there is a pair of NOR and XNOR gates. They both receive the outputs of the two DFFs as their inputs. Therefore, whenever there is a level change in any of the DFF's output, both the NOR and XNOR gates' outputs will toggle. They form the CLK1 and CLK2 signals.

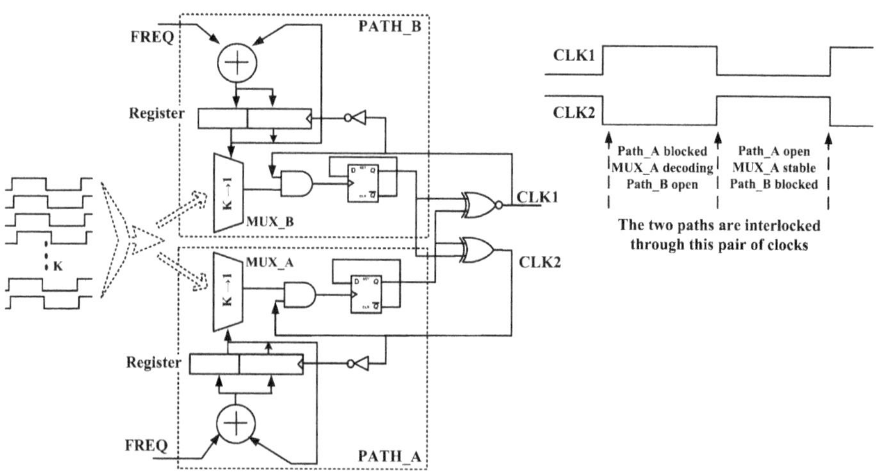

Fig. 4.11. Two paths are used to solve the glitch problem.

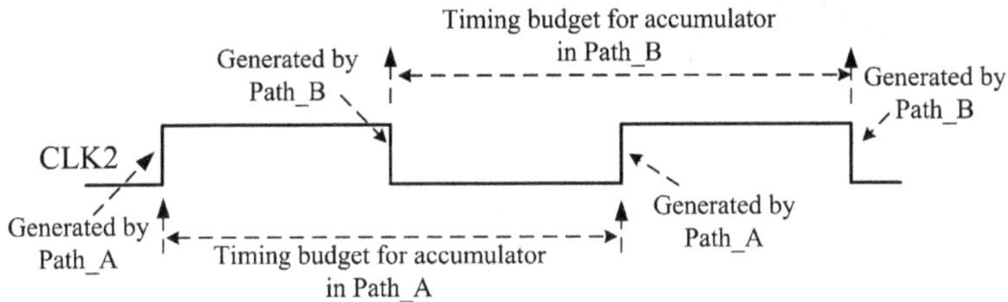

Fig. 4.12. PATH_A and PATH_B are responsible for the output's rising and falling edges, respectively.

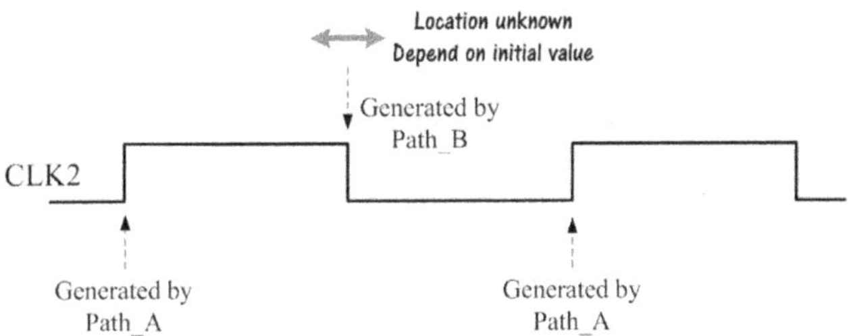

Fig. 4.13. The two paths are unsynchronized. The duty cycle is uncontrollable.

The registers in PATH_A and PATH_B are used to control the accumulators' output. Both the registers are rising-edge triggered. As such, for PATH_A, the accumulator's output reaches MUX_A's decoding circuit only when the MUX_A output is blocked by the associated NAND gate. This is achieved by using CLK2 to control the NAND gate and its inverse (CLK1) to trigger the register; this is the same mechanism for PATH_B. Unlike the case in Figs. 4.2 and 4.4, where the same accumulator register is responsible for generating both the rising and falling edge of the output signal, the rising and falling edges are generated by PATH_A and PATH_B, respectively, in this two-path structure. This is illustrated in Fig. 4.12. As shown, this structure also relaxes the speed constraint on the accumulator. In Figs. 4.2 and 4.4, the accumulator has to run twice the speed of the output frequency since it needs to generate address for both edges. In current circuit, the accumulator in each path is only responsible for one edge.

4.3.2 Synchronize the Two Paths

The structure in Fig. 4.11 successfully solves the glitch problem. It also eases the speed requirement on the accumulator. However, it creates a new problem. As illustrated in Fig. 4.13, the falling edge is unrelated to the rising edge. In other words, the duty cycle is not controllable. This is because the initial values of MUX_A and MUX_B are unrelated. They could be any values after the circuit is powered on. To solve this problem, the structure in Fig. 4.14 is

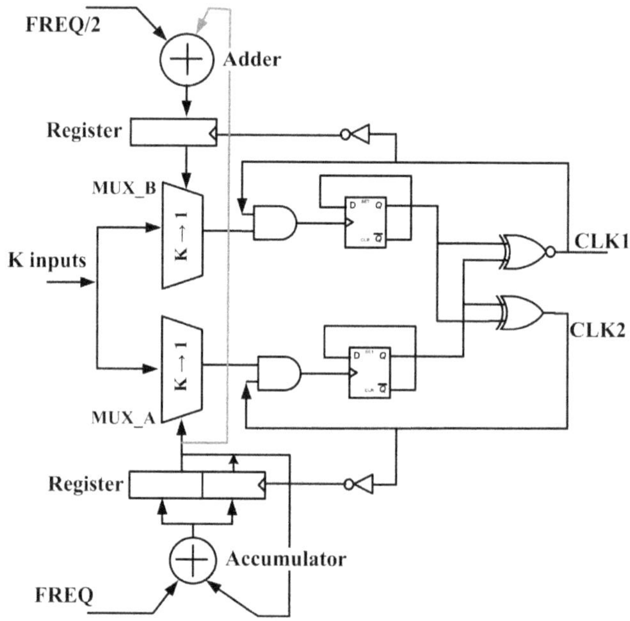

Fig. 4.14. The two paths are synchronized: schematic.

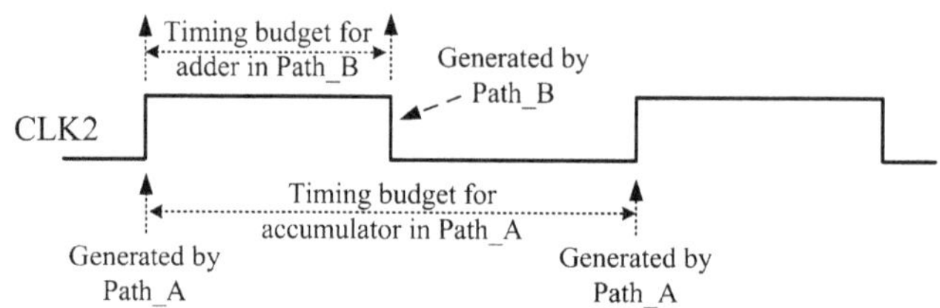

Fig. 4.15. The two paths are synchronized: waveform.

proposed. In this circuit, the MUX_A's address is passed to the PATH_B so that the two MUXs are synchronized. Also, as shown, the frequency control word in PATH_B is half the value of control word in PATH_A. Consequently, the falling edge lies in the middle of the two rising edges. Another minor difference between PATH_A and PATH_B is the size of the accumulator registers. PATH_B only has an integer part. PATH_A has both an integer and a fraction part. This is because the time-average-frequency can be achieved by using just the accumulator in PATH_A. Therefore, the accumulator in PATH_B becomes an adder.

4.3.3 Pipeline for Adder Speed

The circuit in Fig. 4.14 synchronizes the two paths, resulting in a controllable duty cycle. However, as depicted in Fig. 4.15, it tightens the timing budget for

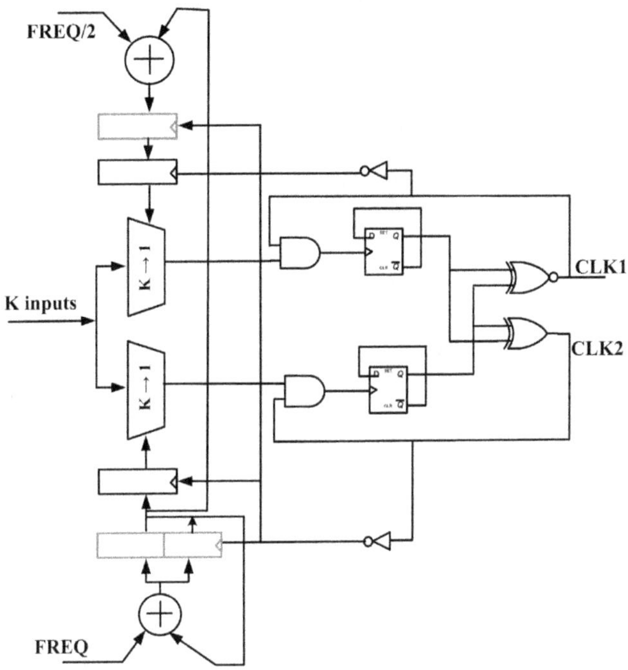

Fig. 4.16. The flying-adder circuit of proof of concept.

the adder in PATH_B. Previously (Fig. 4.11), the accumulators in both paths have a full cycle of time to work. Now, since the address from PATH_A is only available after the rising edge of CLK1 and the adder in PATH_B has to output its address before the falling edge of CLK1, there is only about a half cycle for this adder to work. To deal with this problem, the circuit in Fig. 4.16 is constructed. In this circuit, there is one pipeline stage added. In other words, there are two more registers in the data path. They are all driven by CLK1. Now, the adder in PATH_B has a full cycle of time for its timing budget.

As can be seen from this discussion, from the original circuit of Fig. 4.4, to the circuit of proof of concept in Fig. 4.16, four evolution stages are involved: one path, two paths, synchronized, pipelined. The final circuit bears three characteristics: interlocking between paths, self-clocking, and pipeline. This is the very first flying-adder circuit manufactured (circuit of proof of concept) (Mair and Xiu 2000; Mair et al. 2001).

4.4 THE WORKING CIRCUITRY

The proof-of-concept circuit has been improved to working circuit of Fig. 4.17, primarily for speed advantage. In this circuit, instead of the NAND gates, the interlocking is achieved by a $2 \rightarrow 1$ MUX (MUX_C). The same principle

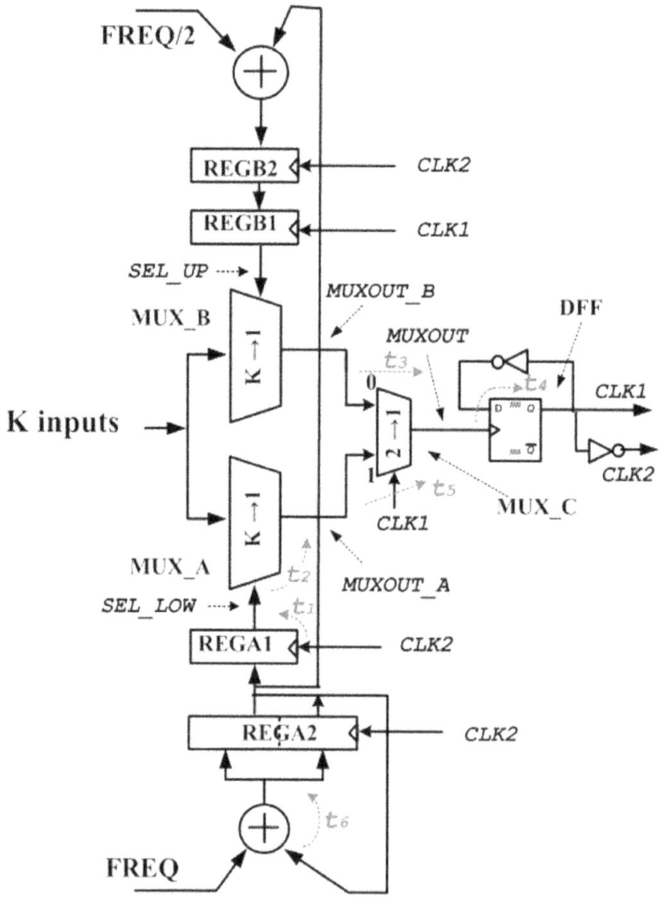

Fig. 4.17. The flying-adder working circuit.

applies in here as well: Their outputs are not used when the two $K \rightarrow 1$ MUXs (MUX_A and MUX_B) are decoding. The pipeline structure is preserved for the same reason of easing the adder speed in PATH_B. The self-clocking mechanism remains the same as in previous structure. The full detail of the circuit will be explained below. The original material is available in Xiu and You (2002); and Xiu and You (2005a).

4.4.1 The Proof of Glitch-Free

There are three MUXs in this circuit, two $K \rightarrow 1$ MUXs and one $2 \rightarrow 1$ MUX. Their outputs all can reach the clock pin of the DFF through the signal of MUXOUT. The proof of glitch-free on this signal will be carried out in two steps: free of random glitch and free of inherent glitch (refer to Fig. 4.6). In the discussion below, it is assumed that all the timing elements (DFF and registers) are rising-edge triggered. Similar discussion can be carried out for falling-edge triggered scenario.

The 2 → 1 MUX does not have intermediate value problem in its address since it only has one address bit. There is the possibility of intermediate values occurring in the address buses of the two $K \rightarrow 1$ MUXs. However, owing to the interlock mechanism discussed in Section 4.3.1, the random glitch has no chance of reaching the DFF. As shown in Fig. 4.17, the two registers (REGA1 and REGB1) that control the two $K \rightarrow 1$ MUXs are driven by *CLK2* and *CLK1*, respectively. The 2 → 1 MUX is controlled by *CLK1*. Thus, when the rising edge of *CLK2* occurs (i.e., *CLK2*'s level will become "high"), MUX_A starts decoding. However, at the same time, MUX_C is selecting the output from MUX_B since *CLK1*'s level is changing to "low" now. This discussion ensures that signal MUX_OUT is free of random glitch.

The next discussion concerns the inherent glitch. For rising-edge-triggered DFF, the trigger event is the signal's low-to-high transition in its CLK pin. This low-to-high transition can be formed by one of the two ways: the transition originating from the VCO (flying-adder input) and the inherent glitch generated from the signal switching of the MUX. The inherent glitch is the one that cannot be passed to the DFF. The necessary condition for a low-to-high inherent glitch is that at the moment of switching, the current signal (IN_A in Fig. 4.6 or in Fig. 4.18) must be low and the future signal (IN_B) must be high. Figure 4.18 is an eight-phase VCO output waveform, which is often used by the flying-adder circuit. Assume that the current signal IN_A for MUX_A is *VCOOUT(0)* (*sel_low*[*2:0*] = "000"). Depending on the next value of *sel_low*[*2:0*], the future signal IN_B could be any of the *VCOOUT*(*0*), *VCOOUT* [1], . . . , *VCOOUT* [7]. By examining those waveforms, none of them satisfies this condition of "IN_A = low and IN_B = high" if IN_B is selected from this current VCO cycle. Assume that next *sel_low*[*2:0*] = "110," which indicates that *VCOOUT* [6] will be scheduled as IN_B (FREQ[3:0] = 6 = "0110"). It can be seen that the only possible rising edge outputted from Z is *End1*, which is originated from the VCO. Since the flying-adder circuit uses two paths, it is possible that the addresses *sel_low* and *sel_up* could hold longer than one VCO cycle (but no more than two VCO cycles). In that case, instead of *End1*, the intended rising edge from *VCOOUT* is *End2* (FREQ[3:0] = 14 = "1110"). The distances between *Start-and-End1* and *Start-and-End2* are 6Δ and 14Δ, respectively. The *End1* will be blocked by the interlock mechanism and will not be passed to Z.

The 2 → 1 MUX (MUX_C) does not generate inherent glitch either. This can be proven by following points:

- The MUX_C's moment of switching is immediately after a low-to-high transition in signal *MUXOUT*.
- The low-to-high transition in signal *MUXOUT* is from *MUXOUT_A*'s or *MUXOUT_B*'s low-to-high transition. This is IN_A. This indicates that IN_A's current level is "high."

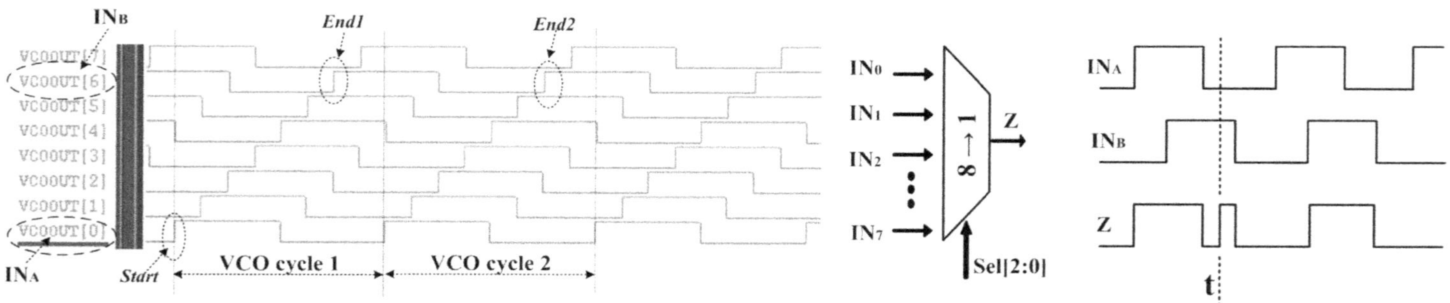

Fig. 4.18. The proof of inherent-glitch-free: flying-adder eight input signals (left), the 8 → 1 MUX (middle), and the MUX output (right).

- IN_A's "high" makes the condition "IN_A = low and IN_B = high" invalid. Therefore, there is no chance of an inherent glitch from MUX_C's switching.

We have proven the flying-adder glitch-free MUX switching from the perspectives of both random glitch and inherent glitch. And we prove it on all the three MUXs. The above discussion is carried out using the $K = 8$ example. It can be generalized to any number.

4.4.2 The Order of the Input Signals

By further investigation it is interesting to note that, in term of flying-adder operation, the order of the K input signals does not matter. In other words, in the left plot of Fig. 4.18, the signals can be labeled 0, 1, . . . , 7 from either top to bottom or from bottom to top. Previous discussion assumes that it is from bottom to top so that inherent glitch will not be produced at the 8 → 1 MUX output. If the order is reversed, then the possibility of glitch does exist. However, following analysis can resolve this issue. The inherent glitch is a short pulse with unintended edges. It can only happen within a small time duration after the address change. The flying-adder's interlock mechanism ensures that the particular MUX's output is immediately blocked after its address change. It is only after a calculated time lapse that it starts to open the path. This interlock circuit does not have "memory." It does not *delay* the signal but *blocks* it. Therefore, all the random glitches and inherent glitches are not seen at the final MUX_C's output.

Besides the glitch problem, the signal order reverse does not have any functional impact since flying-adder operation concerns only the time difference between the selected edges (the number of Δs). Therefore, based on these analyses, flying-adder operation is independent of the input signals' order. This is very convenient for circuit implementation: When a fully symmetric VCO is used, we can add inverters as desired along the delivery path between the VCO and the flying-adder synthesizer to meet the drive strength requirement.

4.4.3 The Analysis of Circuit Speed

The flying-adder structure shown in Fig. 4.17 is a mixed-signal circuit. It must be designed in a high-speed fashion if a high output frequency is desired. The equations listed below are the conditions that must be met for reliable operation.

$$t_1 + t_2 \leqq \min\{T_{c2\to c1}, T_{c1\to c2}\} \tag{4.2}$$

$$t_3 + t_4 + t_5 \leqq \min\{T_{c2\to c1}, T_{c1\to c2}\} \tag{4.3}$$

$$t_6 + t_1 \leqq T_{syn} \tag{4.4}$$

In these equations, T_{syn} is the output signal's period (the desired frequency); $T_{c2\to c1}$ ($T_{c1\to c2}$) is the time elapse from *CLK2*'s (*CLK1*) rising edge to *CLK1*'s

($CLK2$) rising edge. Usually, $T_{c2\to c1} = T_{c1\to c2} = T_{syn}/2$ for a 50% duty cycle when FREQ is an integer and is an even number. When FREQ is an odd number or when a fraction is used in FREQ (TAF clock), $T_{c2\to c1}$ and $T_{c1\to c2}$ have one Δ difference. The variable t_1 is the CLK→ Q time for all the registers (REGA1, REGA2, REGB1, REGB2). The variable t_2 is the MUX_A and MUX_B's decoding time. The variable t_3 is the time required for the signal to pass through MUX_C. The variable t_4 is the DFF's CLK→ Q time. The variable t_5 is the MUX_C decoding time. The variable t_6 is the time required for the accumulator in PATH_A (and the adder in PATH_B) to finish addition. As can be appreciated from these equations, all the variables t_x contribute to the effect of limiting the output frequency. To achieve the highest output frequency, all of them have to be trimmed as much as possible.

4.4.4 The Analysis of Power Consumption

By examining the flying-adder structure in Fig. 4.17, the major power consumers in this circuit are the accumulator/adder and the register. When a large number of bits is used in the accumulator-register pair in the lower path, its power consumption can be significant. Fortunately, the local feedback mechanism (self-clocking) makes the circuit use power in the needed base. As shown in the figure, all the accumulators/adders, registers, and MUXs are driven by the synthesizer's output. As such, the circuit can use power efficiently. Unlike the cases in DDS or in other period synthesizers, where circuits are driven by a fixed high-frequency clock, the flying-adder synthesizer can use less power when outputting low frequency due to this self-clocking.

Moreover, the power consumption is switching activity dependent. Depending on the control word FREQ (thus output frequency), the accumulator/adder-register pair can switch less often or more often. In the case of FREQ = 8, assuming $K = 8$, the accumulator and adder output (and thus the addresses for the two $K \to 1$ MUXs) are constant. The same input signal is always selected, resulting in minimum power usage (refer to Fig. 4.21). In contrast, for the case of FREQ = 9, the accumulator and adder outputs change in every cycle (refer to Fig. 4.20). It might actually use more power than the FREQ = 8 case although its output frequency is lower (9Δ compared to 8Δ).

4.4.5 The Behavioral Simulation

Before being built into a transistor level circuit, the structure in Fig. 4.17 can be modeled in hardware description language (such as VHDL or Verilog) and be simulated at behavioral level. Appendix 4.A contains a complete set of VHDL codes with detailed comments for describing the structure. Figures 4.19–4.21 below are a few simulation examples using this model, on different FREQ settings. In these examples, the VCO frequency is set at 2 GHz ($T_{VCO} = 500$ ps) and it has eight outputs ($K = 8$). Thus, $\Delta = T_{VCO}/8 = 62.5$ ps. The frequency control word FREQ has 32 bits FREQ[31:0]. The MSB 4-bits

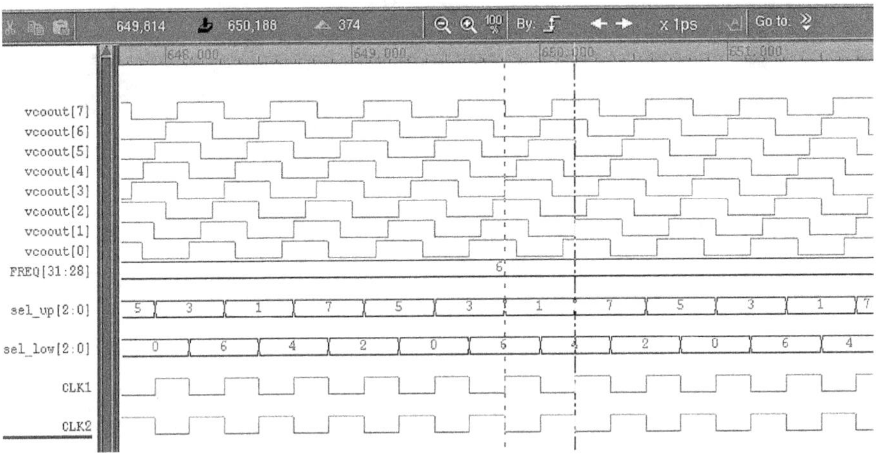

Fig. 4.19. The case of FREQ = 6.0: FREQ[31:28] = 0110b, FREQ[27:0] = 0000000h.

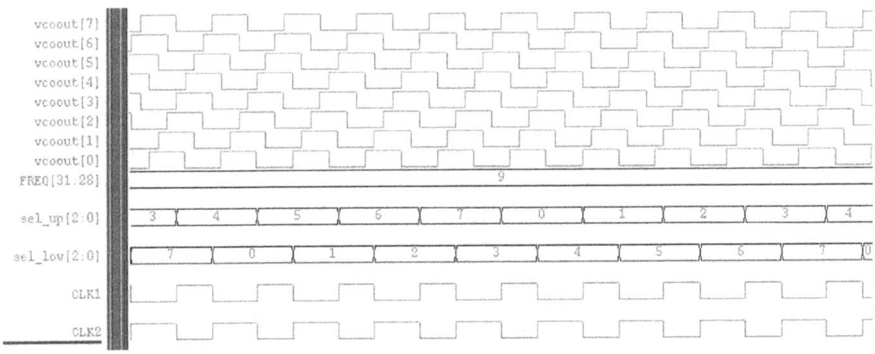

Fig. 4.20. The case of FREQ = 9.0: FREQ[31:28] = 1001b, FREQ[27:0] = 0000000h.

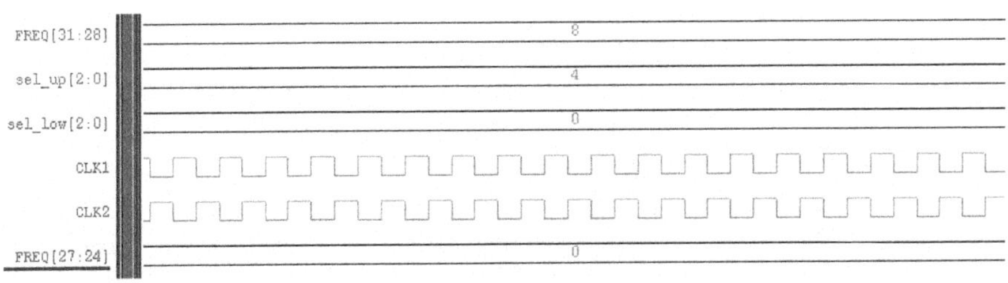

Fig. 4.21. The case of FREQ = 8.0: FREQ[31:28] = 1000b, FREQ[27:0] = 0000000h.

FREQ[31:28] are used for representing the integer: $2 \leqq \text{FREQ} \leqq 2K = 16$. The remaining 28 bits are for fractions.

Figure 4.19 is the case of FREQ = 6.0. The key signals are displayed for illustrating the principle. As can be seen, the *sel_low* sequences at the order of 0, 6, 4, 2, 0, . . . , with the distance of 6Δ at each advance. At the same time,

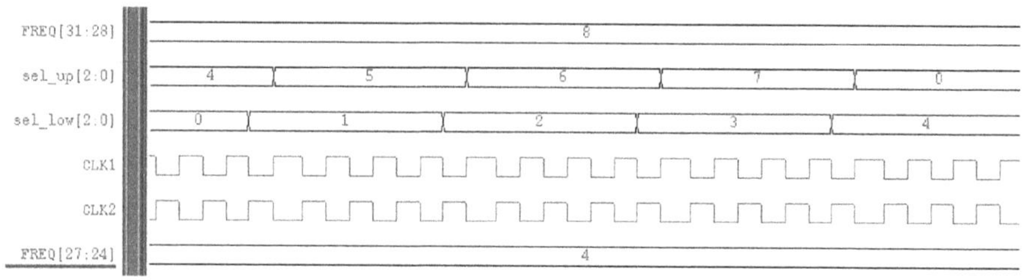

Fig. 4.22. The case of FREQ = 8.25: FREQ[31:28] = 1000b, FREQ[27:0] = 4000000h.

the *sel_up* advances at the order of 3, 1, 7, 5, 3 . . . , also at the distance of 6Δ. It is clear that *sel_low* and *sel_up* are updated alternatively and *sel_up* = *sel_low* + FREQ[31:28]/2. One of them is controlled by *CLK2*, the other by *CLK1*. The output frequency is T_{syn} = FREQ*Δ = 6*62.5 = 375 ps (2.67 GHz). Figure 4.20 shows another example where FREQ = 9.0. The output frequency is 1.778 GHz (562.5 ps). As can be seen, the output duty cycle is not 50/50 but 44/56, because 9 is an odd number.

Figures 4.21 and 4.22 show two interesting cases: FREQ = 8 and FREQ = 8.25. For FREQ = 8, the flying-adder is virtually bypassed since both the *sel_low* and *sel_up* are unchanged (recall that the number of the VCO output is *K* = 8, and the adders are 8-modulus). Clearly, the two addresses stay at 0 and 4, respectively, all the time. In the case of FREQ = 8.25, there is a carry overflow from fraction to integer for every four *add* operations (since the fraction is 0.25). This fact results in a nonperiodic output: for every four output cycles, three of them have lengths of 8Δ and one has a length of 9Δ. The average frequency is 8.25Δ = 515.625 ps (1.939 GHz). Unlike the case in Fig. 4.21, the addresses now change, but they are updated only once after every four output cycles.

For this behavioral model, several important notes are presented below.

1. In a simulation, the initial values on *sel_low* and *sel_up* have to be set before the simulation can start. These values can be any numbers as long as they are not "X." After a few cycles, the circuit will output the correct frequency automatically.
2. For the VCO of *K* outputs, there are virtually 2**K* inputs available for the flying-adder circuit since we use two paths. Therefore, the number of bits required in FREQ to represent the integer part is $\log_2(2*K) + 1 = \log_2(K) + 2$. For example, for an 8-phase VCO, there are 16 inputs available for the flying-adder. Thus FREQ[x+5:x] is needed (where "x" is the number of bits reserved for a fraction). However, since the minimum value for FREQ is 2, the "all-zero" code is invalid. For better efficiency, the circuit is designed in such way that the "all-zero" is used to present the 2**K*. In other words, in the *K* = 8 example, "0000" is used to represent 2*K* = 16 (instead of using "10000"). By doing this, the integer part only

needs $\log_2(2*K)$ bits; one bit is saved in FREQ. Therefore, the general format for FREQ is FREQ[$\log_2(K)$+1+x:0].

3. For the VCO of K outputs, the size of the MUX_A and MUX_B is $\log_2(K)$. The sizes of REGA1, REGB1, REGB2, and the adder in PATH_B are all $\log_2(K)$. The sizes of REGA2 and the adder in PATH_A are $\log_2(K)$+x since factional accumulation is performed here. As can be seen, these registers and adders are all the size of $\log_2(K)$ where only the integer part is concerned. This is one bit less than that required to represent the $2K$ inputs. The circuit is designed in such way that the MSB is implicitly embedded; it is invisible to the circuit designer. For this reason, the suggested method of connecting the FREQ[$\log_2(K)$+1+x:0] to the adders are as follows: FREQ_PATH_A takes FREQ[$\log_2(K)$+x:x] (throw away the MSB bit), and FREQ_PATH_B takes FREQ[$\log_2(K)$+1+x:1+x]. This is equivalent to taking FREQ's MSB $\log_2(K)$ bits and shifting them to the right by one bit position. This operation divides the integer part by 2. Therefore, in previous examples of $K = 8$ and FREQ[31:0], FREQ[30:0] is sent to FREQ_PATH_A, and FREQ[31:29] is passed to FREQ_PATH_B.
4. When the FREQ's integer part is an odd number, the duty cycle will not be 50/50 because the flying-adder output's *high* and *low* will be different by one Δ. The current suggestion of FREQ_PATH_A taking FREQ[$\log_2(K)$+1+x:1+x] (example: FREQ[31:29]) makes the *low* one Δ longer than *high*. When the overflow from the fraction occurs, it will make *low* one more Δ longer. In order to balance this, FREQ_PATH_A can take the value of FREQ[$\log_2(K)$+1+x:1+x] + FREQ[x] [example: FREQ[31:29] + FREQ [28]].
5. The circuit is designed in such way that it will never lock itself, although the self-clocking structure is embedded. This is because in a real environment, all the signals in the *sel_low* and *sel_up* bear certain voltage levels, which are translated into conclusive address codes (not Xs). Therefore, as long as the VCO outputs are available, there is always a rising (or falling) edge coming out of MUX_C (although it might not be the intended edge). This will trigger the DFF and consequently produces transitions in *CLK1* and *CLK2*. When a valid FREQ is presented to the adders, the correct output will be generated after, at most, two cycles.

4.4.6 The Extension to Multipaths

The circuit in Fig. 4.17 has another advantage over the structure in Fig. 4.16: It has scalability. It can be expanded to the multipath to further ease the speed constraint on the adders. Figure 4.23 shows a structure with four paths. The glitch-free is still achieved through interlocking: When any of the $K \rightarrow 1$ MUXs (MUX1–4) is doing the decoding, its output is not used. MUX5 has two address bits, which bear the potential risk of a random glitch. To overcome

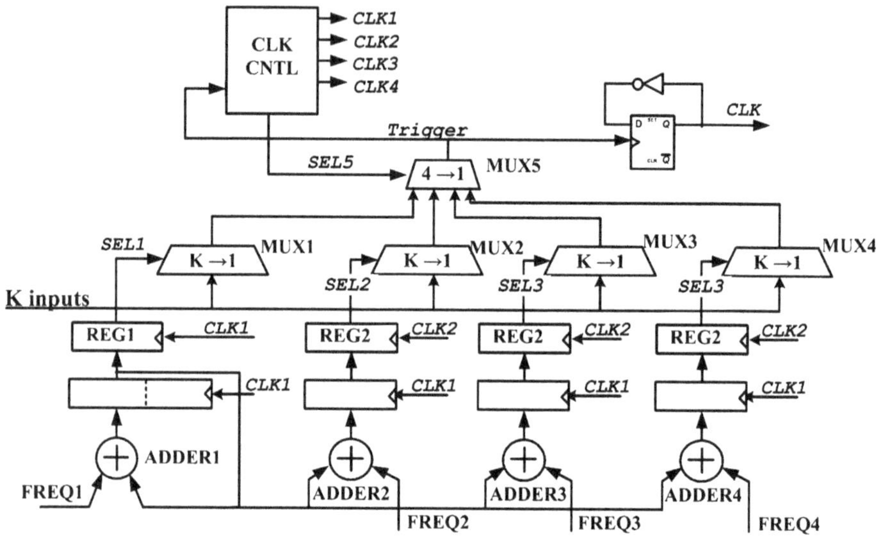

Fig. 4.23. The flying-adder structure of four paths.

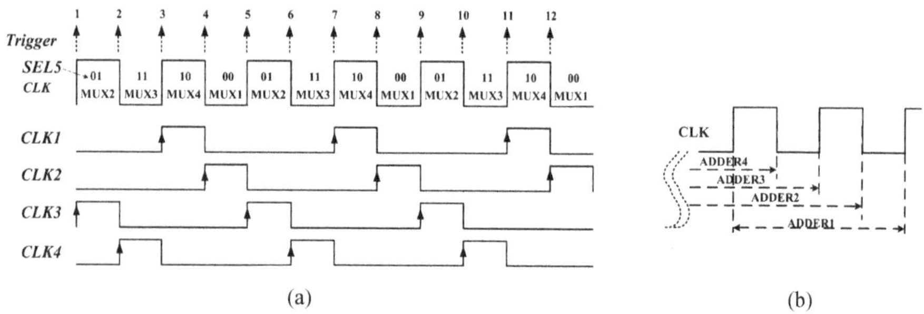

Fig. 4.24. (a) Output signals of CLK_CNTL and (b) the illustration of the adders' timing constraint.

this problem, the Gray code style is used in its address decoding. In other words, at any given moment, there is at most one bit switched. This is possible since the MUX5's sequence of selecting is fixed and known: MUX1 $\rightarrow$ MUX 2 $\rightarrow$ MUX3 $\rightarrow$ MUX4 $\rightarrow$ MUX1 .... For all of the four $K \rightarrow 1$ MUXs (MUX1–4), we cannot do this since their selection sequence is not fixed but depends on FREQ. The CLK_CNTL block is responsible for generating the *CLK1–4* and *trigger* signals. It is also used for producing the decoding signal *SEL5* for MUX5 (in Gary code). This block's outputs are depicted in Fig. 4.24a. The speed of *trigger* is four times faster than that of *CLK1–4*. The Gray code style is visible in *SEL5*'s value. In Fig. 4.24b, it can be seen that each adder has two output cycles to finish its operation. The circuit details can be found in Xiu and You (2005b, 2008).

4.5 FREQUENCY TRANSFER FUNCTION, FREQUENCY RANGE, FREQUENCY RESOLUTION, AND FREQUENCY SWITCHING SPEED

From the previous discussions on the flying-adder working principle, flying-adder circuit structure, and time-average-frequency, the flying-adder synthesizer's frequency transfer function can be derived in Eq. 4.5. Since fractions can be used in the control word FREQ, this transfer function is valid in the time-average-frequency sense. It is obvious that the period transfer function is linear, but the frequency transfer function is in the style of *1/x*. For this reason, flying-adder direct period synthesis is the more appropriate name for this architecture.

$$T_{syn} = FREQ \cdot \Delta \text{ or } f_{syn} = 1/(FREQ \cdot \Delta) = (K/FREQ) \cdot f_{vco} \tag{4.5}$$

From discussion in Section 4.4, it is known that FREQ can take values in the range of [2, 2*K*]. Therefore, the output frequency range is derived as:

$$1/(2 \cdot K \cdot \Delta) \leqq f_{syn} \leqq 1/(2 \cdot \Delta) \tag{4.6}$$

Based on Eq. 4.5, the equation for frequency resolution (Eq. 4.7) can be derived where FREQ has been replaced by *F* for brevity and x is the number of bits used in fraction. The minimum change *F* can take is *dF*. It is decided by the number of fractional bits used in a particular design. The more fractional bits are used, the finer the resolution can be achieved. In other words, its size is one LSB change. For a 20-bits-fraction design, *dF* is $2^{-20} = 0.00000095$. Thus, if $F = 10$, the frequency resolution is 0.095 ppm. It will be 0.006 ppm in the case of 24 bits. Another fact is that the flying-adder's frequency resolution is frequency dependent (it depends on *F*).

$$df_{syn} = \frac{1}{F^2 \cdot \Delta} dF = 2^{-x} \cdot \Delta \cdot f_{syn}^2 \quad \text{and} \quad \frac{df_{syn}}{f_{syn}} = \frac{dF}{F} \tag{4.7}$$

From the circuit in Fig. 4.17, the switching speed can be analyzed. As depicted in Fig. 4.25, whenever there is a change of value in FREQ, it will be

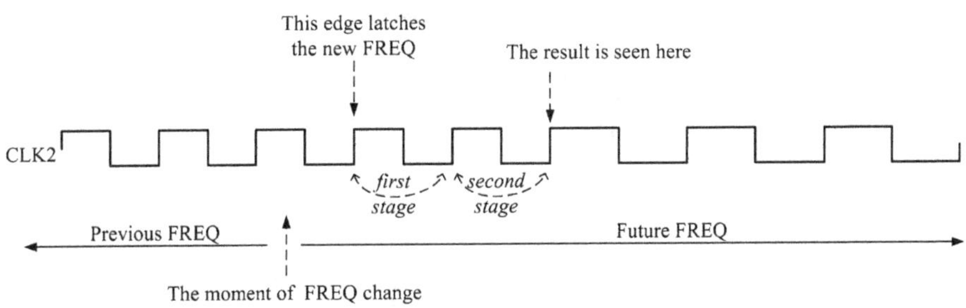

Fig. 4.25. The flying-adder switching speed illustration.

latched into the system by *CLK2* through its rising edge. The result will be seen immediately in the third CLK2 edge. The flying-adder circuit's response speed is virtually instantaneous (thanks to the open loop). It takes only two cycles for its frequency (or period) to change. The two cycles required is due to the two-stage pipeline.

4.6 THE TECHNIQUE OF POST DIVIDER FRACTIONAL BITs RECOVERY

4.6.1 Post Divider Fractional Bits Recovery (PDFR)

When fractional bits are used in a flying-adder synthesizer, the resulting waveform uses two types of cycles: T_A and T_B. When the waveform is examined cycle by cycle, it is nonperiodic and is in time-average-frequency fashion. However, when a post divider is used after the flying-adder synthesizer, the resulting waveform after the divider can become periodic. This is illustrated through an example in Fig. 4.26. In this case, the control word FREQ = I.25, where I is an integer. Consequently, the resulting waveform is made of $T_A = I \cdot \Delta$ and $T_B = (I + 1) \cdot \Delta$. There are three T_A and one T_B for every four cycles, because the fraction is 0.25. If an $M = 4$ post divider is attached after the synthesizer, it can be seen that its output has only one type of cycle: $3T_A + T_B$. In other words, the fraction is recovered. The output is purely periodic now.

This technique has been called post divider fractional bits recovery (PDFR) (Xiu 2006, 2007). In general, for a post divider with divide ratio of M, there are M-1 fractions that are applicable to this technique: 1/M, 2/M, . . . , (M-1)/M. All these fractions can be used safely in a flying-adder synthesizer's frequency control word when the post divider ratio is set at M. The output waveform after the post divider is periodic. PDFR will be discussed in later sections to illustrate its usage for producing TAF-spur-free frequencies.

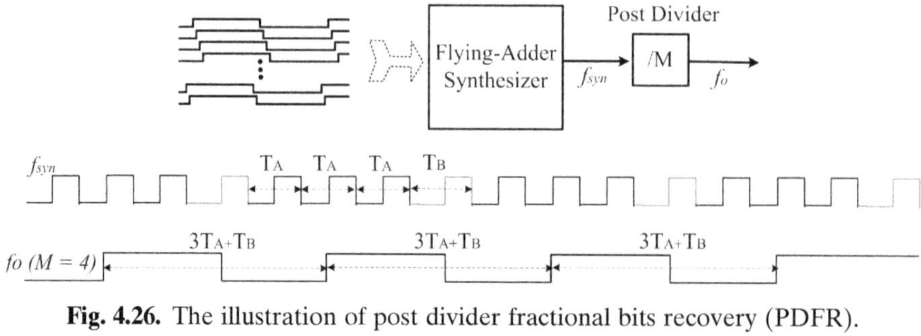

Fig. 4.26. The illustration of post divider fractional bits recovery (PDFR).

4.6.2 PDFR for Virtually Boosting the Number of Inputs *K*

The effect of recovering fractions by PDFR can be viewed from another angle: It virtually increases the number of flying-adder inputs K. Assume that a frequency control word FREQ = $I + r$ is used for generating a frequency $f = 1/T = 1/(\text{FREQ}\cdot\Delta)$. I is the integer in the range of $2 \leqq I \leqq 2K$ and r is the fraction $0 \leqq r < 1$, $r = r_{-1}2^{-1} + r_{-2}2^{-2} + \ldots + r_{-p}2^{-x}$ in an x-bits system. If a post divider of ratio M is attached after the flying-adder circuit, the output period (frequency) becomes $T_{post} = M\cdot T = (M\cdot I + M\cdot r)\Delta$. If we chose M to be $M = 2^m$, T_{post} becomes $(2^m\cdot I + r_{-1}2^{m-1} + r_{-2}2^{m-2} + \ldots + r_{-p}2^{m-x})\Delta$. From the expression of T_{post}, it can be seen that we virtually increase the integer range by the number of m-bits since the first m bits in the original fraction are now moved into the integer part. In other words, the K inputs are virtually boosted to $M\cdot K$ inputs. The following numerical example is useful for further illustrating this point. In this example, there are eight inputs for the flying-adder synthesizer ($K = 8$), and the FREQ uses six bits: FREQ[5:2] for the integer and FREQ[1:0] for the fraction. A post divider of $M = 4$ is attached after the synthesizer.

As can be seen from Table 4.1, the post divider $M = 4 = 2^2$ virtually converts the two fractional bits into an integer. This is equivalent to making the original 8 inputs into 32 inputs. However, in the new "32-inputs" flying-adder synthesizer, instead of starting from $I = 2$, the frequency control word has to start from $2M$. The first few codes (and thus their associated frequencies) are unavailable. The direct result of this approach is the lowered output frequency (since it is after a post divider).

This technique of virtually boosting the number of inputs is very useful in real implementation. It allows the designer to use a less complex structure

TABLE 4.1. PDFR for Virtually Boosting the Number of *K* Inputs

FREQ[5:2] integer part	FREQ[1:0] fraction part	T = FREQ·Δ original	K original	$T_{post} = M\cdot T$ after $M = 4$	FREQ_{new}[5:0] after $M = 4$	K_{new} after $M = 4$
$0010_2 = 2$	00	2Δ	8	8Δ	$001000_2 = 8$	32
$0010_2 = 2$	01	2.25Δ	8	9Δ	$001001_2 = 9$	32
$0010_2 = 2$	10	2.5Δ	8	10Δ	$001010_2 = 10$	32
$0010_2 = 2$	11	2.75Δ	8	11Δ	$001011_2 = 11$	32
$0011_2 = 3$	00	3Δ	8	12Δ	$001100_2 = 12$	32
$0011_2 = 3$	01	3.25Δ	8	13Δ	$001101_2 = 13$	32
$0011_2 = 3$	10	3.5Δ	8	14Δ	$001110_2 = 14$	32
$0011_2 = 3$	11	3.75Δ	8	15Δ	$001111_2 = 15$	32
...	...	...	8	...	...	...
$1111_2 = 15$	00	15Δ	8	60Δ	$111100_2 = 60$	32
$1111_2 = 15$	01	15.25Δ	8	61Δ	$111101_2 = 61$	32
$1111_2 = 15$	10	15.5Δ	8	62Δ	$111110_2 = 62$	32
$1111_2 = 15$	11	15.75Δ	8	63Δ	$111111_2 = 63$	32

when designing the flying-adder synthesizer. A fewer number of inputs can significantly reduce design complexity. A large K uses more resource (such as a larger VCO and a larger multiplex). It also makes the circuit slower. Usually in design, the M in Fig. 4.26 is made programmable so that the first few missing codes (and their corresponding frequencies) can become accessible by changing the M. For example, in Table 4.1, the missing FREQ codes from the range of $2 \leqq \text{FREQ} \leqq 8$ will be available if M is set back to 1. For this reason, in recent flying-adder implementations, more than eight inputs are hardly used for high-performance design. K greater than 8 is often used in low-frequency applications.

4.6.3 The Effective Fraction after Post Divider

When it is considered as part of a flying-adder synthesizer, the post divider affects the synthesizer's frequency control word. If the original control word is expressed as $F_{org} = I + r$, the equivalent control word after the post divider of M is $F_{new} = M{\cdot}(I + r) = M{\cdot}I + M{\cdot}r$. The integer part becomes $I_{new} = \text{floor}(M{\cdot}I + M{\cdot}r)$; the resulting new fraction is $r_{new} = F_{new} - I_{new} = M{\cdot}r - \text{floor}(M{\cdot}r)$. However, regardless the value of M, the flying-adder synthesizer's basic timing unit is still Δ. The divider is an edge selector, and while it discards certain clock edges, it does not change time resolution.

4.7 FLYING-ADDER PLL: FAPLL

In many practical cases, the multiple inputs needed by the flying-adder synthesizer are generated from an integer-N PLL as depicted in Fig. 4.27. This structure is so commonly used that it is given the name of flying-adder PLL (FAPLL) (Xiu 2007; http://focus.ti.com/lit/ug/sprugx9/sprugx9.pdf; http://focus.ti.com/lit/ug/sprugx7/sprugx7.pdf; http://focus.ti.com/lit/ug/sprugx8/sprugx8.pdf). The usage of PLL in this structure provides several advantages:

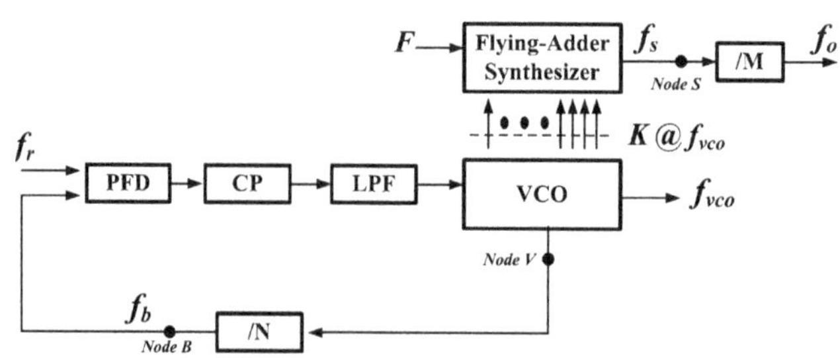

Fig. 4.27. The block diagram of flying-adder PLL.

- The multiple inputs, which are usually high frequency, can be generated from a relatively low-frequency f_r.
- The multiple inputs can be conveniently generated from a multiple-delay-stage VCO, which is very common in modern CMOS VCO design.
- The frequency of the multiple inputs can be adjusted through the PLL (i.e., the size of Δ can be adjusted). When combined with the flying-adder synthesizer, it creates a very powerful frequency generation machine.

From a reference frequency of f_r, the VCO frequency can be calculated as $f_{VCO} = N \cdot f_r$. Therefore, $\Delta = T_{VCO}/K = 1/(K \cdot N \cdot f_r)$. From the flying-adder transfer function of Eq. 4.5, it is obtained that $f_s = (K \cdot N \cdot f_r)/F$. Taking the M post divider into consideration, the FAPLL's output–input frequency relationship can be derived in Eq. 4.8.

$$f_o = \frac{K \cdot N}{F \cdot M} f_r \tag{4.8}$$

4.8 FLYING-ADDER FRACTIONAL DIVIDER

A frequency divider is used to divide a high-frequency f_i into low-frequency f_o. One low-frequency cycle contains a number of high-frequency cycles. The number is determined by the divider's divide ratio. In this operation, the base unit is one f_i cycle. When viewed from this angle, the flying-adder synthesizer can also be viewed as a divider. Moreover, this divider can reach a resolution in time that is finer than one f_i cycle. This can be illustrated with the assistance of Fig. 4.28. From Section 4.5, Eq. 4.9 can be easily derived. It is clear that this divider's time resolution is Δ, or T_{VCO}/K, which is finer than T_{VCO}. In this sense, this divider is termed a fractional divider since it can reach inside the input signal's cycle. Similar to the fact that a frequency divider is often called a frequency counter, the flying-adder fractional divider can also be called a phase counter. In a frequency counter case, the base unit is one cycle. In a phase counter, the base unit is one cycle divided by K. The right side of Fig. 4.28 shows the difference between the two. From a base VCO frequency, a

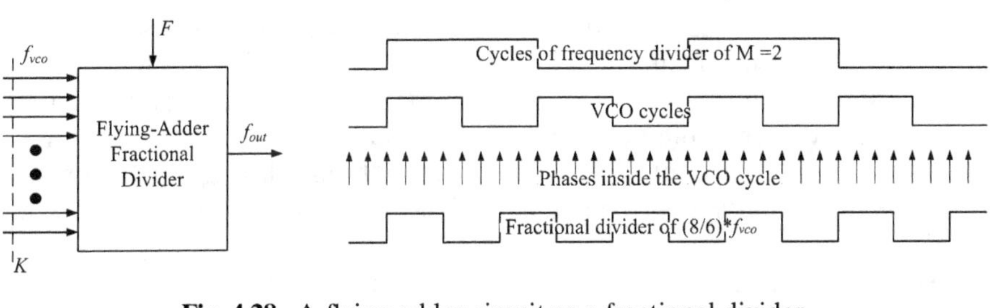

Fig. 4.28. A flying-adder circuit as a fractional divider.

frequency divider of $M = 2$ has its output waveform displayed at the top. Every cycle in its waveform bears the length-in-time that is twice of one VCO cycle. On the other hand, the fractional divider has a cycle length of 6/8 of one VCO cycle ($K = 8$ and $F = 6$ in Eq. 4.9) since it can reach the inside of a VCO cycle through individual phases. This fractional divider, or phase counter, concept will be used in the following integer-flying-adder discussion.

$$f_{out} = \frac{K}{F} f_{vco} \quad \text{or} \quad T_{out} = F \cdot \Delta = \frac{F}{K} T_{vco} \tag{4.9}$$

One important observation from Eq. 4.9 is that, when $F < K$, the flying-adder synthesizer will produce a frequency that is higher than the input frequency. This unique feature is only available in the flying-adder synthesizer. In PLL-based techniques, the highest frequency point is always at its VCO.

4.9 INTEGER-FLYING-ADDER ARCHITECTURE

In the previous flying-adder discussion, the frequency control word FREQ (or F) includes both integer and fraction. As a result, time-average-frequency is outputted when the fractional part is not zero. Inspected from cycle to cycle, the output is nonperiodic. In this section, a flying-adder synthesizer of periodic output will be discussed; only integers and PDFR-compatible fractions are allowed to be used. Thus, this subset of the flying-adder circuit is termed as integer-flying-adder architecture, which can be useful for applications where a spectrally pure clock is required.

4.9.1 Integer-Only FAPLL: How Close Can It Reach an Integer?

Figure 4.27 in Section 4.7 is the generic FAPLL structure. In Eq. 4.8, the number of inputs K is a constant when an implementation plan is fixed. The input reference f_r usually cannot be changed either after design plan is finalized. If we take the post divider M out of the consideration (M is treated as a fixed number not a variable), the output frequency f_o can be expressed in Eq. 4.10, where γ is a constant. The output frequency f_o can be any number since it is user dependent. Therefore, γf_o is a real number. N is the divide ratio of the frequency divider, which is an integer. If only integers are allowed in F, Eq. 4.10 cannot be satisfied for some given f_o. In certain cases, approximation has to be made.

Equation 4.10 represents a known mathematic problem: using two integers to approximate a real number. For any asked real number, the solution theoretically exists. The two integers N and F can be found through the algorithm of *continued fraction approximation** IF no constraint is attached to the

* See this website for a useful tool: http://www.math.mtu.edu/mathlab/COURSES/holt/dnt/pell1.html.

integers that can be used. However, in the real world, N is limited to range of $[N_L, N_H]$, and F is confined in $[2, 2K]$. Now this problem is transformed into the following one: Given certain allow-to-use ranges on the two integers, how accurately can we approximate the desired real number? In other words, what is the bound of the error?

$$\gamma \cdot f_o = \frac{N}{F} \tag{4.10}$$

If we rearrange Eq. 4.10 into Eq. 4.11, the question can be asked: Given a real number η and a group of integers in a certain range, by multiplying the real number and an integer chosen from this group, how close can it reach another integer? This is an open problem. One partial solution is offered in Appendix 4.B.

$$F = N/(\gamma \cdot f_o) = \eta \cdot N \tag{4.11}$$

Unlike the generic FAPLL, where virtually any frequency can be generated as long as enough fractional bits are used, the concern with the integer-only FAPLL is that some frequencies cannot be reached. They have to be approximated by the nearby synthesizable frequency. This leads to the issue of frequency error. Assume that we want to generate a frequency f_w, but the synthesizable one is f_s. The frequency error can be defined as $(f_w - f_s)/f_s = (T_s - T_w)/T_w$. From transfer function Eq. 4.5, $T_w = F_w \cdot \Delta = (I \pm r) \cdot \Delta$ and $T_s = F_s \cdot \Delta = I \cdot \Delta$, where I is the integer part and r is the fractional part. Therefore, $|T_s - T_w|/T_w = r \cdot \Delta / T_w$. From the discussion in Section 4.7, the flying-adder synthesizer's output is $f_w = K \cdot N \cdot f_r / F_w$. Hence, $F_w = I \pm r = (K \cdot f_r \cdot T_w) \cdot N$. From this we can see that $r \leqq (1/2) \cdot K \cdot f_r \cdot T_w$ since we can chose it between two adjacent N values. Based on these arguments, a frequency error upper bound can be found in Eq. 4.12, where N_L is the lower limit of the N's valid range. Equation 4.12 shows that when integer-only FAPLL is concerned, the guaranteed frequency accuracy is $1/(2N_L)$ for any frequency requested (within the designed range). For $N_L = 200$, this gives 0.25%. Appendix 4.B offers a method of finding improved error upper bound.

$$\frac{|f_w - f_s|}{f_w} = \frac{|T_s - T_w|}{T_w} = \frac{r \cdot \Delta}{T_w} \leq \frac{1}{2} K \cdot f_r \cdot \Delta = \frac{1}{2} K \cdot f_r \cdot \frac{T_{VCO}}{K} = \frac{1}{2 \cdot N} \leq \frac{1}{2 \cdot N_L} \tag{4.12}$$

The above frequency error upper bound analysis is useful when the desired output frequency is treated as a variable. In real design cases, if the desired frequency f_o is known, a simple algorithm can be used to sweep all the possible F and N values. The pairs that give the minimum frequency error will be the solution of choice (Xiu and You 2003).

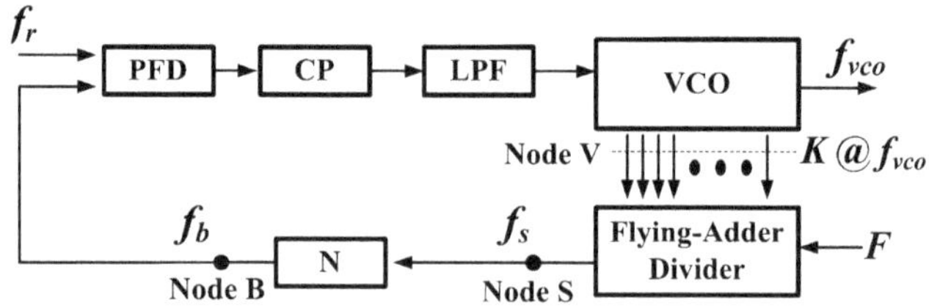

Fig. 4.29. Using a flying-adder fractional divider inside a PLL loop.

4.9.2 Incorporating Flying-Adder Fractional Divider Inside Integer-*N* PLL

Another route of the integer-flying-adder approach is to use the flying-adder circuit as a fractional divider inside the PLL loop. In integer-N PLL architecture, the divider N inside the loop can only take an integer. Therefore, its output's period is adjustable only in a step of one VCO cycle. Since a flying-adder fractional divider can reach a finer resolution than one VCO cycle, it is beneficial to use it inside the PLL loop. This idea is illustrated in Fig. 4.29. From Eq. 4.9, at node S, f_s can be expressed as $f_s = (K/F) \cdot f_{vco}$ (a fractional divider as discussed in Section 4.8). At node B, $f_b = f_s/N$ (a conventional frequency divider). When the PLL reaches lock, we have $f_b = f_r$. Therefore, the frequency relationship between f_{vco} and f_r can be expressed in Eq. 4.13. Clearly, compared to integer-N PLL, frequency resolution can potentially be improved from f_r to f_r/K. Since the PDFR technique introduced in Section 4.6 can be employed between the pair of $\{F, N\}$, the resolution at f_{vco} can be made f_r/K with this technique. The can easily be seen when F takes two PDFR-compatible values $\mathrm{I} + a/N$ and $\mathrm{I} + (a+1)/N$, where a is an integer $a < N$. Using Eq. 4.13, the frequency difference between the two resulting frequencies is f_r/K. One important point is that the output at node B is periodic. There is no phase jump at PFD as in the case of fractional-N PLL.

$$f_{vco} = \frac{F \cdot N}{K} f_r = (F \cdot N) \cdot \frac{f_r}{K} \tag{4.13}$$

In terms of available frequency points, Fig. 4.30 graphically demonstrates the comparison between integer-N PLL with and without a flying-adder divider. In this example, $f_r = 26$ MHz. The N is limited to the range of [50; 100]. The flying-adder divider uses eight inputs $K = 8$; its control word can take any integer between $2 \leqq F \leqq 16$. As shown, in this VCO range of 1.3 GHz to 2.6 GHz, the integer-N PLL can produce 51 frequencies (in 26 MHz steps). With the flying-adder divider, 284 frequencies can be generated (Fig. 4.30, top plot). After removing the redundant frequencies and sorting them in ascending order, the flying-adder divider has 193 unique frequencies (Fig. 4.30, bottom plot). The average frequency step is 6.7 MHz.

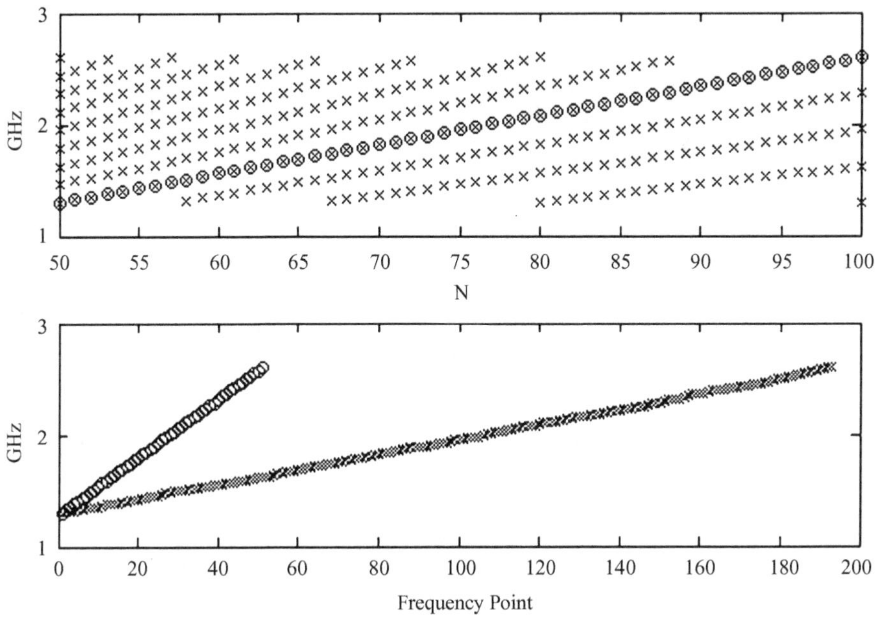

Fig. 4.30. The available frequency points without (o) and with (x) a flying-adder divider: distribution when N varies (top), redundant frequencies removed, and the result sorted (bottom).

Figure 4.31 is the result when the PDFR technique is used. With the help of PDFR-compatible fractions, the top plot in Fig. 4.31 shows the frequency set. As expected, for each given N, the vertical line (corresponding to F change) contains many more frequencies compared to that of Fig. 4.30. The points are so densely packed that overall it looks like a line. After redundancy removing and sorting, the frequency distribution is shown in bottom plot of Fig. 4.31. In this range of 1.3 GHz to 2.6 GHz, there are 401 frequencies with steps of 3.25 MHz. The 3.25 MHz agrees with the predication of Eq. 4.13: $f_r/K = 26/8$. The other two lines in the bottom plot of Fig. 4.31 are the same lines in Fig. 4.30. They are plotted together for visual comparison.

4.9.3 Integer-Flying-Adder Architecture

An architectural extension from the structure in Fig. 4.29 is to add another flying-adder synthesizer outside the PLL as shown in Fig. 4.32. Because of its importance, this structure is given a name of integer-flying-adder PLL (IFAPLL). When the PLL is in a lock state, from Eq. 4.13 we have $f_{vco} = ((F_0 \cdot N)/K) \cdot f_r$. From Eq. 4.9, the synthesized frequency f_{s1} is $f_{s1} = (K/F_1) \cdot f_{vco}$. Therefore, the final output f_o can be derived as in Eq. 4.14.

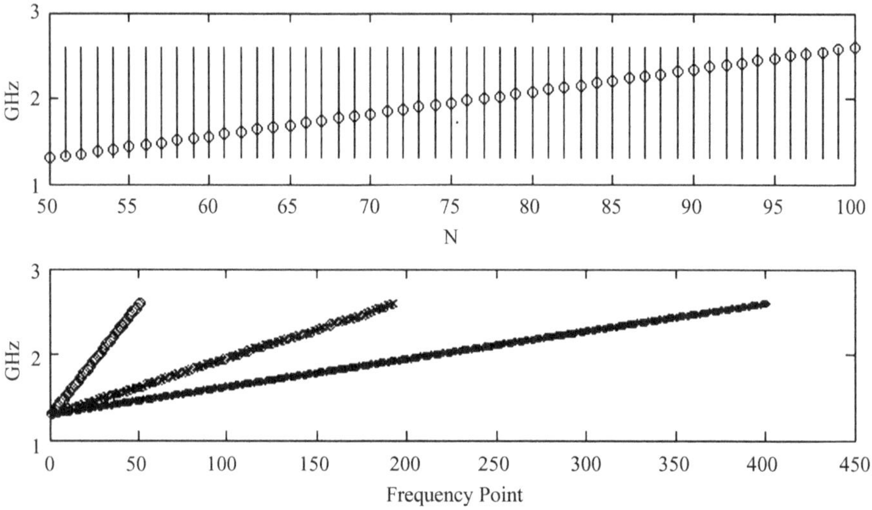

Fig. 4.31. The available frequency points with a flying-adder divider plus PDFR: distribution when N varies (top), redundant frequencies removed, and the result sorted (bottom).

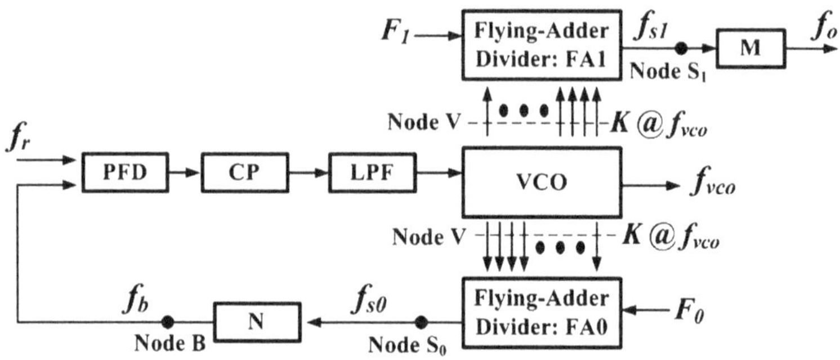

Fig. 4.32. The integer-flying-adder architecture.

$$f_o = \frac{F_0 \cdot N}{F_1 \cdot M} f_r \tag{4.14}$$

In Eq. 4.14, N and M are integers. In implementation, they can be designed as programmable dividers having a certain range: $N_L \le N \le N_H$ and $M_L \le M \le M_H$. F_1 and F_0 are also integers in the range of $2 \le F_1, F_0 \le 2K$. Furthermore, F_1 and M is a pair; PDFR can be applied between them. The same is true for F_0 and N. At node S_1, using the same argument as we used in analyzing Eq. 4.13, the frequency resolution is f_r/F_1. Since F_1 can take values up

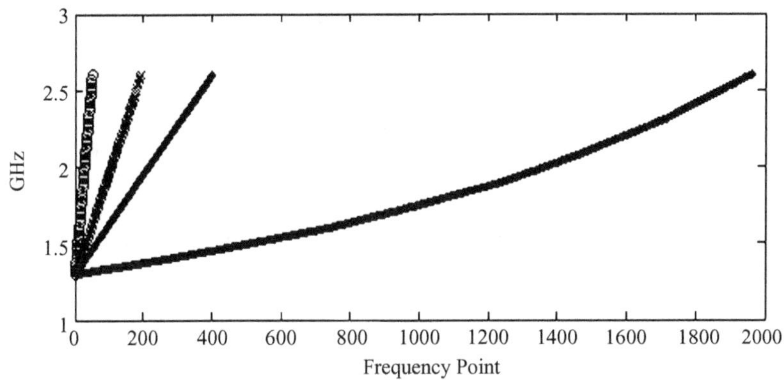

Fig. 4.33. The available frequency points from integer-flying-adder PLL.

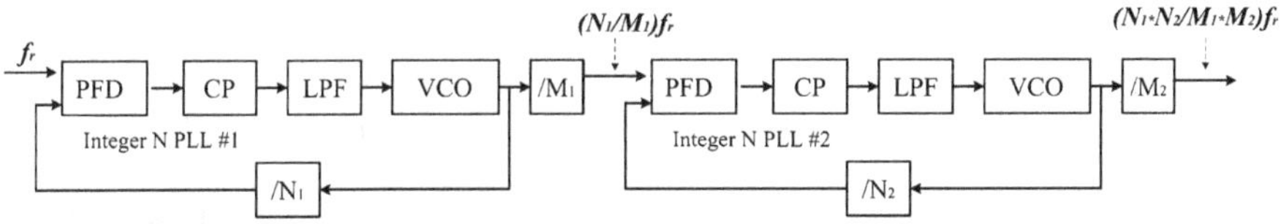

Fig. 4.34. Two integer-N PLLs cascaded together to produce frequency.

to $2K$, the finest resolution is therefore improved to $(f_r/K)/2$. If M is brought into the picture, the resolution can be further improved.

Figure 4.33 is obtained from Eq. 4.14 with PDFR applied on both $\{F_1, M\}$ and $\{F_0, N\}$. Between 1.3 GHz and 2.6 GHz, there are 1,959 frequencies (the leftmost line). The average step is about 0.66 MHz (since M is involved). These frequencies are not uniformly distributed. More frequencies are available in the lower end. The other three lines are the same lines from Fig. 4.31. They are plotted in this figure for visual comparison.

It is interesting to compare Eq. 4.14 with that of integer-N PLL: $f_o = (N/M) \cdot f_r$. It can be immediately appreciated that there are two more adjustable variables in the IFAPLL and they are incorporated in the equation with the powerful multiply operator. As a result, it is understandable that the IFAPLL's frequency generating capability is greatly boosted. In fact, in terms of frequency generation capability, one IFAPLL is more than two integer-N PLLs connected together. If two integer-N PLLs, each with its own loop divider N and post divider M, are cascaded together as shown in Fig. 4.34, the final output is $f_o = ((N_1 \cdot N_2)/(M_1 \cdot M_2)) \cdot f_r$. This expression is mathematically equivalent to Eq. 4.14. However, in the IFAPLL case, F_1 and F_2 have an additional capability of using fractions, and both the $\{F_0, N\}$ and $\{F_1, M\}$ pairs can utilize the PDFR feature. For integer-flying-adder architecture, a more detailed analysis on synthesizable frequencies is presented in Appendix 4.C.

4.10 THE ALGORITHM TO SEARCH OPTIMUM PARAMETERS

The IFAPLL's frequency function in Eqs. 4.13 and 4.14 is very powerful in producing frequencies. As shown, there can be four control parameters $\{N, F_1, F_0, M\}$ that can be adjusted to achieve a requested frequency. If a prescaler is employed at the input reference, there will be five adjustable parameters. This provides great flexibility. However, care should be taken when these parameters are used. In real practice, K is a fixed number after the design plan is finalized. The most popular numbers for K are 8, 16, or 32. After K is chosen, the following is the general consideration for $\{N, F_1, F_0, M\}$ (refer to Fig. 4.32).

- F_1 and $F_0 \rightarrow$ within a range of $[2, 2K]$
- F_0 and $N \rightarrow f_{vco}$ falls in an appropriate range
- $F_1 \rightarrow f_{s1}$ is within the flying-adder synthesizer's speed limit
- $F_0 \rightarrow f_{s0}$ is within the flying-adder synthesizer's speed limit
- $M \rightarrow$ prefers being an even number (duty cycle)
- $N, M \rightarrow$ prefer being 2's power for easily implementation of PDFR

For an asked frequency, owing to the power of Eq. 4.14 and the PDFR, it is usually difficult to identify the best-fit parameters $\{N, F_1, F_0, M\}$ by mind calculation. Therefore, a blind search algorithm is suggested for assistance. Following are the predefined parameters, followed by the algorithm pseudocode. It is very likely that, for a given f_{req}, there would be many groups of $\{F_1, F_0, N, M\}$ that can directly generate this f_{req} (error = 0). Therefore, the algorithm saves them into a file for users to chose the one which best fits their requirements.

f_{req}:	requested frequency
f_{ref}:	reference frequency
K:	number of VCO output phases
N_L, N_H:	the low and high limit for the N divider
M_L, M_H:	the low and high limit for the M divider
f_{vcol}, f_{vcoh}:	the low and high limit of the VCO working range
f_{slim}:	the flying-adder synthesizer's high limit

```
error_min = a very large number ;
for ( N_L ≤ N ≤ N_H, N++ ) {
  for ( 2 ≤ F_1 ≤ 2K, F_1++) {
    for ( 2 ≤ F_0 ≤ 2K, F_0++) {
      for (M_L ≤ M ≤ M_H, M++) {
        for (i = 0, N-1) {
          FREQ_0 = F_0 + i/N #include fractions for PDFR
          of {F_0, N}
```

```
        For (j = 0, M-1) {
          FREQ1 = F1 + i/M #include fractions for PDFR
          of {F1, M}
          fout = (FREQ0*N*fref)/(FREQ1*M)
          error = abs(freq - fout)
          fvco = (FREQ0*N*fref)/K
          fs1 = K*fvco/FREQ1
          fs0 = K*fvco/FREQ0
if (error_min ≧ error AND fvcol ≤ fvco ≤ fvcoh AND
fs1 ≤ fslim AND fs0 ≤ fslim) {
          error_min = error
          F1best = FREQ1, F0best = FREQ0, Nbest = N, Mbest = M
          If (error == 0) {save F1best, F0best, Nbest, Mbest
          into file}
}}}}}}}
```

4.11 THE CONSTRUCTION OF THE ACCUMULATOR

As discussed in Section 4.4.3, the flying-adder circuit's working speed depends heavily on the speed of the adder. This fact is directly expressed in the timing constraint of Eq. 4.4. By examining Fig. 4.17, it is clear that the accumulator in PATH_A is the speed bottleneck since it not only has the integer portion but also has fractional part. In most cases, this fractional part uses a large number of bits (such as 20, 28, 36, . . .) for fine frequency resolution, as apparent from Eq. 4.7. However, the drawback with a larger size is the higher speed requirement on the adder. Figure 4.35 shows the schematic of a conceptual multiple-bits adder with its output registered. The circuit at left is the conventional adder with two inputs of augend and addend. As shown, the multiple-bits adder is made of multiple one-bit full adders. These one-bit adders are serially connected through their carry bits. The adder at the right is configured as an accumulator (one of its two inputs comes from its registered output). In this drawing, a decimal point is inserted to indicate that the operation has both integer and fraction parts. This accumulator is exactly what we need in a flying-adder circuit. The value represented by the bits to the left of the decimal point is the integer part, which is used for controlling the two $K \rightarrow 1$ MUXs. The value corresponding to the right-hand side of the decimal point is the fractional part, which is used for accumulating the "error." For producing correct addition result, the operation has to be finished in one *CLK* cycle. In other words, all the carries have to be propagated to the leftmost position in one cycle. This is obviously not an easy task when the adder size is large.

Being the core element of complex arithmetic circuits, the adder has been studied intensively in its VLSI implementation. There are numerous architectures available to meet various performance and resource requirements, such as ripple carry adder (RCA), Manchester carry chain adder (MCC), carry skip adder (CSK and VSK), carry select adder (CSL), carry lookahead adder

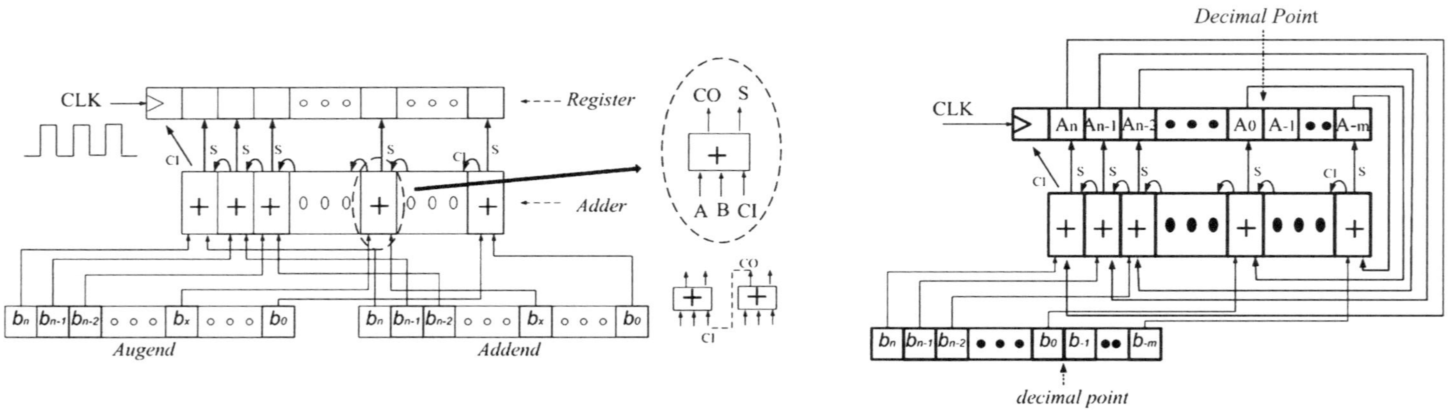

Fig. 4.35. The multiple-bits adder (left) and accumulator (right).

(CLA) and carry save adder (CSA) (Nagendra et al. 1996). The differences among these architectures lie in the ways the carry bits are generated and propagated. There are also several logic styles to implement the basic full adder cell: complementary CMOS, complementary pass transistor logic, transmission function full adder, transmission-gate adder, 14-transistors adder, and 10-transistors adder (Alioto and Palumbo 2002; Bui et al. 2002; Goel et al. 2006; Lin et al. 2007; Shams et al. 2002). All these different logic styles have their cost-performance tradeoffs. The selection criteria concerned include supply voltage range, voltage swing, speed, power-delay product, output skew, driving capability and area.

All the aforementioned adder-related techniques can be used for implementing the accumulation function required in flying-adder architecture. However, based on the definition of time-average-frequency presented in Section 3.3, a higher level (system level in this case) of innovation can provide new opportunities in circuit level implementation. The spirit of time-average-frequency is to ensure that a predetermined number of operations are guaranteed to occur within a fixed time window (such as 1 second). Based on this understanding and the discussion around Fig. 4.5, the following can be said:

- We actually do not care ***when*** the carries are propagated from the fraction part to the integer part, nor do we care about their sequence.
- We only care ***how many*** carries are propagated in a fixed time window.

From this observation, a different style of accumulator, called the XIU-accumulator (Xiu 2009, 2010), is proposed. In Fig. 4.36, the circuit on the left is the conventional accumulator (CON-accumulator). A 1-bit full adder with A, B, CI as inputs, and S, CO as outputs is depicted at the top. CI is the carry-in and CO is the carry-out. In the bottom drawing, a multi-bits version is shown

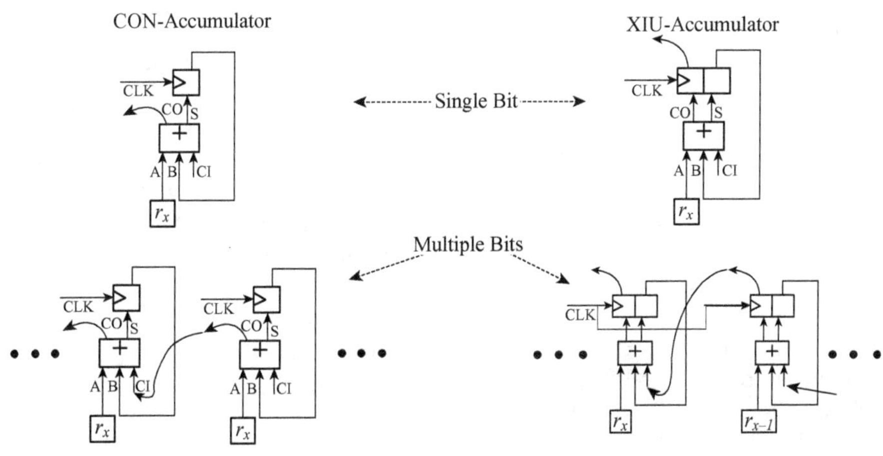

Fig. 4.36. The CON-accumulator (left) and the XIU-accumulator (right).

where the CO of the previous stage is fed into the CI of the current stage. On the right-hand side, the XIU-accumulator is illustrated. In the 1-bit version, its difference from the conventional full adder is the use of one extra flip-flop, which is for storing the CO. In other words, unlike the case of the CON-accumulator, where only the sum S is stored, both the S and CO are registered in the XIU-accumulator. When the multi-bits structure is constructed, the stored CO is forwarded into the CI of the current stage as shown at the bottom right. Circuit-wise, the significance of this modification is that now the carries are not required to propagate from last stage to first stage in one cycle. From the carry-propagation perspective, the *multiple-stages* operation has been separated into multiple *single-stage* operations.

With the XIU-accumulator, the accumulator's speed is no longer related to the accumulator's size. However, an important question has to be asked: Within a given time window, do the CON-accumulator and the XIU-accumulator produce the same number of carries? Appendix 4.D provides the formal proof that they indeed generate the same number of carries (Xiu 2009). This question can also be validated through simulation, either at the transistor level, or the behavior level. Figure 4.37 shows the behavior-level simulation results for a 6-bits accumulator by using Matlab®. In the plots at left, the input to the accumulators is 000001_2. As can be seen, for the CON-accumulator, the accumulation result increases linearly with each operation (controlled by *CLK* cycles). As expected, the accumulation result from the XIU-accumulator is different. However, the numbers of carries generated are the same, although they do not occur at the same times. The plots in the middle correspond to another case of input being 001001_2. Again, the accumulation results are different, but the numbers of carries in a given window are the same. In the second and fifth subwindows, the "window of b^m" is 2^6 cycles. The plots at right shows the total numbers of carries generated from 700 *CLK* cycles for the same 001001_2 input.

Tables 4.2 and 4.3 list some transistor-level simulation results when the XIU-accumulator is used in the flying-adder circuit. The design is implemented in a 90-nm 1v process. The CON-accumulator is built by using Synopsys's Design Compiler®. The design constraint for the accumulators is set at 0.4 ns. Table 4.2 shows the result of four different width accumulators: 24-, 32-, 48- and 64-bits. As shown, when width increases, the circuit becomes slower for the CON-accumulator since the carry-propagation path is longer. For the XIU-accumulator, the speed is independent of width due to the local isolation of the carry-propagation path. For both accumulators, the circuit size increases with the width, as expected. The number reported is the total gate count in units of NAND2. The first number inside the parenthesis is the size of combination logic; the second number is for sequential logic. It is apparent that, for sequential logic, the XIU-accumulator uses roughly twice as much area as the CON-accumulator. This is understandable since it uses one extra storage unit for each stage. The ratio is not exactly 2 because different types of flip-flops are used. Overall, the CON-accumulator uses about twice the area of the

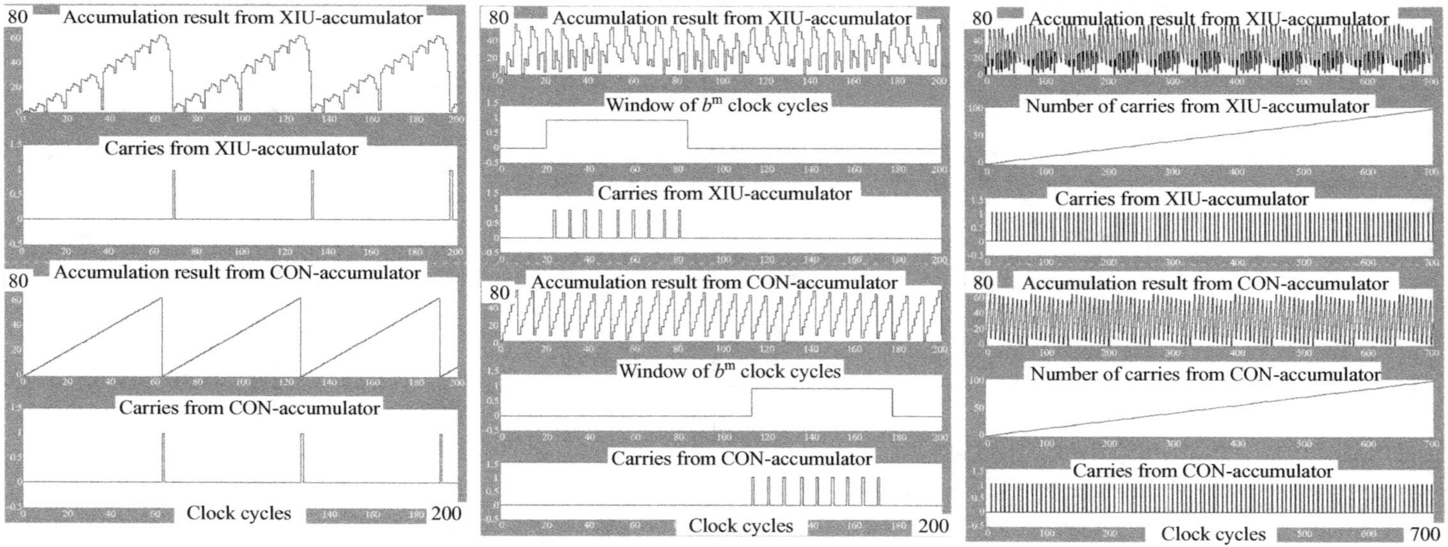

Fig. 4.37. Behavior-level simulation for validating the XIU-accumulator: case of 000001_2 (left), case of 001001_2 (middle), case of 001001_2 (right).

TABLE 4.2. The Speed and Area Comparisons on Various Accumulator Widths

Size	CON-accumulator's speed (ns)	XIU-accumulator's speed (ns)	CON-accumulator's size (in units of NAND2)	XIU-accumulator's size (in units of NAND2)
24-bit	0.61	0.43	622.75[a] (516[b], 106.75[c])	315.5 (135.5, 180)
32-bit	0.63	0.43	887.75 (743.75, 144)	417.5 (173.5, 244)
48-bit	0.72	0.43	1295.5 (1085.5, 210)	621.5 (249.5, 372)
64-bit	0.72	0.43	1914.5 (1627.5, 287)	825.5 (325.5, 500)

[a]Total area.
[b]Area of combination logic.
[c]Area of sequential logic.

TABLE 4.3. The Comparison of Power Consumption on Various Accumulator Widths

Accumulator size	CON-accumulator (mA)			XIU-accumulator (mA)		
	1 GHz	500 M	100 M	1 GHz	500 M	100 M
24-bit	3.33	1.69	0.36	1.75	0.88	0.18
32-bit	4.51	2.27	0.47	2.20	1.13	0.23
48-bit	6.22	3.13	0.67	3.35	1.68	0.35
64-bit	9.76	4.96	1.04	4.41	2.18	0.46

XIU-accumulator. Table 4.3 is the power consumption comparison for the two types of accumulators under three different speeds: 100 MHz, 500 MHz, and 1 GHz. It can be seen that the XIU-accumulator uses about half of the power consumed by the CON-accumulator.

4.12 THE CONSTRUCTION OF THE HIGH SPEED MULTIPLEX

From the flying-adder circuit of Fig. 4.17, it is clear that the $K \rightarrow 1$ multiplex is another key component that has great impact on circuit speed. It limits the highest synthesizable frequency. This fact is reflected in the constraint of Eq. 4.2, where t_2 is the multiplex's decoding time. From the moment a new address is received, we want the multiplex to switch to the new input path as soon as possible. The time required is defined as the decoding time. In most flying-adder implementations, this multiplex is constructed from the transmission gate. Figure 4.38 is the schematic of $2 \rightarrow 1$ multiplex (MUX2). The two paths associated with inputs A and B are controlled by two transmission gates. Their control signals SB and SS originate from the selection signal S so that the decoding times from $A \rightarrow B$ and $B \rightarrow A$ are roughly the same. The delay of SB is balanced with that of SS through an always-closed transmission gate. The

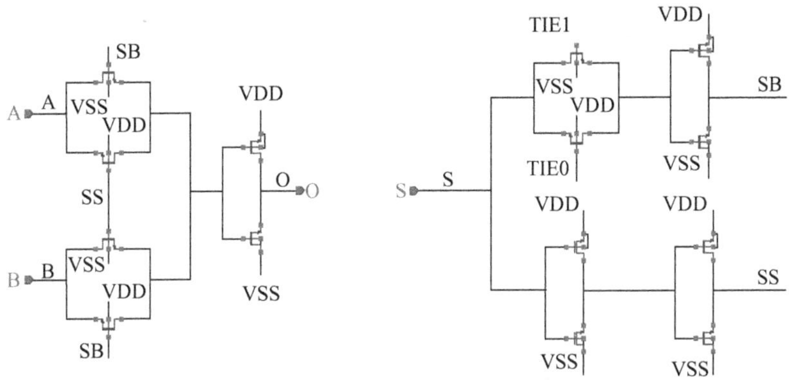

Fig. 4.38. MUX2 of inputs A, B, S, and output O.

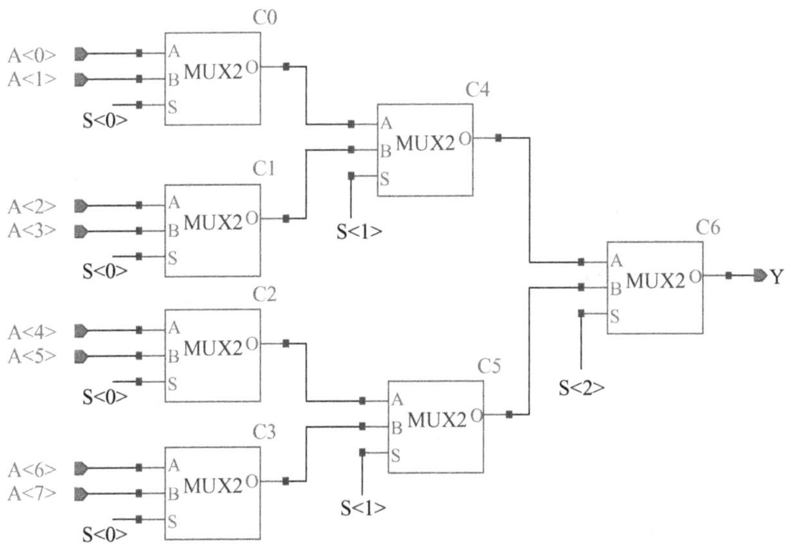

Fig. 4.39. MUX8 made of three stages of MUX2s.

key requirements are: (1) the delays of A → O and B → O need to be balanced, and (2) the decoding time (from S change to A ↔ B switching) needs to be as short as possible.

Figure 4.39 is the schematic of an 8 → 1 multiplex, which is made of the MUX2s described above. As shown, the path-decoding operation has been separated into three stages. The requirements remain the same as that of MUX2: balancing all the paths from input to output and making the address decoding as fast as possible. In this case, there are three individual address signals. The most demanding one is the LSB (S<0> in Fig. 4.39) since it drives the largest load. Hence, buffering should be considered for this signal (and

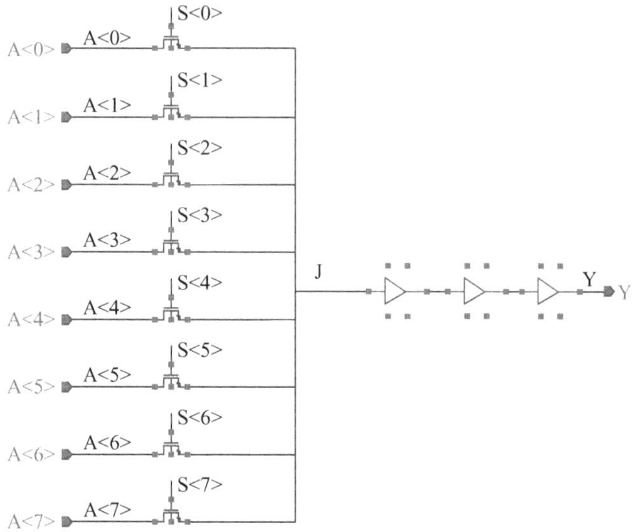

Fig. 4.40. MUX8 made of pass transistors.

maybe for others as well). When new address signals arrive, all the three stages start decoding at the same time. However, the overall decoding time is dominated by the first stage (from left). Buffering the S<0> improves the transition edge, which helps the transmission gates' on–off switching. But buffers require extra delays, which might negate the result. Therefore, buffering must be done with great care. This is especially true for large size multiplexes.

Figure 4.40 shows another design style for the $K \rightarrow 1$ multiplex ($K = 8$ in this case). In this case, the decoding is done in one stage. The eight pass transistors are controlled by eight individual signals (instead of three as in the previous case). These eight signals are in "one-hot" fashion. In other words, at any given time, there is one and only one signal that is "high." The advantage of this structure is the faster decoding speed since each address signal controls only one transistor. The drawback is the potential problem at node J, where all outputs from the pass transistors join. The pass transistor that is currently active has to drive the large capacitive load presented at this node. This could be a problem for large multiplexes. Moreover, a new issue with this scheme is the coding of the address. The outputs from the flying-adder adder and accumulator are coded in binary with a width of $\log_2(K)$. This multiplex requires K individual signals. Hence, a decoding circuit for $\log_2(K) \rightarrow K$ is required. Fortunately, we can insert the logics associated with this decoder between the registers REGA1 and REGA2, and REGB1 and REGB2 (refer to Fig. 4.17).

A more serious issue is the prevention of "all-zero" code. If "all-zero" occurs in these signals, the flying-adder circuit will lock itself (no output) since there will be no "low-to-high" transition available from the K inputs to trigger any event inside the flying-adder circuit. This is not the case for the circuit in Fig. 4.39 because there is always an active path no matter what the value that

$\log_2(K)$-bits produces. Therefore, for the multiplex of Fig. 4.40, an extra circuit is needed to ensure that "all-zero" is mapped to other value (any "one-hot" value). And, more importantly, this circuit must be physically located between the register and the $K \rightarrow 1$ MUX to take into account the possibility of "all-zero" from the register's initial value. The scenario of "more than one of one" code is also a problem, but less severely. When properly designed, the decoder will not output code with "high" on more than one bit. However, this could happen at the initial power-up stage due to the value stored in the register. When this occurs, two or more paths are active and their respective signals will "fight" at node J. But, eventually, a "low-to-high" transition will appear, and this is enough to put the flying-adder circuit into the right track. Overall, when properly designed, the structure of Fig. 4.40 could produce an extremely fast circuit that can be useful for high frequency applications. For even higher frequency applications, current mode logic (CML) design can be used.

Another way to solve the "all-zero" and "more than one of one" code problems is to use an extra signal *RESET*. As illustrated in Fig. 4.41, a *RESET* can be used to set the two key signals *SEL_LOW* and *SEL_UP* to predetermined values. Started from these legal values, it is guaranteed by design that the illegal values are not accessed by the flying-adder circuit.

4.13 NON-2'S POWER FLYING-ADDER CIRCUIT

Up to now, all the discussions around flying-adder architecture have been carried out under the condition that the number of inputs is in 2's power. In principle, any number of inputs can be used. This fact can be very useful for

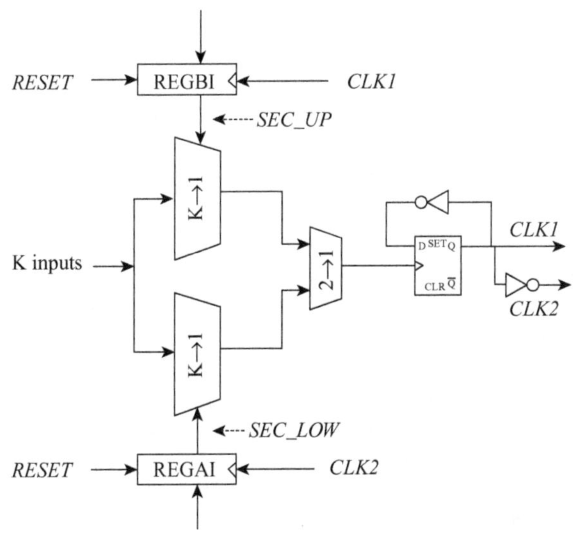

Fig. 4.41. Using RESET to avoid circuit lock.

certain applications where the number of phases from a source is non-2's power (e.g., 10 or 20). As a matter of fact, the first proof-of-concept flying-adder circuit (Mair and Xiu 2000) is based on a group of 31 inputs. When non-2's power inputs are used, three issues need attention: (1) the accumulator design, (2) the $K \rightarrow 1$ multiplex design, and (3) the frequency control word adjustment.

The adders used in a flying-adder circuit are based on the binary system. Naturally, it is in 2's power. When K is not in 2's power, a modulus-K adder has to be built. The size of the adder can be chosen as $\text{int}[\log_2(K)] + 1$. However, in operation, any code value that is equal to or greater than K has to be skipped (rollover). This can be easily achieved in HDL coding through HDL-based design methodology. For the accumulator in PATH_A, the modulus-K operation only needs to be done in the integer part since only the integer controls the $K \rightarrow 1$ multiplex. The multiplex style in Fig. 4.39 is naturally in 2's power. Therefore, the unused input ports have to be either unconnected or connected to one of the available inputs (such as input0). In normal operation, these unused ports will never be referenced by the adders' output (due to the modulus-K operation). But they could be accessed in the initial power-up stage. If these ports are connected to one of the active inputs, the circuit can tune itself into the right track automatically. If these ports are unconnected, an extra safeguard circuit must be inserted between the register and the multiplex to ensure that the illegal codes are mapped to one of the valid codes. This is similar to the prevention of the "all-zero" code in Section 4.12. Also, the circuit in Fig. 4.41 can be used to prevent unused ports access if *RESET* is preferred. With this, the codes corresponding to those unused ports are never produced from the flying-adder circuit.

Frequency control word adjustment can be explained through an example. Assume that we need to design a 20-inputs flying-adder synthesizer. Since $K = 20$, $\text{int}[\log_2(20)] + 1 = 5$ bits are required for the address of the $K \rightarrow 1$ multiplex. For the two integer adders inside the PATH_A and PATH_B (refer to Fig. 4.17), whenever the 5-bits adders' output is equal to or greater than 20 (10100_2), it needs to roll over from 0. In other words, out of the 32 possible values, 12 of them are invalid. To utilize the two-paths flying-adder circuit, one more bit is needed for the frequency control word; thus FREQ[5:0]. If a FREQ $\leqq$ 19 (10011_2) is the current control word, it can be directly plugged into the synthesizer. No modification is needed. In the range of $20 \leqq$ FREQ $\leqq 39$, an adjustment of FREQ(valid) = FREQ + 12 is needed. For FREQ = 40, FREQ(valid) = FREQ + 24 = 64 (000000_2). As discussed in Section 4.4.3 and shown in Appendix 4.A, FREQ[4:0] can be directly fed into the PATH_A as FREQ_LOW. FREQ[5:1] is sent to PATH_B as FREQ_UP[4:0] = FREQ[5:1] (recall that this actually divides the size of FREQ by half). Further, to balance the duty cycle, we would prefer to make FREQ_UP[4:0] = FREQ[5:1] + FREQ(0). For the $K = 20$ case, FREQ_UP(valid) = FREQ_UP + 12 if FREQ_UP $\geqq 20$. Using the flying-adder Eq. 4.9 of $f_o = (K/F)*f_{in}$, four numerical examples are listed in Table 4.4 below. VHDL code can be modified from the example in Appendix 4.A for this 20-inputs flying-adder synthesizer.

TABLE 4.4. Numerical Examples of a 20-Input Flying-Adder Synthesizer (Fractional Divider)

f_o/f_{in} ratio[a]	FREQ[5:0]	FREQ[5:0] (valid)	Note	FREQ_ LOW[4:0]	FREQ_ UP[4:0]
20/7	7	$7 = 000111_2$	Direct use since FREQ < 20	$00111_2 = 7$	$00100_2 = 4$
4/5	25	$37 = 100101_2$	FREQ (valid) = FREQ + 12	$00101_2 = 5$	$01101_2 = 13$
2/3	30	$42 = 101010_2$	FREQ (valid) = FREQ + 12	$01010_2 = 9$	$01111_2 = 15$
1/2	40	$64 = 000000_2$	FREQ (valid) = FREQ + 24	$00000_2 = 0$	$00000_2 = 0$

[a] f_o/f_{in} = K/F = 20/F.

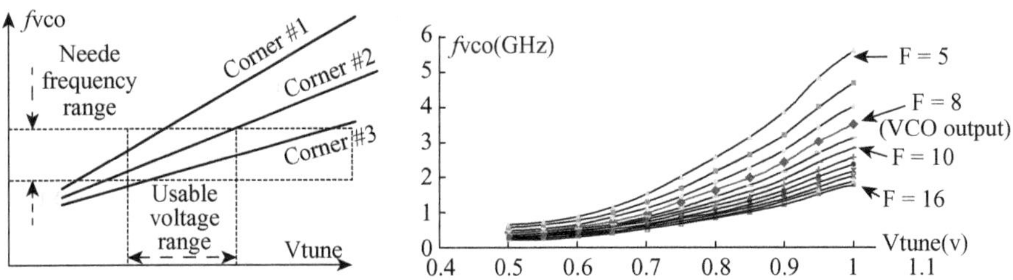

Fig. 4.42. VCO tuning curves of three corners (left), using a flying-adder to expand the VCO range (right).

4.14 EXPANDING VCO FREQUENCY RANGE IN NANOMETER CMOS PROCESSES

One of the most distinguished features of modern nanometer CMOS processes is their ever-decreasing transistor channel length. Consequently, the transistors switch faster and faster. However, at the same time, the switching speed becomes more sensitive to the surrounding environment (process corner, temperature, and voltage). Another important characteristic of nanometer processes is that the supply voltage becomes lower and lower (around 1 v now). Therefore, the usable voltage range for function is much reduced. Together, these facts have made ring-VCO design in these processes a challenging task. As shown in the left drawing of Fig. 4.42, a particular VCO is designed in an advanced CMOS process. The VCO tuning curves for three different corners spread greatly in frequency range. As a result, the needed frequency range cannot be covered under all the conditions.

From Eq. 4.9 of $f_s = (K/F){\cdot}f_{vco}$, it can be seen that the flying-adder synthesizer can help expand the VCO frequency range. For every VCO frequency f_{vco}, theoretically there are $2K$-1 extra frequencies that can be produced without using a fraction (no TAF clock). This is because of the $2K$ integer values available to F. One of them is the original f_{vco} when $F = K$. Therefore, for each VCO curve, several more associated curves are available if the flying-adder synthesizer is used together with the VCO. The plot on the right side of

Fig. 4.42 illustrates this point. In this case of $K = 8$, the curve of $F = 8$ is one of the VCO curves in the left. Additionally, there are several more curves ranging from $F = 5$ to $F = 16$. They expand the original frequency range considerably. For each of the three curves on the left, such flying-adder-assisted expansion can be carried out. As a result, the needed frequency range will have greater chance to be covered.

Another potential usage of flying-adder-assisted frequency expansion is to help reduce VCO gain. Due to the very limited usable voltage range, ring-VCOs designed in modern nanometer processes usually bear large gain (over GHz/v) in order to cover relatively large frequency ranges. This makes the VCO very sensitive to any disturbance that occurs on its voltage control line, which could result in significant jitter. With a flying-adder synthesizer, the VCO gain can be intentionally designed smaller for better jitter performance. The lost frequency range can be compensated by the synthesizer.

The technique of using a flying-adder synthesizer inside the PLL (Section 4.9.2) can help us reach more frequency points from the VCO. The discussion presented here (a flying-adder synthesizer used outside the loop) can expand the VCO range. Together, as shown in Fig. 4.32, flying-adder synthesizers can greatly enhance the frequency generation capability of a PLL/VCO design.

4.15 MULTIPLE FLYING-ADDER SYNTHESIZERS

The flying-adder's power of generating frequency is demonstrated by Eqs. 4.8 and 4.14, and Figs. 4.27 and 4.32. This capability can be further empowered by attaching multiple flying-adder synthesizers to the same multiphases input. Usually, this is implemented together with an integer-N PLL as depicted in Fig. 4.43. One flying-adder synthesizer is incorporated inside the PLL to reach more frequencies from the VCO. From outside the loop, four more synthesizers are attached to the same VCO's K outputs. Each synthesizer has its own independent control, resulting in multiple independent clock outputs. Based on the discussions of previous sections, the f_{vco} and f_{ox} (where x = 1, 2, 3, 4) can be calculated in Eqs. 4.15 and 4.16, respectively.

$$f_{vco} = \frac{F_o \cdot N}{P \cdot K} f_r \tag{4.15}$$

$$f_{ox} = \frac{F_o \cdot N}{P \cdot F_x \cdot M_x} f_r \quad where \quad x = 1, 2, 3, 4 \tag{4.16}$$

Out of the five clock outputs f_{vco} is completely independent; f_{ox} (x = 1, 2, 3, 4) is partially independent since it relies on f_{vco}. Furthermore, if needed, the output from the internal flying-adder synthesizer (controlled by F_0) can also be a clock output for supporting additional loads. In principle, more synthesizers can be attached to the VCO and make this clock generator more powerful.

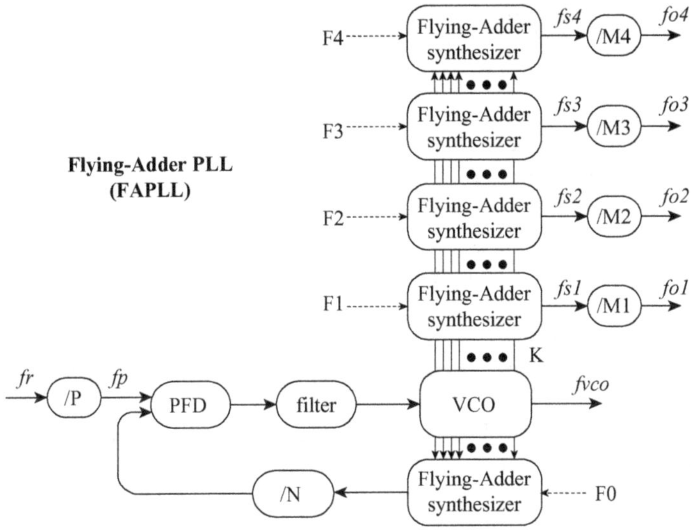

Fig. 4.43. Flying-adder PLL with multiple synthesizers.

However, care must be taken since this will increase the difficulty of implementation. Another point worth mentioning is that PDFR can be applied between the pair of $\{F_0, N\}$, and also between $\{F_X, M_X\}$. Lastly, non-PDFR compatible fractions can be utilized in F_X when time-average-frequency is appropriate for the application of interest.

The structure of multiple flying-adder synthesizers on an integer-N PLL has been used in commercial products for many years. It is especially suitable for reducing the number of PLLs in large SoC. The latest examples can be found in http://focus.ti.com/lit/ug/sprugx9/sprugx9.pdf, http://focus.ti.com/lit/ug/sprugx7/sprugx7.pdf, and http://focus.ti.com/lit/ug/sprugx8/sprugx8.pdf.

4.16 FLYING-ADDER IMPLEMENTATION STYLES

For a particular application, an appropriate implementation style can be chosen to best fit the current situation with the lowest cost. Overall, three styles are available for different applications: low-cost, middle range, and high-end.

In the low-cost domain, a flying-adder synthesizer can be realized in an all-digital fashion (Chau and Chen 2008; Chau et al. 2006; Gharaee and Tathesari 2006; Sung et al. 2010). In other words, the entire synthesizer can be built on standard cells. The multi-inputs can be generated from a standard-cell-based ring oscillator or by a chain of flip-flops that are driven by a higher frequency clock (refer to Fig. 4.10). The MUXs can be constructed from standard cells as well. This fashion of flying-adder can be designed in a very short time and at a very low cost. It can be useful in many low-frequency applications where frequency flexibility is important and jitter performance is secondary. In extreme cases, the flying-adder synthesizer can even be realized in field programmable gate array (FPGA) for quick turnaround.

In the middle-range category, the multi-inputs are usually generated from an integer-N PLL. The VCO can be built from a ring-type CMOS circuit. The VCO, the filter, and other components in the PLL loop, and the MUX used in the synthesizer, are all designed and laid out in analog fashion. In terms of frequency range and jitter performance, this style of implementation is appropriate for constructing the clock generator of most ASIC chips. Nowadays, the highest output frequency from a flying-adder synthesizer is around 2~3 GHz with *rms* period jitter of a few ps (55 nm). Better jitter performance and higher output frequency can be achieved with the help of special circuit techniques and more advanced processes.

For high-end applications where the frequency synthesizer's phase noise is of primary concern (such as for LOs in RF application, for LOs in TV silicon turners, for high data rate serial links, for clocking high-precision ADC/DACs, etc.), the multi-inputs can be generated from LC-VCO-based PLL. The rotary traveling wave oscillator (RTWO) described in Section 4.2.3 is one of the potential options. The circuit structure in Fig. 4.9 is another option. The current mode logic (CML) technique can be used for constructing other high-speed components (such as the divider and the MUXs). Integer-flying-adder PLL can be employed to efficiently utilize the frequencies available from the high-quality but frequency-range-limited LC VCO.

4.17 SIMULATION APPROACHES

Flying-adder circuits can be simulated at two levels: the behavior level and the transistor level. Behavior-level modeling can be carried out by using HDL (such as VDHL and Verilog; Appendix 4.A is the example of VHDL). The circuit can also be modeled in higher-level modeling tools, such as Simulink® from MathWork, *C* language, or *Perl* scripting language. Behavior level simulation is good for understanding the circuit. Its main purpose, however, is for studying the output clock's time domain and frequency domain behaviors. It is also often used for modeling the higher-level system that uses the flying-adder synthesizer as its building block. Transistor-level simulation is largely used for verifying the circuit and evaluating the performance when the transistor-level circuit design is finished.

In all these simulations, one key issue is the generation of the multi-inputs. In VHDL, this is done by using "TRANSPORT *signal-X* after" (see Appendix 4.A). In Simulink, the K inputs can be generated by using a group of K *pulse generators* with phase delay of *period/K* to each other. Similarly, in transistor-level simulations, a group of signal sources (*vpulse*) is often used to model the K inputs. The following is an example of a SPICE subckt for modeling a four-differential-stage VCO with eight outputs. This model is often used as inputs for flying-adder synthesizers to assist in real transistor-level designs. The VCO frequency is programmable through a variable f_{vco}. The *vpulse* generators are separated by a delay of $1/(f_{vco} \cdot 8)$ from each other.

```
.subckt VCO_MODEL vco<7> vco<6> vco<5> vco<4> vco<3> vco<2> vco<1> vco<0>
v7 vco<7> 0 DC=0 PULSE 0 1.1 '(7/fvco)/8' 5e-12 5e-12 '500e-3/fvco' '1/fvco'
v6 vco<6> 0 DC=0 PULSE 0 1.1 '(6/fvco)/8' 5e-12 5e-12 '500e-3/fvco' '1/fvco'
v5 vco<5> 0 DC=0 PULSE 0 1.1 '(5/fvco)/8' 5e-12 5e-12 '500e-3/fvco' '1/fvco'
v4 vco<4> 0 DC=0 PULSE 0 1.1 '(4/fvco)/8' 5e-12 5e-12 '500e-3/fvco' '1/fvco'
v3 vco<3> 0 DC=0 PULSE 0 1.1 '(3/fvco)/8' 5e-12 5e-12 '500e-3/fvco' '1/fvco'
v2 vco<2> 0 DC=0 PULSE 0 1.1 '(2/fvco)/8' 5e-12 5e-12 '500e-3/fvco' '1/fvco'
v1 vco<1> 0 DC=0 PULSE 0 1.1 '(1/fvco)/8' 5e-12 5e-12 '500e-3/fvco' '1/fvco'
v0 vco<0> 0 DC=0 PULSE 0 1.1 0 5e-12 5e-12 '500e-3/fvco' '1/fvco'
.ends VCO_MODEL
```

4.18 THE IMPACT OF INPUT MISMATCH ON OUTPUT JITTER

As is apparent from Figs. 4.1 and 4.2, the K input signals are the driving force of the flying-adder synthesizer. Consequently, any time error caused by mismatch among these signals could be transferred to the flying-adder output and induce jitter (time domain) and spurs (frequency domain). This issue is so important for direct period synthesis that a serious effort will be spent in this section to address this subject in detail.

4.18.1 The Cause of Mismatch and Its Characteristics

In most cases, the flying-adder synthesizer is incorporated with an integer-N PLL as depicted in Fig. 4.27. The VCO used in the PLL usually is constructed from a multiple-delay-stages ring oscillator. Multiple outputs are naturally available from this type of VCO. Figure 4.44 is an example of a VCO of the type that is often used in flying-adder implementations. This VCO will be used

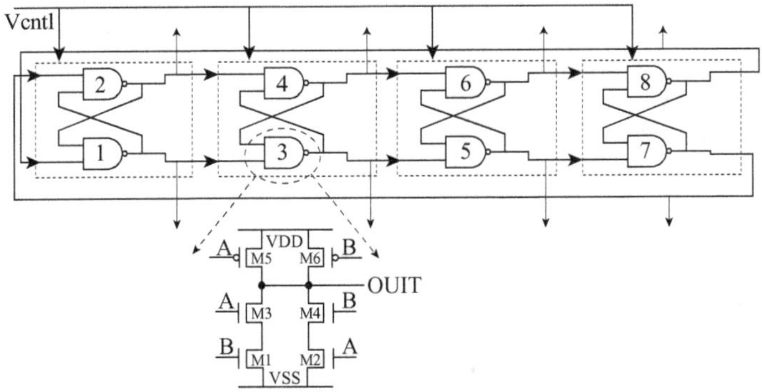

Fig. 4.44. The four-differential-stages CMOS ring VCO used for a flying-adder synthesizer.

in our input mismatch study. The results obtained from this study can be applied to more general mismatch scenarios.

This VCO is made of four differential stages. Each stage consists of two cross-coupled NAND gates as shown in Fig. 4.44. Its delay can be varied by adjusting its supply voltage V_{cntl}. There are only six transistors in each of the NAND gates. The design and layout of the stages are fully symmetrical to match the rise and fall times (to reduce the phase noise caused by rise/fall time mismatch [Hajimiri and Lee 1998]) and the time offsets among all the phases. In terms of matching, compared to other structures, this VCO offers following benefits: (1) fewer number of transistors for less chance of mismatching, (2) no local bias circuit (within each stage) for less chance of mismatching, (3) simple structure for low supply voltage, and (4) easy oscillation, which is very important for a clock generator in SoC (it always oscillates when V_{cntl} reaches a certain level). Although it is simple and fully symmetric, this VCO still bears the possibility of producing unbalanced outputs among its eight outputs. This could be caused by the transistors' V_t mismatch, size mismatch, and layout mismatch (the imbalanced local metal connections of wiring the transistors). Regardless of the root cause, the final result of these physical imperfections is the unbalanced output waveforms that result in time errors in the base unit Δ.

When a single NAND gate is investigated, it is apparent that transistors M2, M3, and M5 can be grouped together since they are all tied to the same input A. Any deviation from its normal state (V_t, width, length, etc.) on any of these three transistors will produce the same end result: causing a rise or fall time error on signal OUT. From outside the NAND, it does not matter which transistor is the source of the problem. For this reason, it is possible to use only one transistor, such as M5, to model the mismatch associated with signal A. Further, we only need to vary the width of the M5 to study the mismatch. By varying the width in both directions, the length effect is also included because the transistor's behavior is determined by the ratio *W/L*. Another reason to use M5, instead of M2 or M3, is that M5 is responsible for the low-to-high transition that is the active edge used in a flying-adder synthesizer. A similar argument holds true for M6 (input B). In summary, to make a particular NAND different from the rest of the normal NANDs (termed "bad-cell" hereafter), the following variations can cover all the possibilities. Twenty-five percent of width variation is chosen based on manufacture reality (the VCO is very small in size and all the transistors sit closely together). It also includes the contributions from the other factors: the length variation, the M2, M3 (or M1, M4).

- Increase M5's width by 25%, called M5+25
- Decrease M5's width by 25%, called M5-25
- Increase M6's width by 25%, called M6+25
- Decrease M6's width by 25%, called M6-25
- Increase both M5's and M6's widths by 25%, called M56+25.
- Decrease both M5's and M6's widths by 25%, called M56-25

TABLE 4.5. The Scenarios of Bad-Cell Occurrence

Number of bad-cells	Patterns of bad-cell location	Possibility
One	3[a] or any other	High
Two	3-4, 3-5, 3-6, 3-7, 3-8	Medium
Three	3-4-6, 3-4-5, 3-5-6, 3-8-7, 3-4-7, 3-4-8	Low
Four	3-4-5-6, 3-4-7-8, 1-3-5-7, 1-4-6-8, 1-3-6-8, 1-4-5-7, 1-4-5-8.	Very low

[a]This is the cell index. Refer to Fig. 4.44.

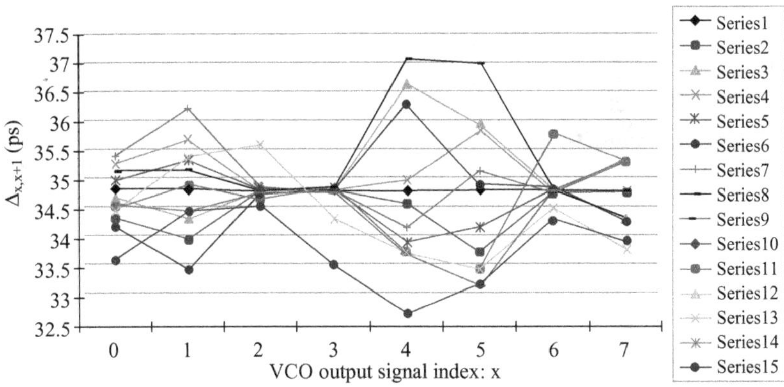

Fig. 4.45. VCO mismatch simulation. Each curve is made of eight $\Delta_{x,\ x+1}$s that correspond to one VCO configuration.

When studying at a higher level (the ring level with all 48 transistors considered), there are scenarios of one, two, three, and four bad-cells that could exist in the ring. Seven, six, and five are equivalent to one, two, and three (symmetrical counterpart). Table 4.5 lists all the scenarios. Refer to Fig. 4.44 for the cell index. There are 19 patterns in total. Within each pattern, for each NAND, there are six transistor variations, as stated before. This gives us a total of 19*6 = 114 variation patterns. To make the situation even more complicated, the variations can take different directions among the cells when there are more than one bad-cell in the pattern. For example, in the pattern of 3-4, cell3 takes the variation of M5+25. At the same time, cell4 could take any of the six variations: M5+25, M5-25, M6+25, M6-25, M56+25, or M56-25.

Figure 4.45 shows the transistor-level simulation result of VCO mismatches. In this simulation, the VCO of no -mismatch is set at 3.6 GHz (278.58 ps). Thus $\Delta = 278.58/8 = 34.82$ ps. Then 15 cases of one bad-cell, two bad-cells, three bad-cells, and four bad-cells are simulated. Under each case, the VCO period T_{VCO} is first measured. Then Δ is defined as $\Delta = T_{VCO}/8$. For any two adjacent VCO outputs (they are listed in the x-axis), the time difference between their rising edges (called $\Delta_{x,x+1}$) is also measured. These eight values are plotted in the figure for each case. Each curve corresponds to one VCO mismatch configuration. When no bad-cell is included, the maximum variation among the eight $\Delta_{x,x+1}$ is within 0.1%. From all the simulated cases of various mismatch configurations, the maximum variation is under 5%.

The VCO mismatches bear at least two characteristics: it is deterministic and it is bounded. The mismatch does not change with time once the manufactured imperfections have occurred. They are also bounded because the variation of $\Delta_{x,x+1}$ cannot be large than one Δ (otherwise, signal order will be changed). Furthermore, the effect of these mismatches is not accumulative, owing to the flying-adder operation. The synthesizer's transfer function is expressed as $T_s = \text{FREQ}\cdot\Delta$, where T_s is the desired output period. Its ideal period is FREQ of Δ. The actual period is expressed in Eq. 4.17, where i is the index for the input signals (the address value of the 8 → 1 MUX) and j is the signal that is responsible for the rising edge of the current flying-adder output cycle. As can be seen, the actual synthesizer output is made of FREQ of $\Delta_{x,x+1}$s. This operation is carried out continually. Thus, whenever it crosses one VCO period (eight Δs in this particular VCO design), all mismatches are canceled since we return to the same signal. Therefore, the absolute error upper bound on the output signal's period is $\Delta/T_s = 1/\text{FREQ}$.

$$T_s(j) = \sum_{i=j}^{FREQ+j} \Delta_{x,x+1}(i) \tag{4.17}$$

4.18.2 The Mismatch Modeling

From the above discussion, it is reasonable to make the following assumptions regarding $\Delta_{x,x+1}$: (1) statistically, the variation of the size of $\Delta_{x,x+1}$ is a random process, and (2) the maximum variation of $\Delta_{x,x+1}$ is limited to ±5%. To take into account the metal connection variation and to make some extra margin, the maximum variation of $\Delta_{x,x+1}$ can be increased to ±10%. Therefore, a model with a mismatch effect included can be created to represent the real VCO. Figure 4.46 shows this model. There are eight independent voltage sources (*vpulse* in SPICE; refer to the discussion in Section 4.17) that produce pulse trains of a certain frequency. All of them have the same frequency f, but each has its own unique phase Φ_x. The time difference between the rising edges of any two adjacent signals is defined as $\Delta_{x,x+1} = \Phi_x - \Phi_{x+1}$. The ideal base unit is defined as $\Delta = T/8 = 1/(8f)$. Therefore, the following constraints are applied to $\Delta_{x,x+1}$:

1. $\Delta_{x,x+1}$ is a random variable.
2. $0.9\Delta \leqq \Delta_{x,x+1} \leqq 1.1\Delta$.
3. $\sum_{x=1}^{8} \Delta_{x,x+1} = 8\Delta$ (4.18)

4.18.3 The Mismatch and the Frequency Control Word

Consider the flying-adder output as an infinitely long pulse train; each pulse is made of some number of $\Delta_{x,x+1}$s. The FREQ controls the number of $\Delta_{x,x+1}$s to use. After power-on, the flying-adder circuit starts from a particular VCO signal that is selected by the initial value on the lower 8 → 1 MUX (refer to Fig. 4.17). From there, this MUX address is increased by FREQ at every output clock cycle (CLK1/CLK2). This addition operation is carried out in modulo-8 fashion. The rising edge of the currently selected VCO signal is responsible for generating the rising edge of the output signal. Table 4.6 shows the signal selection sequence for different FREQ settings. The first column is the index of flying-adder output cycles. The numbers in the table represent the VCO signals selected. In other words, these numbers are the addresses for the lower 8 → 1 MUX. For all the FREQ settings, we assume that the flying-adder circuit starts with VCO signal0 (could be any other signal).

Several important observations can be made from this table.

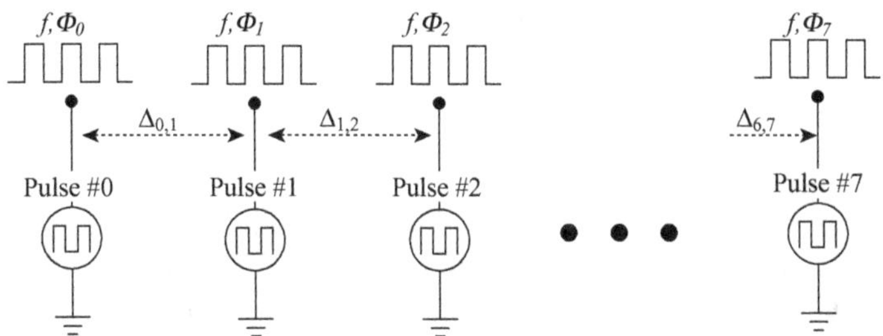

Fig. 4.46. Model of representing the flying-adder's input mismatch.

TABLE 4.6. VCO Signals Selection Sequence of Various FREQs

FREQ	4	5	6	7	8	9	10	11	12	13	14	15	16
1st cyc.	0	0	0	0	0	0	0	0	0	0	0	0	0
2nd cyc.	4	5	6	7	0	1	2	3	4	5	6	7	0
3rd cyc.	0	2	4	6	0	2	4	6	0	2	4	6	0
4th cyc.	4	7	2	5	0	3	6	1	4	7	2	5	0
5th cyc.	0	4	0	4	0	4	0	4	0	4	0	4	0
6th cyc.	4	1	6	3	0	5	2	7	4	1	6	3	0
7th cyc.	0	6	4	2	0	6	4	2	0	6	4	2	0
8th cyc.	4	3	2	1	0	7	6	5	4	3	2	1	0
9th cyc.	0	0	0	0	0	0	0	0	0	0	0	0	0
Ω[a]	***2***	***8***	***4***	***8***	***1***	***8***	***4***	***8***	***2***	***8***	***4***	***8***	***1***
Dist.[b]	***4***	***3***	***2***	***1***	***0***	***1***	***2***	***3***	***4***	***3***	***2***	***1***	***0***

[a]This is periodicity.

[b]This is the corresponding distance to 8 or 16.

- For all the FREQ settings, the address patterns will repeat after at most eight flying-adder output cycles.
- There are four address patterns. They can be classified as P1 (8, 16), P2 (4, 12), P4 (6, 10, 14), and P8 (3, 5, 7, 9, 11, 13, 15).
- P1 has only one address sequence. P2 has two address sequences (e.g., 0→4, 4→0). P4 has four address sequences. P8 bears eight address sequences. The number of address sequences is defined as periodicity Ω.
- For a given FREQ and a given VCO mismatch configuration, each address sequence corresponds to a unique output period (refer to Eq. 4.17).
- For group P1, all mismatches are canceled (refer to Eq. 4.18).

The mismatch reset point is signal0. Whenever the address crosses 0, the previously accumulated mismatch effect vanishes. For any given FREQ, its distance to 8 or 16 (whichever is smaller) determines its mismatch accumulation factor. The larger the distance, the more chance a mismatch will accumulate. Using this argument, between FREQ = 8 and 16, it is predicated that 11, 12, and 13 shall have the potential for the largest period variation, with 12 having the greatest. However, this is only true statistically for a large number of VCO mismatch configurations and different starting addresses. Since 12 belongs to group P2, which has only two address sequences, it might actually experience a small period variation for a given VCO mismatch configuration and a given initial address.

When a post divider of ratio M is used after the flying-adder synthesizer, its cycle length is increased to $M{\cdot}T_s$ (Section 4.6.2). However, the absolute magnitude of the mismatch is unchanged since the mismatch resets after one VCO period. Therefore, the output period's error upper bound is improved from 1/FREQ to 1/(M·FREQ). In certain cases, a divider can help completely eliminate the mismatch effect. From a circuit perspective, a divider is an edge selector and it discards certain edges from the source signal. At the same time, it discards the mismatch associated with the edges as well. For example, for the case of FREQ = 12 of address sequence 0→4→0→4 . . . , a divider of $M = 2$ makes the address sequence become 0→0→0→0 . . . , which is integer multiples of the VCO period. Hence, mismatch will not be seen at the divider output.

4.18.4 The Mismatch's Impact on Output Period

Using the model created in Section 4.18.2, behavior-level simulations are performed to study the output signal's time domain characteristics. Three parameters are observed from a large number of simulations: the output periods' distribution, the output periods' time trend, and the output periods' maximum deviation.

From Table 4.6, it is clear that FREQ from 8 to 12 can cover all the mismatch scenarios from best to worst case when the frequency control word is concerned. They include cases from all the four address patterns P1, P2, P4,

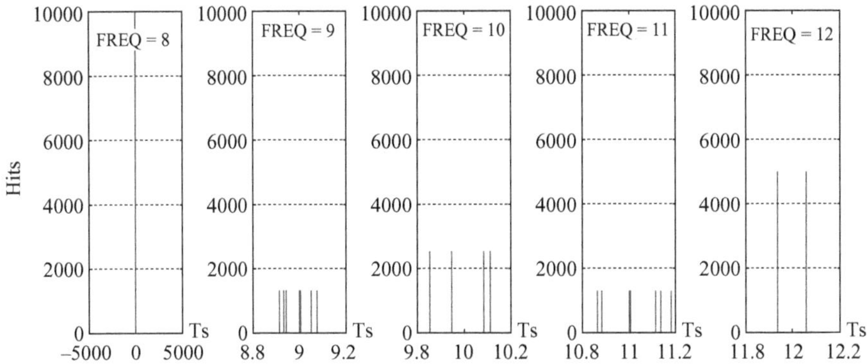

Fig. 4.47. The flying-adder output's 10K-cycles-period distribution for five different FREQ: 8, 9, 10, 11, and 12.

and P8. From the developed behavior-level simulator, Fig. 4.47 shows the simulation result of the output period distribution under these FREQ settings. In each setting, there are 10,000 flying-adder cycles simulated. The x-axis is the flying-adder output period in units of Δ. The span is 0.4Δ for all the cases except where FREQ = 8. The y-axis is the number of samples. These simulations are performed for a particular VCO mismatch configuration. It is interesting to recognize the following:

- The number of bins agrees with the number address sequence, or periodicity Ω (some bins are collapsed in plotting; there is a need to zoom in to view). This is because each address sequence corresponds to one unique period.
- Under one FREQ setting, all the bins have same strength (same number of samples). For example, the number of samples for all eight bins in FREQ = 9 case is $10{,}000/\Omega = 1250$. This is because all the address sequences occur sequentially with same possibility.
- The mean of each distribution is at its center (the value of FREQ). This is because the address sequence has a periodicity of Ω. The mismatch effect resets after Ω flying-adder cycles.

As expected, all the mismatches are canceled in the case of FREQ = 8 (there is only one bin, and it is at the expected position). The most important observation from this figure is that once the mismatch configuration is given, the period distribution is deterministic. No randomness is observed.

Figures 4.48 and 4.49 plot the time trend of the flying-adder output's periods. In Fig. 4.48, for a given particular VCO mismatch configuration, four different FREQ settings are used for simulation. From top to bottom, the FREQs are 9, 10, 11 to 12. Figure 4.49 shows four different VCO mismatch configurations under the same setting of FREQ = 13. The important observations from these

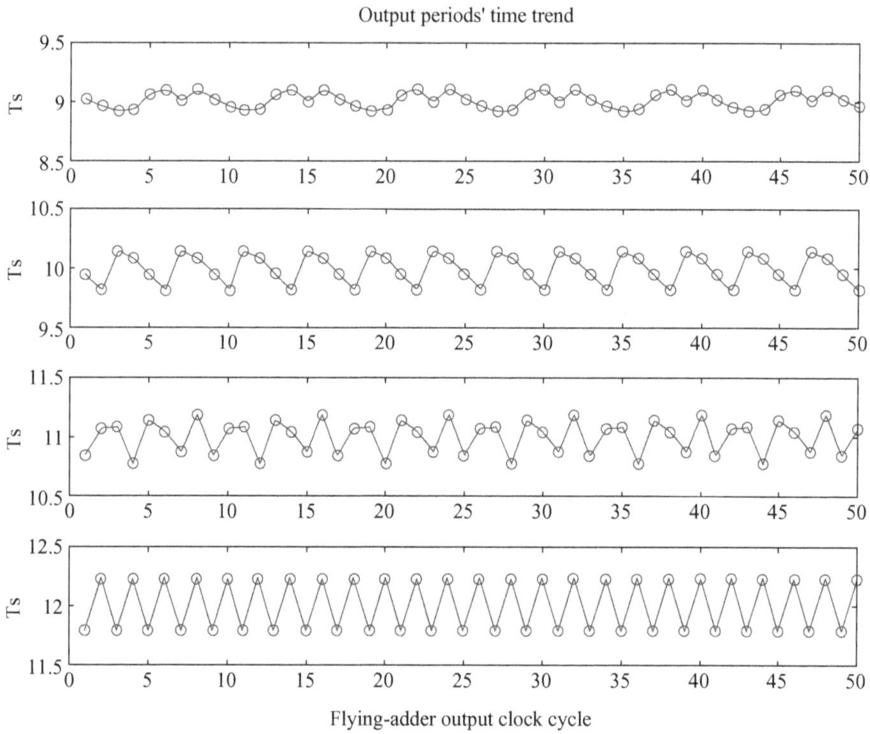

Fig. 4.48. The flying-adder output's period time trend for different FREQ settings under one particular VCO mismatch configuration.

plots are that (1) the mismatch-induced period variation is periodic, (2) the periodicity found in the variations equals to Ω, (3) the VCO mismatch configuration has no impact on periodicity (however, it influences the size of the variation), and (4) the period variation is bounded. It is not accumulative. For Ω flying-adder cycles, the average period equals to the ideal period (FREQ). In other words, everything resets after Ω cycles.

Previous study investigates the mismatch-induced period variation for individual VCO mismatch configurations. Using the model of Section 4.18.2, a large number of simulations are carried out to reveal the statistical relationship between the VCO mismatch and the frequency control word. We want to find the maximum period deviation from its ideal value under all the VCO mismatch scenarios. This number will directly correspond to the *bound* of the output signal's peak-to-peak jitter. Table 4.7 lists the simulation result. The simulations have been carried out under two scenarios of 5,000 and 50,000 VCO mismatch configurations. It shows that, statistically, the further the FREQ is away from 8, the larger the peak-to-peak jitter would be (since the mismatch effect can accumulate more). The period's maximum deviation is found to be 1–2% of the output period for most of the FREQ settings.

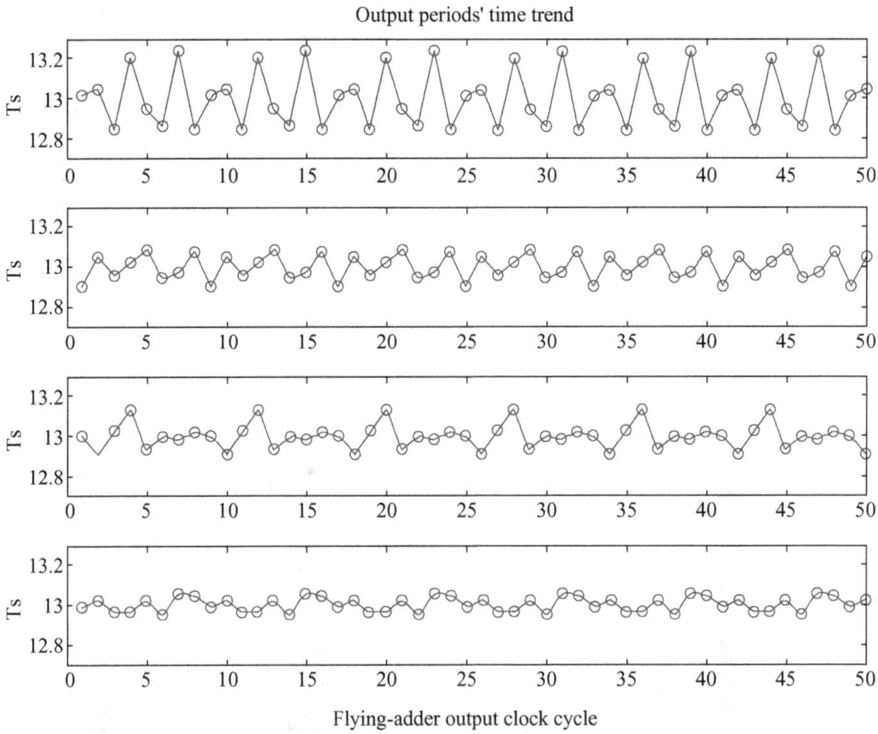

Fig. 4.49. The flying-adder output's period time trend for four different VCO mismatch configurations under the same FREQ setting of 13.

TABLE 4.7. Simulation Result for Periods' Maximum Deviation

	5000 VCO mismatch configurations		50,000 VCO mismatch configurations	
FREQ	Max. Devi.[a]	Percentage[b]	Max. Devi.	Percentage
4	0.312	±3.9	0.323	±4.04
5	0.276	±2.76	0.290	±2.9
6	0.199	±1.66	0.200	±1.67
7	0.100	±0.72	0.100	±0.72
8	0	0	0	0
9	0.100	±0.56	0.100	±0.56
10	0.198	±0.99	0.200	±1.00
11	0.272	±1.24	0.294	±1.34
12	0.322	±1.34	0.320	±1.33
13	0.291	±1.12	0.290	±1.12
14	0.199	±0.71	0.199	±0.72
15	0.100	±0.33	0.100	±0.33
16	0	0	0	0

[a] In units of Δ.
[b] Calculated as maximum deviation divided by FREQ.

Figure 4.50 reveals this fact graphically. In the simulations, 10,000 VCO mismatch configurations are used for each FREQ setting. Then the periods are all displayed with the x-axis span of one Δ. It is seen that the case of FREQ = 12, although with a small Ω of 2, has the greatest span in its period distribution (greatest *potential* peak-peak jitter). This is because it bears the greatest distance to 8 (or 16). The cases of FREQ = 7, 9, and 15 have relatively small spans since they are close to 8 or 16.

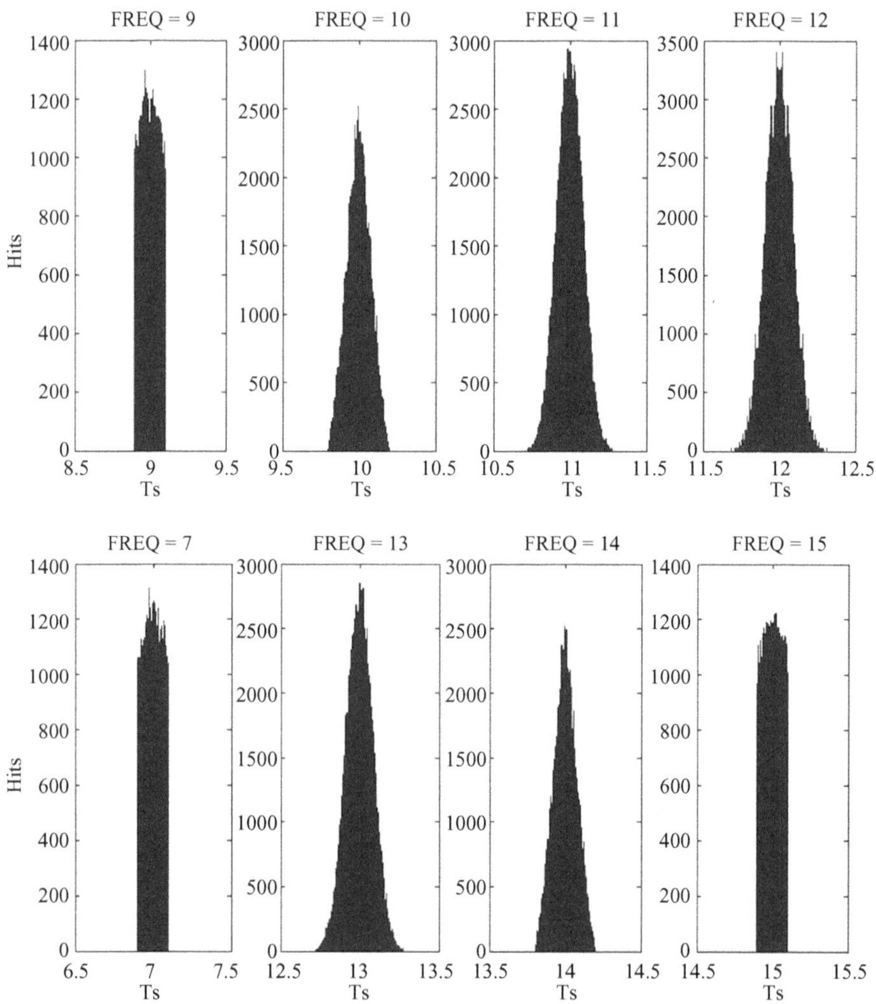

Fig. 4.50. The flying-adder output's period distribution from 10,000 VCO mismatch configurations.

4.18.5 The Mismatch's Impact on Output Spectrum

Figure 4.51 is the result of FFT performed on 1048576 periods of flying-adder output for the cases of FREQ = 9, 10, 11, 12, and 13, respectively. Clearly, their periodicities agree with the Ω found in Table 4.6. They also align with the observations obtained from Fig. 4.48. Figure 4.52 illustrates the point that, although the stems' locations are determined by Ω, their strengths vary with the VCO mismatch configuration. This plot is another way of illustrating the same fact revealed by Fig. 4.49.

Transistor-level simulations are carried out to further evaluate the predication obtained from previous analysis. Figure 4.53 is the flying-adder output periods' time trend plot (counterpart of Fig. 4.49). The VCO is running at 2 GHz (Δ = 62.5 ps). From top to bottom, FREQ are 9, 10, 11, 12, 13, 14, and

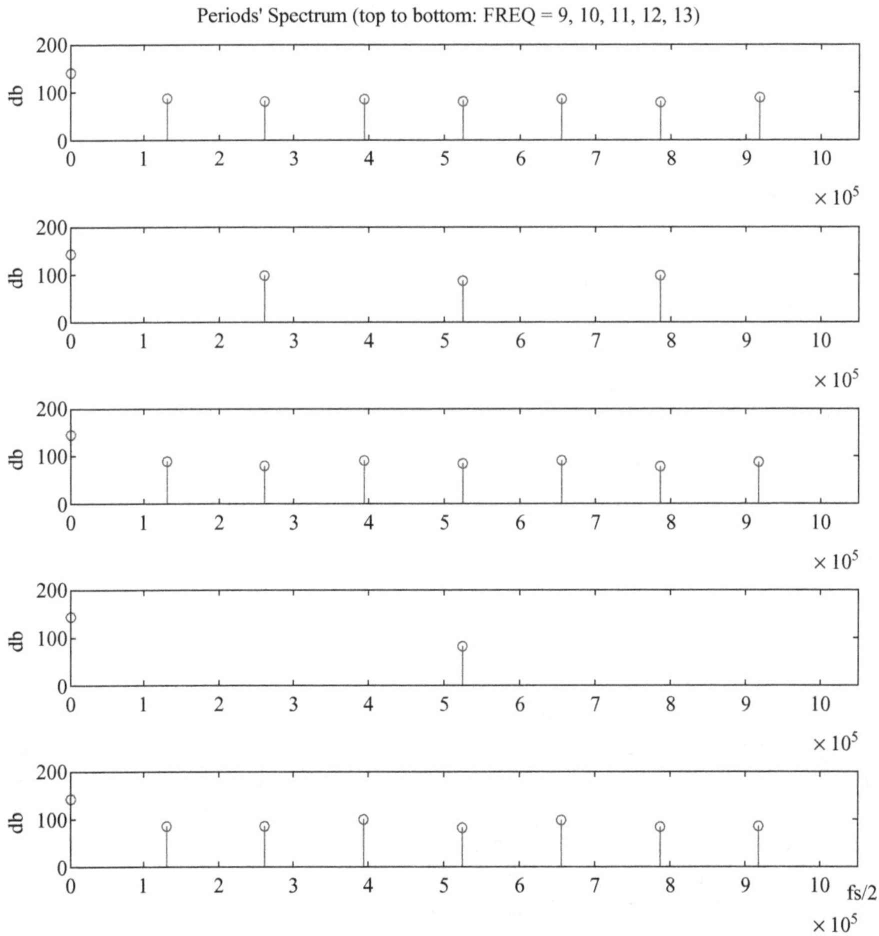

Fig. 4.51. Periods' spectrums of various FREQs.

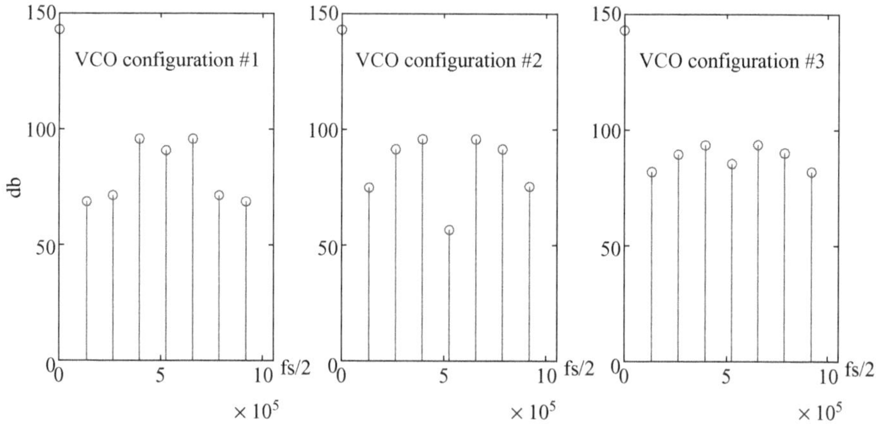

Fig. 4.52. Periods' spectrums of different VCO mismatch configurations (FREQ = 13).

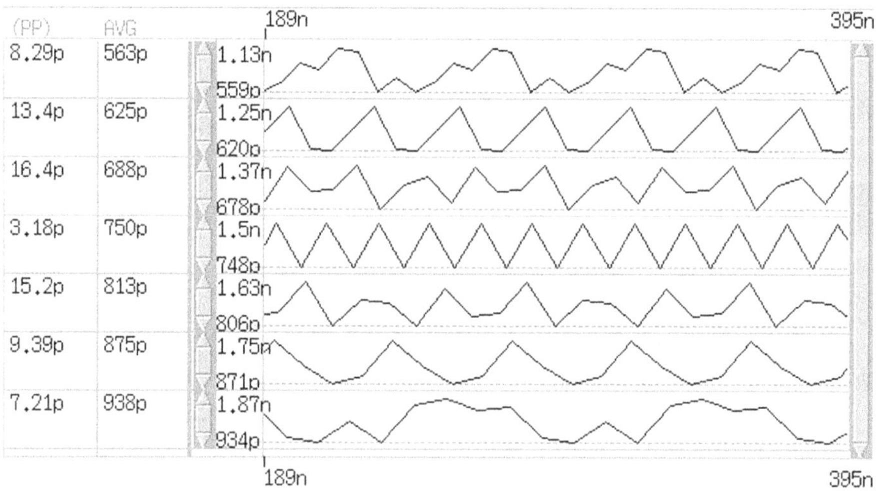

Fig. 4.53. Transistor-level simulation of a flying-adder output's period time trend under one VCO mismatch configuration (from top to bottom, FREQ = 9, 10, 11, 12, 13, 14, 15).

15 respectively. The number in the first column is the periods' peak-peak value of their variation. The one in the second column is the average period (frequency). It can be seen that the averages align with the corresponding calculated frequencies for each FREQ, which indicates that the mismatch effect is not cumulative. Instead, it resets regularly. It is also clear from the waveform that the periodicity embedded in the periods agrees with Ω. For the case of FREQ = 10 (the second one from the top), $\Omega = 4$. Although it seems like there are only three discrete periods, there are actually four. Two of them are very close to each other. Figure 4.54 shows the corresponding clock pulses'

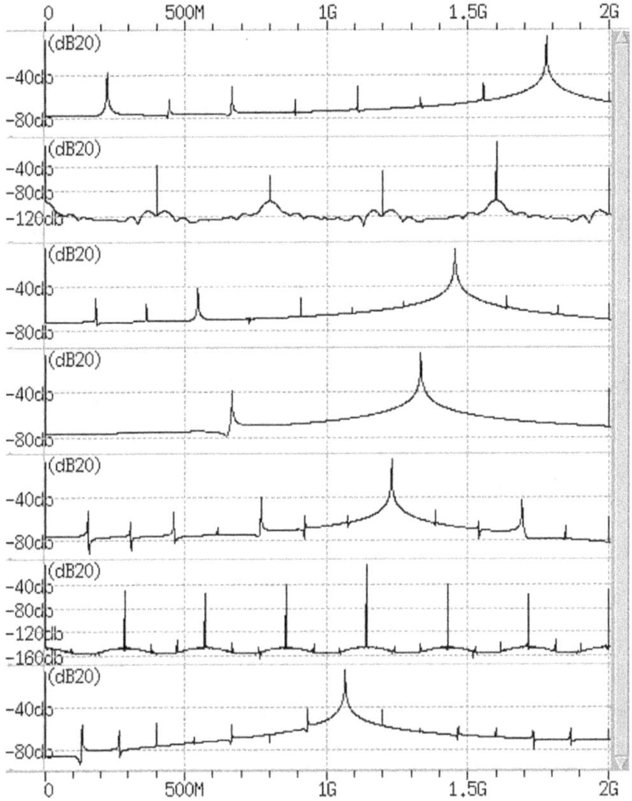

Fig. 4.54. Transistor-level simulation of a flying-adder clock pulse spectrum under one VCO mismatch configuration (from top to bottom FREQ = 9, 10, 11, 12, 13, 14, 15).

spectrums. The output frequencies are all at their expected locations. Not surprisingly, the mismatch-induced spurs are spaced at f_s/Ω, which is already predicted from the previous study.

Figures 4.55–4.57 show the transistor-level simulation result for three different VCO mismatch configurations. In these simulations, $f_{vco} = 2$ GHz and $F = 10$ ($f_s = 1.6$ GHz, 625 ps). Figure 4.45 shows the periods' time trend. Figure 4.46 is the period distributions. These plots, although significantly different, all show the periodicity of 4 since $\Omega = 4$ for $F = 10$. The spectrums for all three cases have their spurs located at the same locations. They all spaced at $f_s/\Omega = 400$ MHz. However, the spurs' strengths are slightly different, as expected.

4.18.6 Summary on Mismatch's Impact (Xiu 2011)

Based on the previous analysis, listed below are summaries of the mismatches' impacts. Although they are obtained from this special case of $K = 8$, it is

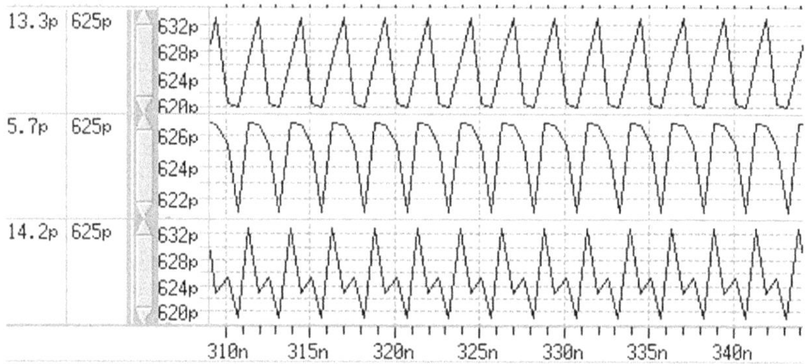

Fig. 4.55. Periods' time trend for three VCO mismatch configurations ($F = 10$, $f_{vco} = 2$ GHz).

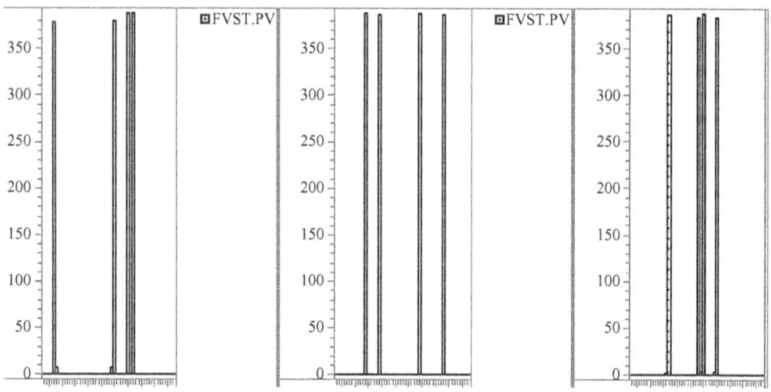

Fig. 4.56. Periods' distribution for three VCO mismatch configurations ($F = 10$, $f_{vco} = 2$ GHz).

believed that they are valid in a general sense. These observations will be compared against the experiment data in Section 4.21.10.

- The impact of design-layout-mismatch-induced time error on flying-adder output is deterministic. It is not cumulative. Its influence on the flying-adder output's peak-peak jitter is bounded.
- The transference of input time error to the flying-adder output's period jitter depends on the frequency control word used (frequency dependent).
- The input time error causes the output's period to vary periodically. The periodicity equals to the number of unique address sequences.
- In the frequency spectrum, the input time error induces spurs spaced at f_s/Ω, where f_s is the output frequency and Ω is the periodicity embedded in the address sequence.

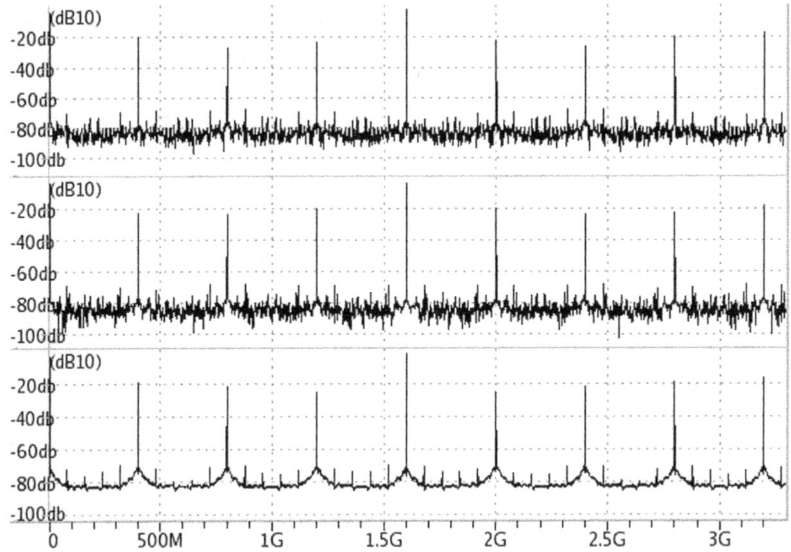

Fig. 4.57. A flying-adder output's spectrum for three VCO mismatch configurations ($F = 10, f_{vco} = 2$ GHz).

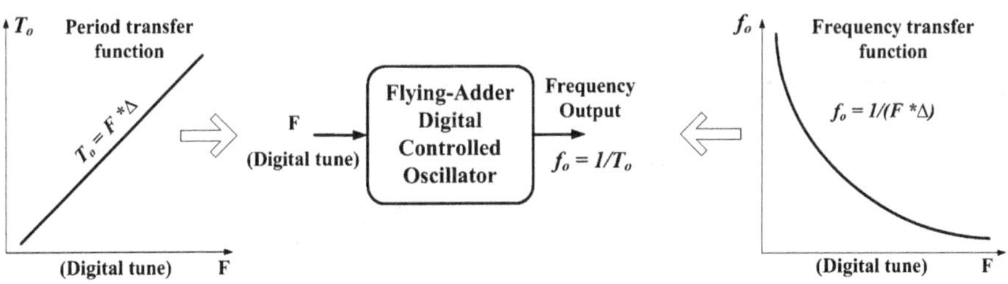

Fig. 4.58. A flying-adder circuit as a digital controlled oscillator.

4.19 FLYING-ADDER CIRCUIT AS DIGITAL CONTROLLED OSCILLATOR

The discussion in Section 4.5 reveals two flying-adder distinguished features: *arbitrary frequency generation* and *instantaneous response*. These features make it suitable for a flying-adder circuit to function as a digital controlled oscillator (DCO) (FADCO). From the flying-adder frequency transfer function Eq. 4.5, its period and frequency tuning curves are plotted in Fig. 4.58. The tuning variable is the digital frequency control word FREQ (*F* for short). As can be seen, its period increases linearly with *F*; its frequency is inversely proportional to *F*. More importantly, these curves are single-value and monotonic. This guarantees that *one and only one* frequency is produced for an input digital code. This is a necessary condition for an oscillator.

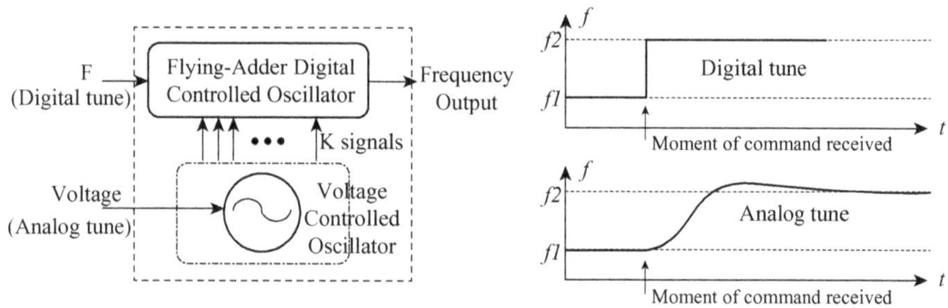

Fig. 4.59. A flying-adder oscillator with both digital and analog tunes.

If the K inputs are generated from a VCO and the VCO control voltage is regarded as another control parameter, there will be two frequency tuning knobs in the oscillator of Fig. 4.59. In this configuration, the digital and analog tunes are independent of each other. The digital tune has a fast response, owing to the flying-adder's instantaneous response. The analog tune is much slower since it usually uses a feedback mechanism to compare its output with a known reference. The FADCO will be used in Chapter 6 when we discuss system-level innovations.

4.20 FLYING-ADDER TERMINOLOGY

Along the path of flying-adder technology development, the same baseline flying-adder circuit has been used in various applications. The term "flying-adder" has been used in different environments for different purposes. For the convenience of future discussions, more clearly defined terms on its usage are preferred.

- **FADPS.** Flying-adder direct period synthesizer. This term refers to the circuit in Fig. 4.17.
- **FAPLL.** A flying-adder circuit is used as an on-chip frequency generator. In this application, since an integer-N PLL is usually incorporated with the synthesizer, the circuit is often called FAPLL (Fig. 4.27).
- **IFAPLL.** Integer-flying-adder PLL; one flying-adder synthesizer is incorporated inside the PLL loop as a fractional divider. For all the other synthesizers outside the PLL loop, only integers and PDFR-compatible fractions are used (Fig. 4.32).
- **FADCO.** Flying-adder DCO; a flying-adder circuit is used as a digital controlled oscillator (Fig. 4.59).
- **FADFLL.** Flying-adder digital FLL; FADCO is used as an oscillator in a digital frequency locked loop (refer to Chapter 6, Section 6.10).

- **FADPLL.** Flying-adder digital PLL; FADCO is used as an oscillator and a delay generator in a digital phase locked loop (refer to Chapter 6, Section 6.11).
- **FA Divider.** A flying-adder circuit is used as a fractional divider that can reach a time resolution that is finer than its input reference's period (Fig. 4.28).
- **FAFSK.** Flying-adder FSK modulator; a flying-adder circuit is used as a frequency shift keying modulator (refer to Chapter 6, Section 6.17).
- **FAPWM.** Flying-adder PWM modulator; a flying-adder circuit is used as a pulse width modulator (refer to section 6.18).

4.21 FLYING-ADDER SYNTHESIZER AND TIME-AVERAGE-FREQUENCY: THE EXPERIMENTAL EVIDENCE

As the workhorse for time-average-frequency, a flying-adder synthesizer is uniquely equipped with capability of implementing this new frequency concept in circuit. The two most distinguished flying-adder features are stated below.

1. Arbitrary frequency generation. As long as enough bits are used for the frequency control word's fractional part, any frequency resolution can be virtually achieved since frequency is generated by simple open-loop counting.
2. Instantaneous response speed. The period (frequency) can be changed in next two cycles after a command is received.

In this section, the result from a FAPLL implemented in a 55-nm process will be used to provide experimental evidence for supporting the above statements.

4.21.1 The FAPLL Structure

The FAPLL of Fig. 4.27 is designed and manufactured in a 55-nm process. The integer-N PLL is a charge pump PLL commonly used in industry. Its block diagram is shown in the left drawing of Fig. 4.60. In this particular implementation, four differential stages are used in the VCO ($K = 8$). Its frequency tuning control is the supply voltage. The PLL uses a 125-MHz frequency source as reference. The VCO can operate in a range from 800 MHz to 2.75 GHz.

Figure 4.61 show this VCO's measured period jitter at 2 GHz. The *rms* jitter is 2.97 ps and the peak-peak jitter is 23.8 ps with 3.7-million samples (measured with 16 GHz, 100 Gs/s Tektronix DSA71640C). Figure 4.62 includes the spectrum plot and phase noise plot at 2.5 GHz and 2 GHz, respectively (Agilent Spectrum Analyzer E4440A PSA, 25 GHz). The left plot in Fig. 4.62 is the frequency spectrum of $f_{vco} = 2.5$ GHz. The plot at right is the phase noise plot at

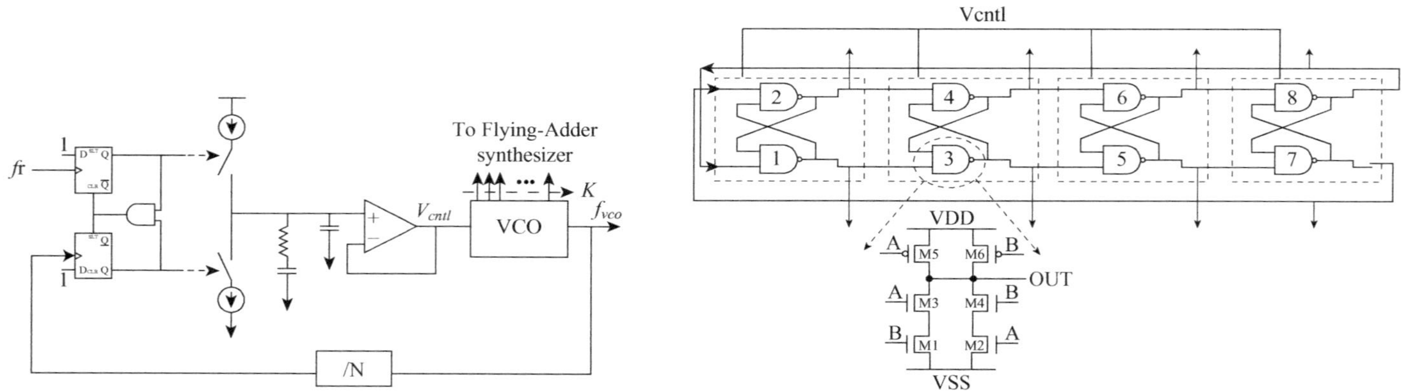

Fig. 4.60. The PLL (left) and the VCO (right, $K = 8$) for generating the flying-adder K inputs.

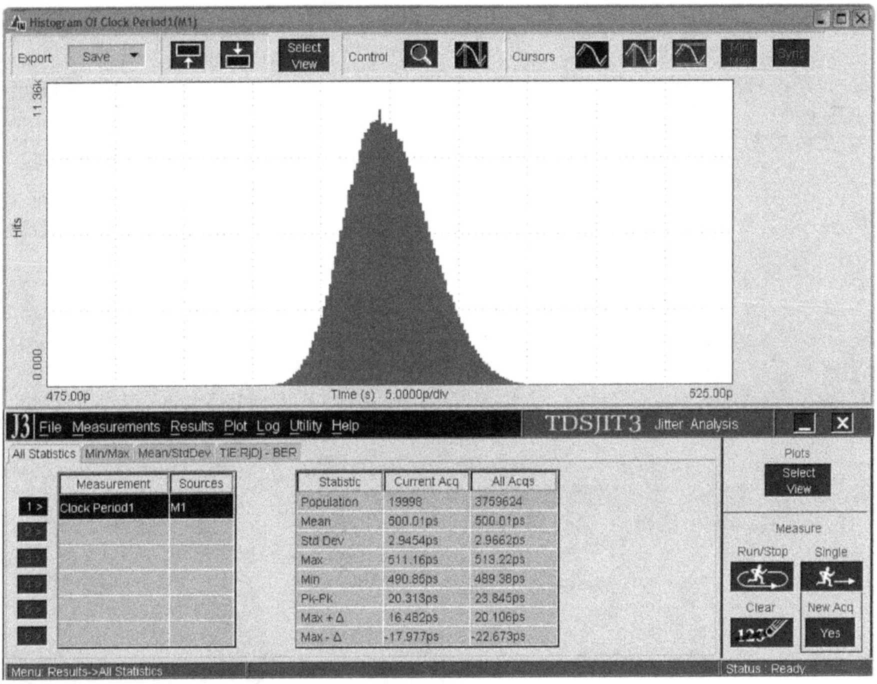

Fig. 4.61. VCO output at 2 GHz. Period jitter: *rms* 2.97 ps, peak-peak 23.8 ps, 3.7 million samples.

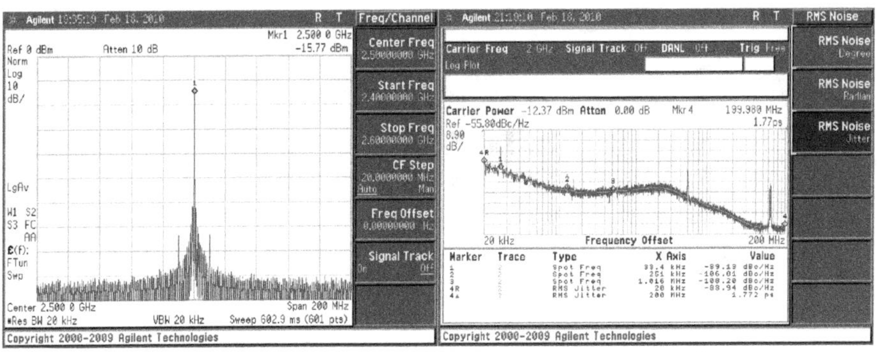

Fig. 4.62. The spectrum plot of a VCO output at 2.5 GHz (left). The phase noise plot of VCO at 2 GHz.

$f_{vco} = 2$ GHz. At the 1-MHz offset, the noise number is –108 dBc/Hz. The integrated *rms* jitter from 20 KHz to 200 MHz is 1.772 ps. In the following flying-adder-related measurements, the eight outputs from this VCO will be used.

The flying-adder synthesizer used in this experiment is the circuit in Fig. 4.17. There are 24 bits in its control word FREQ. FREQ[23:20] is the integer part; FREQ[19:0] is the fraction.

TABLE 4.8. Measured Period Jitter from a Flying-Adder Synthesizer

FREQ	Frequency/period (GHz, ps)	*rms* jitter (ps)	pk-pk jitter (ps)	Percentage of pk-pk/ period	Number of acquisitions
8	2 GHz, 500 ps	2.75	21.93	±2.19%	4,579,542
9	1.78 GHz, 562.5 ps	6.2	36.72	±3.26%	3,288,560
10	1.6 GHz, 625 ps	3.68	28.93	±2.31%	2,495,688
11	1.46 GHz, 687.5 ps	5.80	36.71	±2.67%	1,527,120
12	1.33 GHz, 750 ps	2.91	22.32	±1.49%	1,519,848
13	1.23 GHz, 812.5 ps	5.67	37.42	±2.30%	1,279,824
14	1.14 GHz, 875 ps	3.59	27.61	±1.58%	742,820
15	1.07 GHz, 937.5 ps	4.96	35.59	±1.90%	885,278
16	1 GHz, 1 ns	3.70	22.4	±1.12%	1,019,796

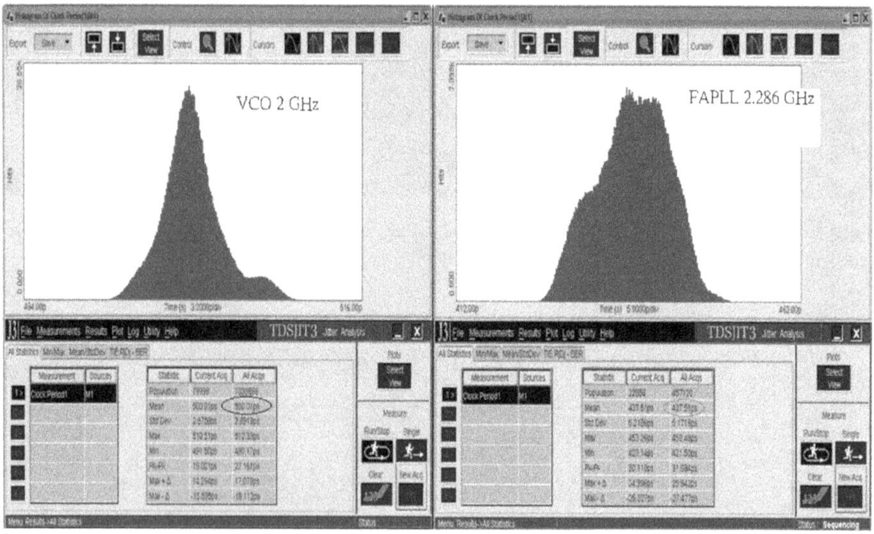

Fig. 4.63. FAPLL output when FREQ = 7 (right, 457 K samples) and FREQ = 8 (left, 1.04 million samples). FAPLL can produce a higher frequency than f_{vco}.

4.21.2 Jitter Performance

The FAPLL's frequency calculation function is expressed in Eq. 4.8: $f_o = ([K \cdot N]/[F \cdot M])f_r$. In this implementation, $K = 8$ and $f_r = 125$ MHz. Table 4.8 lists the measured jitter under the setting of $N = 16$ ($f_{vco} = 2$ GHz) and M = 1. Figure 4.63 shows the FAPLL output period distribution when FREQ = 7 (2.286 GHz, 437.5 ps). This is a very unique case. It illustrates the fact that, when $F < K$, FAPLL can produce a higher frequency than that of its input. In the conventional PLL-based indirect frequency synthesis approach, the highest frequency is always at the VCO.

4.21.3 Frequency Generation Capability

The function of $f_o = ([K \cdot N]/[F \cdot M])f_r$ is in $1/x$ fashion when F is the variable, as illustrated in Fig. 4.58. Figure 4.64 shows the measurement result in this regard. In these plots, $K = 8$, $N = 8$ ($f_{vco} = 1$ GHz) and $M = 1$. The plots at left shows the measurement when F takes the values from 4 to 15. The trace is made of the calculated values from Eq. 4.8 based on $\{K = 8,\ N = 8$ and $M = 1,$ $f_r = 125$ MHz$\}$. The dots are the actual measured results. The plot at right is the zoom-in view around the area of $F = 8$. Again, the curve represents the expected values. The dots are the measured results. The errors are in the range of 10^{-8} for all the points when error is defined as *error = (calculated – measured)/calculated.* This frequency measurement is carried out by using the Agilent HP53131A frequency counter with 12-digit accuracy. Since its highest frequency capability is 225 MHz, the measurement is actually done with setting of M = 8 and then the results are converted back to M = 1. Figure 4.64 confirms the FAPLL's $1/x$ transfer function style (left plot). It also shows that, in a small region, the transfer function becomes linear (right plot). Furthermore, the plot at left provides more hard evidence that a flying-adder synthesizer can produce frequencies that are higher than its input frequency (recall that $f_{vco} = 1$ GHz in this measurement).

4.21.4 Frequency Resolution

Table 4.9 lists the measurement result regarding the frequency resolution (Agilent HP53131A). The input reference is $f_r = 125$ MHz and $N = 8$ ($f_{vco} = 1$ GHz). The settings of FREQ and M are all listed in the table. In this table, every three FREQ settings are clustered as one group. FREQ[23:20] is the integer part; FREQ[19:0] is the fractional part. The difference between the two adjacent FREQs is one LSB. In other words, a 2^{-20} is added to FREQ[23:0] each time. The calculated f_o (according to Eq. 4.8) and measured f_o are listed side by side for comparison. The frequency resolution is defined as current frequency minus the next frequency (from measured values). From Eq. 4.7, the resolution can be roughly predicted from $df_s/f_s = -dF/F$, which is equivalent to $df_o/f_o = -dF/F$ (after the divider). In the last two columns, the df_o/f_o from measured values is listed alongside dF/F.

4.21.5 Frequency Spectrum

Figure 4.65 shows the measured FAPLL's output spectrum and phase noise by Agilent Spectrum Analyzer E4440A PSA. In this measurement, $N = 16$ and $F = 10$, which gives $f_{vco} = 2$ GHz and $f_s = 1.6$ GHz. The 10-MHz and 25-MHz spurs are caused by other sources presented in the system, not related to FAPLL. In the phase noise plot (right), the 125-MHz PLL reference is visible. The noise at the 1-MHz offset is -110 dBc/Hz, which is at the same level as the VCO output (Fig. 4.62). Comparing the two phase noise plots of Figs. 4.62

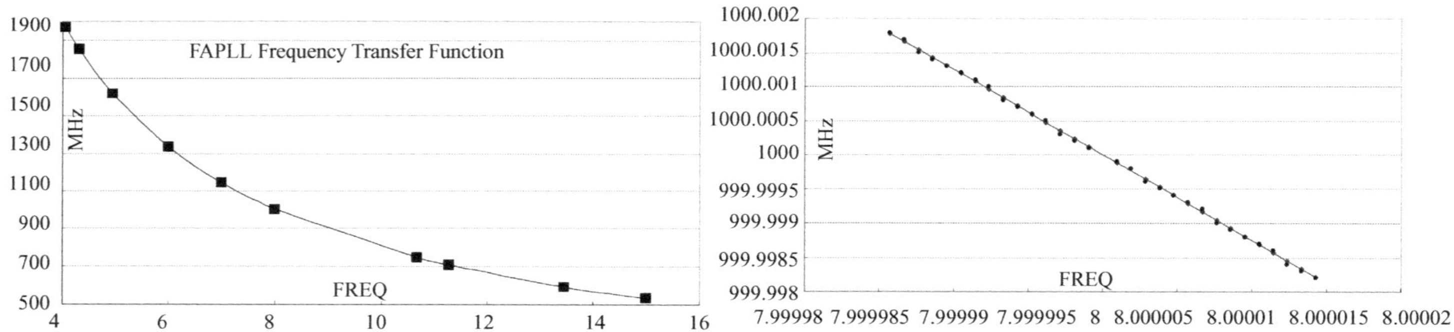

Fig. 4.64. Measured FAPLL frequency function (left) and zoom-in around $F = 8$ (right).

TABLE 4.9. Frequency Resolution Measurement

FREQ (F)	F[23:0]	M	f_o (MHz)	f_o Measured	Resolution (Hz)[a]	df_o/f_o	dF/F
14.941174507	$EF0F0D_{16}$	250	2.141732565	2.141681475	0.136703792	0.0630	0.0669
14.941175461	$EF0F0E_{16}$	250	2.141732428	2.14168134	0.136703775	0.0621	0.0669
14.941176414	$EF0F0F_{16}$	250	2.141732292	2.141681207	N/A	N/A	N/A
8.941174507	$8F0F0D_{16}$	250	3.578948154	3.57886273	0.381734029	0.1006	0.1118
8.941175461	$8F0F0E_{16}$	250	3.578947773	3.57886237	0.381733948	0.1090	0.1118
8.941176414	$8F0F0F_{16}$	250	3.578947391	3.57886198	N/A	N/A	N/A
10.99999905	$AFFFFF_{16}$	250	2.909091161	2.909021735	0.252211411	0.0911	0.0909
11	$B00000_{16}$	250	2.909090909	2.90902147	0.252211368	0.0883	0.0909
11.00000095	$B00001_{16}$	250	2.909090657	2.909021213	N/A	N/A	N/A
6.941174507	$6F0F0D_{16}$	250	4.610170796	4.610060732	0.633408781	0.1397	0.1441
6.941175461	$6F0F0E_{16}$	250	4.610170162	4.610060088	0.633408606	0.1356	0.1441
6.941176414	$6F0F0F_{16}$	250	4.610169529	4.610059463	N/A	N/A	N/A

[a] Resolution is calculated as current frequency minus the next frequency (measured values).

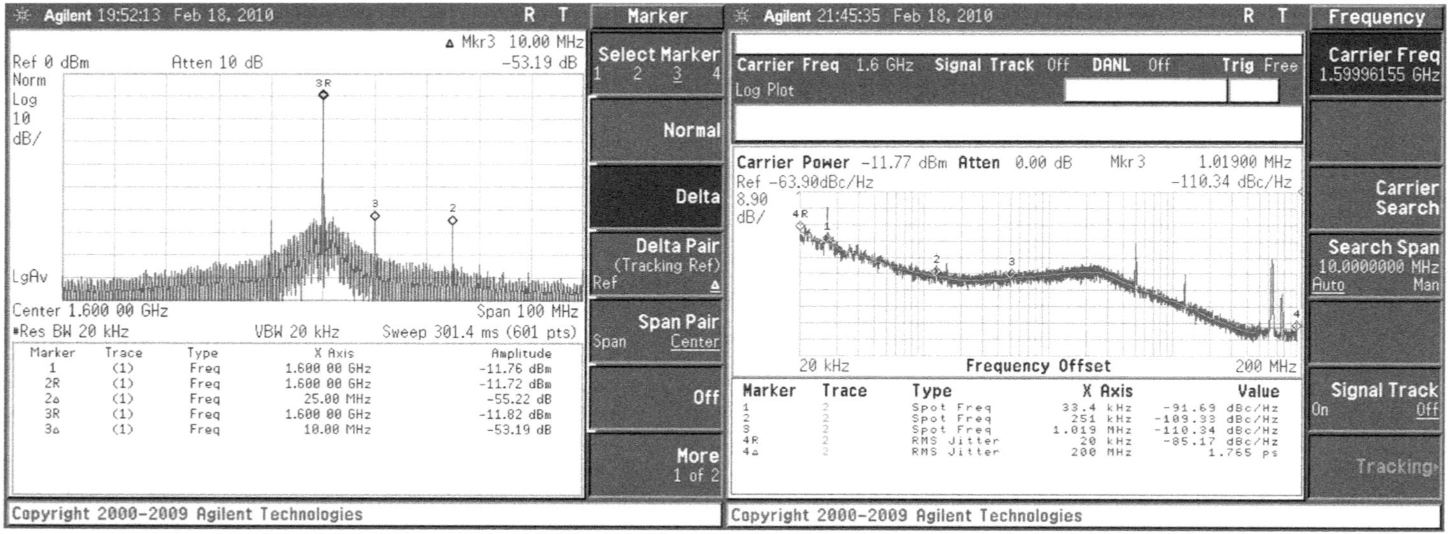

Fig. 4.65. Measured FAPLL output spectrum (left) and phase noise (right), when $F = 10$ (1.6 GHz).

and 4.65 (one is from the VCO, the other is from the synthesizer), it can be seen that the *rms* jitters are both about 1.7 ps. In other words, the flying-adder synthesizer does not add random jitter. This is due to the fact that the flying-adder synthesizer is a driven system; it is not an oscillator in itself (autonomous system).

4.21.6 Instantaneous Switching Demonstration

One of the distinguished features of the flying-adder synthesizer is its capability to switch frequency quickly. As explained in Section 4.5, its output frequency (period) can be changed in the next two cycles after the command is received. Furthermore, the switching is achieved in seamless fashion without a glitch. Figure 4.66 is the demonstration of this feature (LeCory Wavemaster 8500 oscilloscope). In this measurement, $N = 8$ ($f_{vco} = 1$ GHz). In the left plot, the bottom trace is a signal of 1 MHz that is used to control the FREQ. The FREQ alternates between FREQ = 8 ($f_s = 1$ GHz) and FREQ = 12 ($f_s = 667$ MHz). The upper trace is the flying-adder synthesizer's output (after divider M = 8). As can be seen, it alternates between 125 MHz and 83.3 MHz quickly. The plot at right is the magnified view around the area of FREQ as it changes from 8 to 12.

Figure 4.67 shows the instantaneous switching feature in high speed fashion (Tektronix DSA71640C). In this measurement, the f_{vco} is still 1 GHz. The FREQ is forced to switch among three values: 4, 8, and 12, which results in three frequencies of 2 GHz, 1 GHz, and 667 MHz. The FREQ update is controlled by a 50-MHz signal (20 ns). A high-speed differential IO cell is incorporated in the design to bring the high-speed clock signal directly out of the chip for observation. As is apparent from the figure, at 2 GHz, the signal level is significantly reduced due to the high frequency. This is because the amplitude of the output buffer's swing is heavily frequency dependent. As can be seen from the plot, in these 20-ns spans, the output switches quickly among the three frequencies, as expected.

Using another example, Fig. 4.68 shows in detail the seamless characteristic when switching between frequencies. Figure 4.69 plots the trend of the flying-adder output frequency, using the advanced features of the Tektronix DSA71640C. In this case, the f_{vco} is 1 GHz. FREQ alternates among 6, 8, and 10. Consequently, the three frequencies are 1.33 GHz, 1 GHz, and 800 MHz. The FREQ is controlled by a signal of 50 MHz (20 ns). The top plot is the frequency trend versus time (230-ns span). The bottom plot is frequency trend versus cycles (the span is about 4,000 cycles).

4.21.7 Time-Average-Frequency Demonstration

Time-average-frequency is a revolutionary new concept. The flying-adder direct period synthesizer is squarely suitable to generate this type of frequency, owing to its open-loop and direct-synthesis style. This section will provide

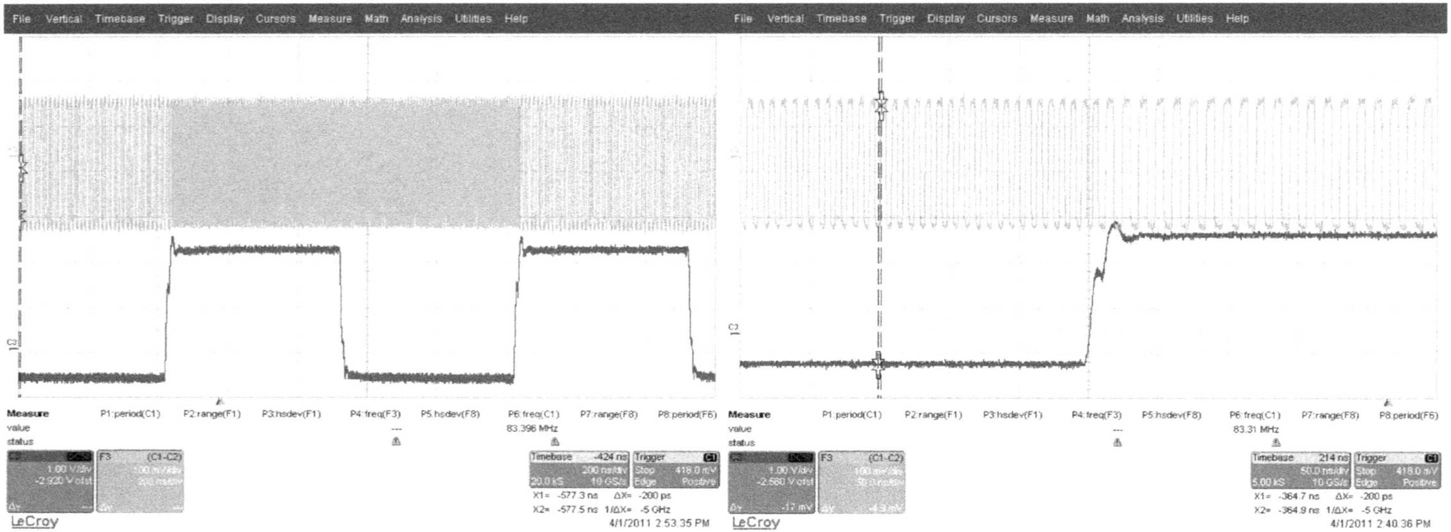

Fig. 4.66. The flying-adder output waveform alternates between 83.3 MHz and 125 MHz.

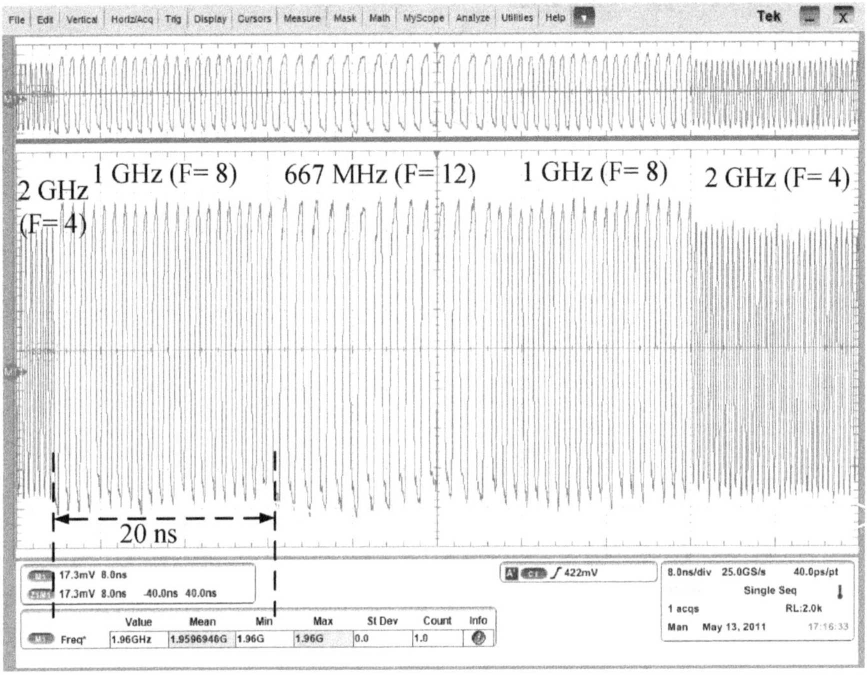

Fig. 4.67. The flying-adder output waveform alternates among 667 MHz, 1 GHz and 2 GHz.

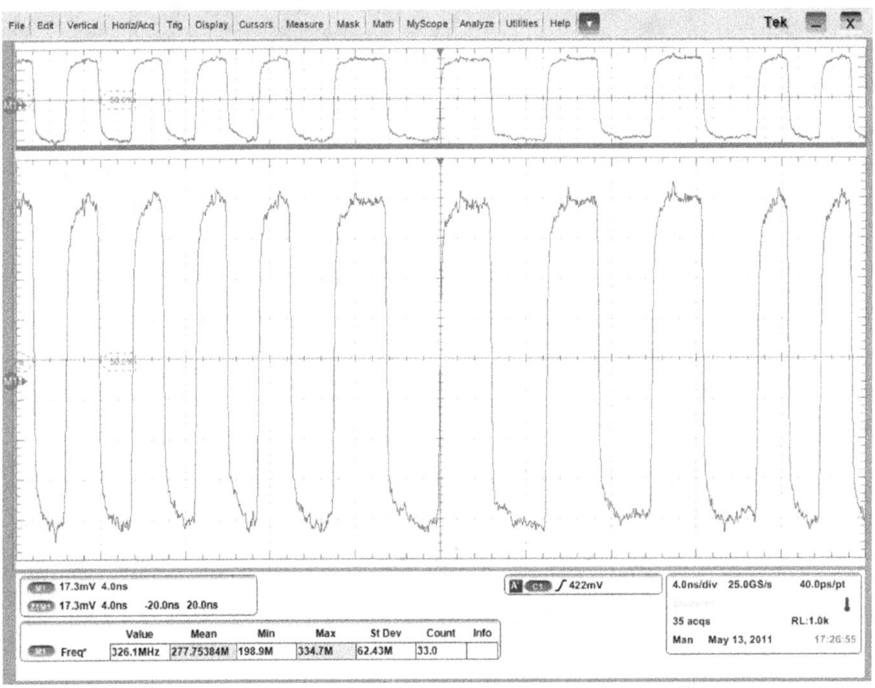

Fig. 4.68. The flying-adder output's seamless switching between frequencies.

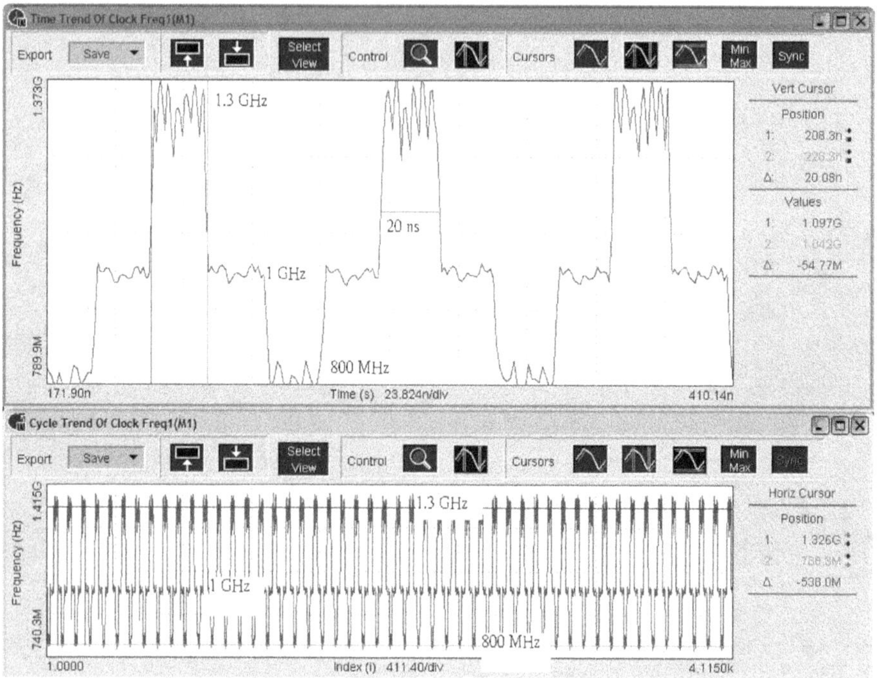

Fig. 4.69. The frequency trend plot: frequency vs. time (top) and frequency vs. cycle (bottom).

some hard evidence on this breakthrough concept. In the following measurement plots, the VCO is set at 2 GHz ($\Delta = 62.5$ ps). The control word FREQ's integer part is 8. Its fractional part is varied to demonstrate the time-average-frequency. The corresponding frequency can be calculated using Eq. 4.8: $f_o = ([K \cdot N]/[F \cdot M]) \cdot f_r$, where $f_r = 125$ MHz, $N = 16$ ($f_{vco} = 2$ GHz, $\Delta = 62.5$ ps), $K = 8$, and $M = 1$.

Figure 4.70 is the case for FREQ = 8. Since no fraction is present in FREQ, the resulting waveform is created in the conventional frequency fashion. The plot at the upper right is the spectrum on the output's period. Clearly visible are 125-MHz reference-induced spurs.

Figure 4.71 shows the case of FREQ = 8.5. Under this setting, the resulting output is 1.88 GHz (531.25 ps) according to Eq. 4.8. In the upper-left plot of period distribution, there are clearly two peaks (distinguishable unique periods). They are the T_A and T_B cycles introduced in Chapter 3. The T_A is generated by FREQ = 8 and T_B by FREQ = 9. They are separated by one Δ of 62.5 ps. Since the fraction is 0.5, the weight is 50% for each. The time-average-frequency is exactly at 531.25 ps as expected. The significant difference between these distributions' shape is due to the input mismatch. As discussed in Section 4.18, all input mismatches are canceled in FREQ = 8

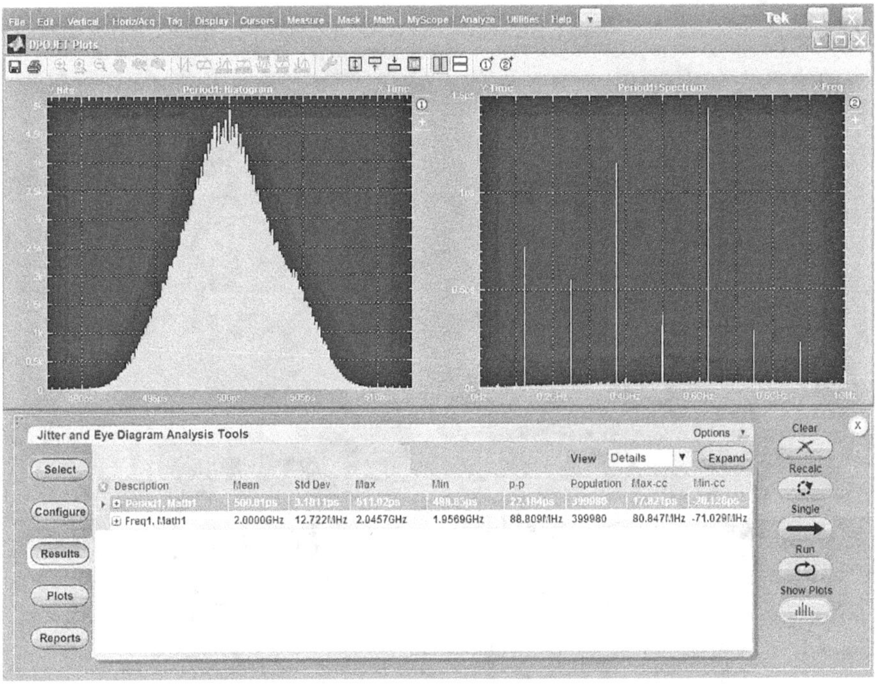

Fig. 4.70. FREQ = 8, conventional frequency.

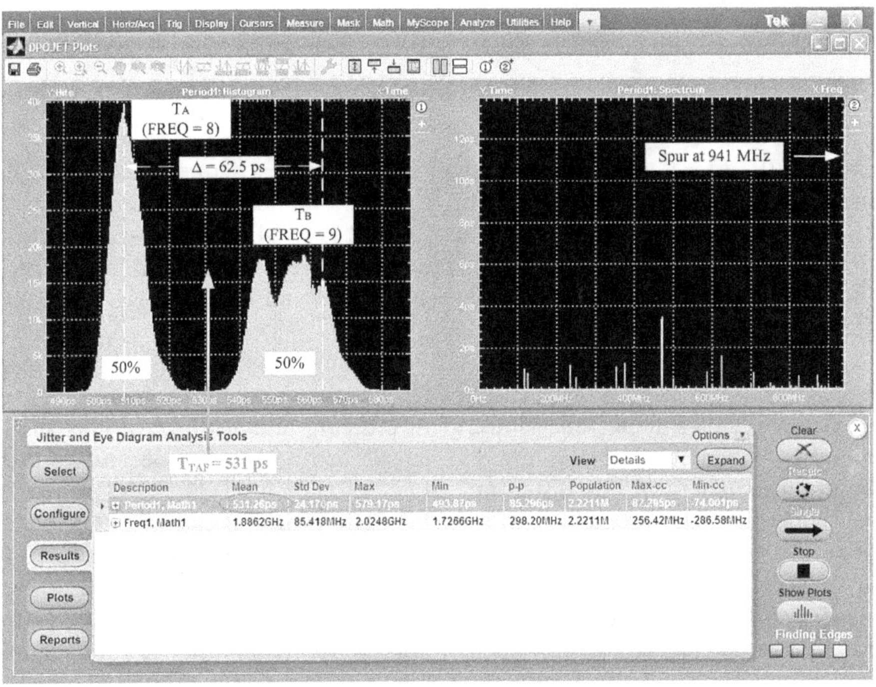

Fig. 4.71. FREQ = 8.5, time-average-frequency.

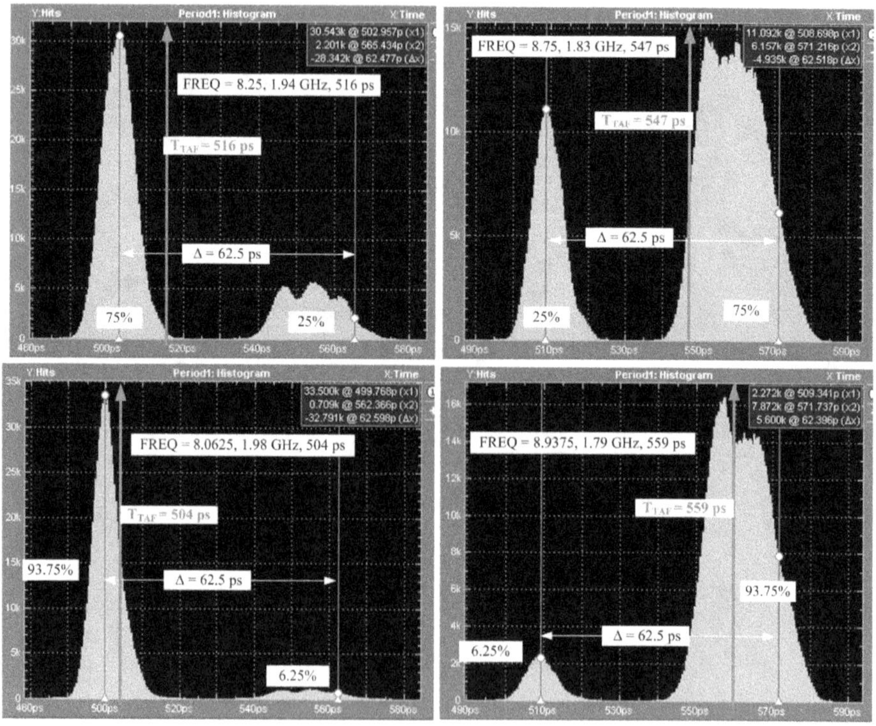

Fig. 4.72. Time-average-frequency of FREQ = 8.25, 8.75, 8.0625, and 8.9375.

case. In FREQ = 9 case, there are eight unique address sequences that result in eight slightly different periods. The plot at the upper right is the spectrum of the output periods. The spur located at $f_o/2$ (941 MHz) is caused by fraction 0.5 (and also by the input mismatch). The remaining weak stems are the input-mismatch-induced spurs spaced at $f_o/8$ and the 125-MHz-induced spurs.

Figure 4.72 shows a few more cases of time-average-frequency at settings of FREQ = 8.25, 8.75, 8.0625 and 8.9375. All the frequencies (periods) are generated at their expected values. Ideally, the T_A and T_B distributions shall be mirror symmetric for 8.25 and 8.75, 8.0625 and 8.9375. But because of the input mismatch, they look different.

Figure 4.73 shows an extreme case of FREQ = $8.01000_{16} = 8 + 2^{-8} = 8 + 1/256 = 8.00390625$. In this case, the T_A to T_B ratio is 255:1. The output frequency is 1.999 GHz (500.25 ps). In the period's spectrum plot, the spurs caused by the fraction 1/256 can be seen (the stronger spurs in the plot are 125-MHz-reference induced). Figure 4.74 further plots the trend of output periods versus cycles. As can be seen from this measurement, for every 256 output cycles, there are 255 T_A cycles and one T_B cycle.

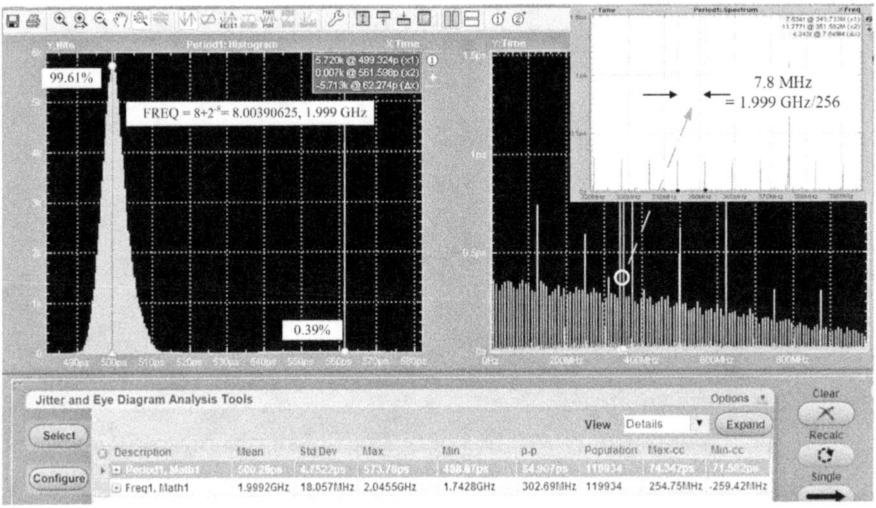

Fig. 4.73. Time-average-frequency of FREQ = $8.01000_{16} = 8+2^{-8} = 8+ 1/256 =$ 8.00390625.

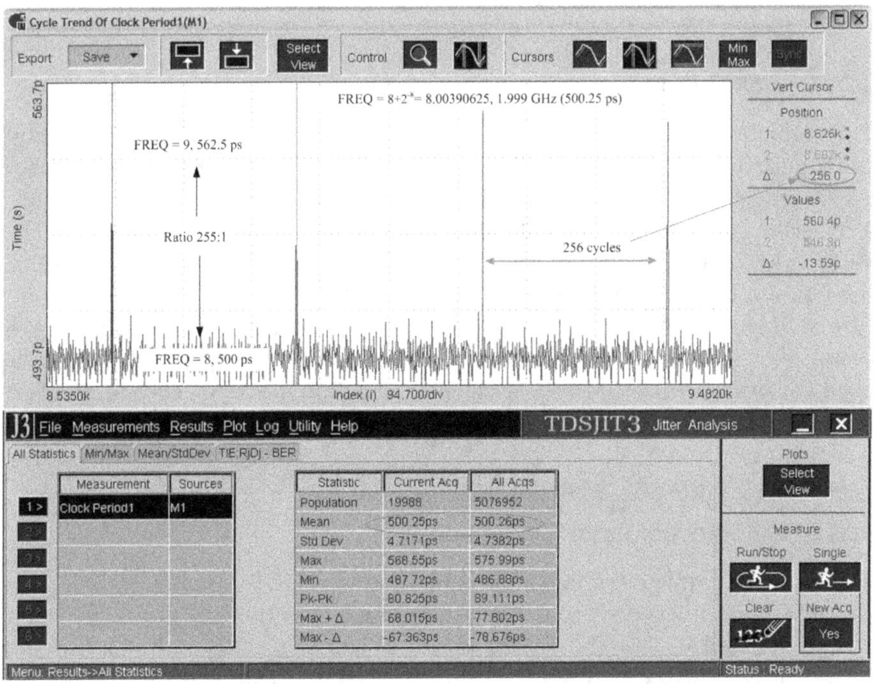

Fig. 4.74. Period's cycle trend of FREQ = 8.00390625.

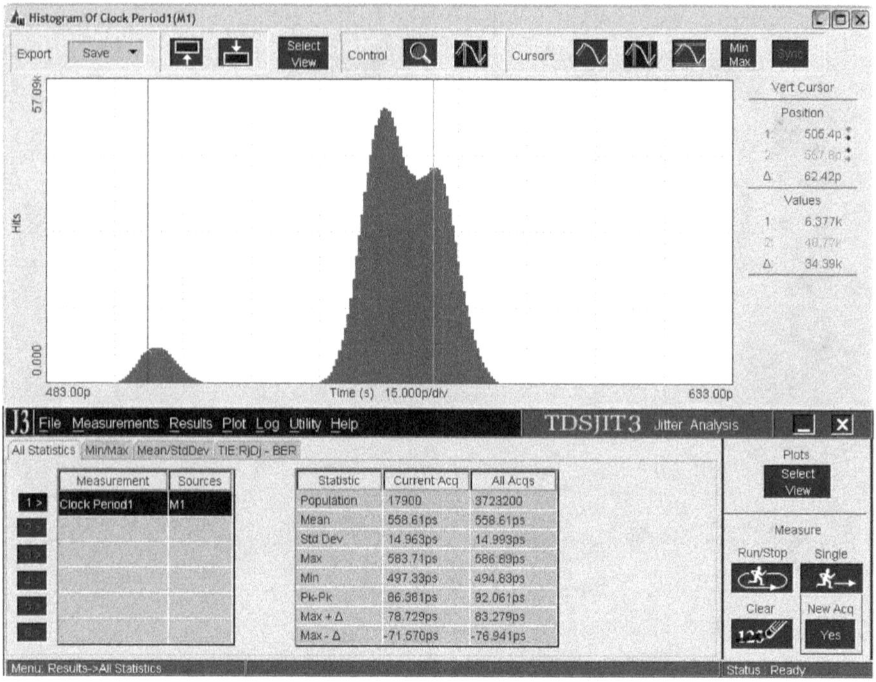

Fig. 4.75. Flying-adder output at FREQ = 8. $F0000_{16}$ = 8 + 15/16 = 8.9375.

4.21.8 PDFR Demonstration

The technique of post divider fractional bit recovery is introduced in Section 4.6. Figure 4.75 is the flying-adder output at FREQ = $8.F0000_{16}$ = 8 + 15/16 = 8.9375. The f_{vco} = 2 GHz (Δ = 62.5 ps) results in f_o = 1.79 GHz (559 ps). Figure 4.76 is the measurement result after a post divider of M = 16 (111.9 MHz, 8.938 ns). Clearly, T_A and T_B have merged into one period of 8.938 ns since the fraction 15/16 can be recovered by the divider of M = 16. Figure 4.77 is the case of M = 8. Clearly, there are still two distinguishable periods because M = 8 is not able to recover the fraction 15/16. However, the effective fraction is changed from 0.9375 to 0.5 as discussed in Section 4.6.3. As a result, the two peaks in Fig. 4.77 both have 50% as their weights compared to Fig. 4.75, with 6.25% and 93.75%.

4.21.9 XIU-Accumulator Evaluation

The XIU-accumulator is introduced in Section 4.11. It effectively separates the accumulator's speed from its size. Figure 4.78 is the measurement result when FREQ = $8 + 2^{-20}$ and f_{vco} = 2 GHz (Δ = 62.5 ps). The plot at left is the case of FREQ = 8. Without a fraction, the peak-peak jitter is 18 ps (only T_A, no T_B). The plot at the right is the case with 2^{-20} as the fraction (it is the LSB

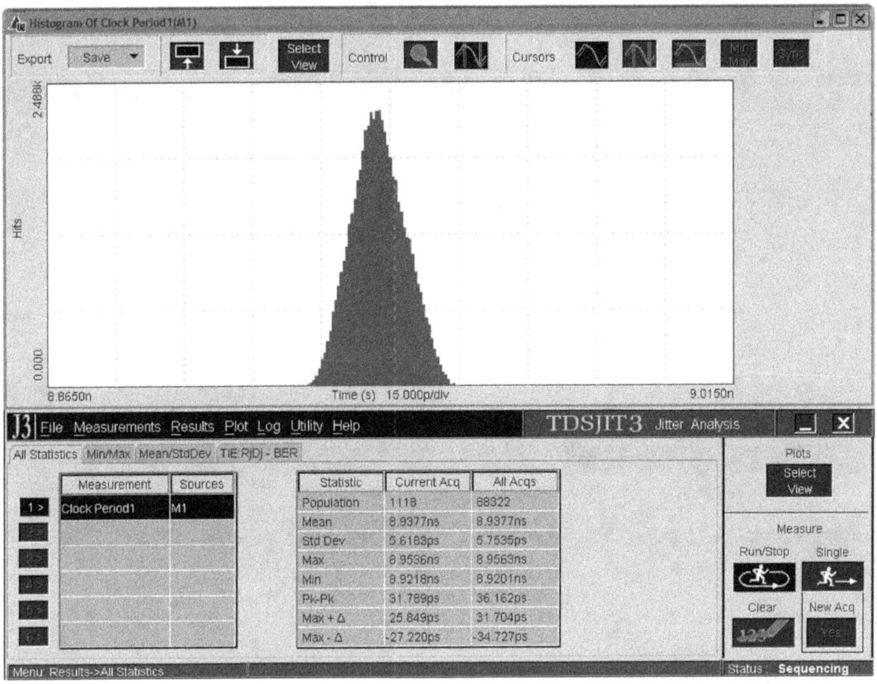

Fig. 4.76. Flying-adder output at FREQ = 8.9375 after $M = 16$, effective fraction = 0.

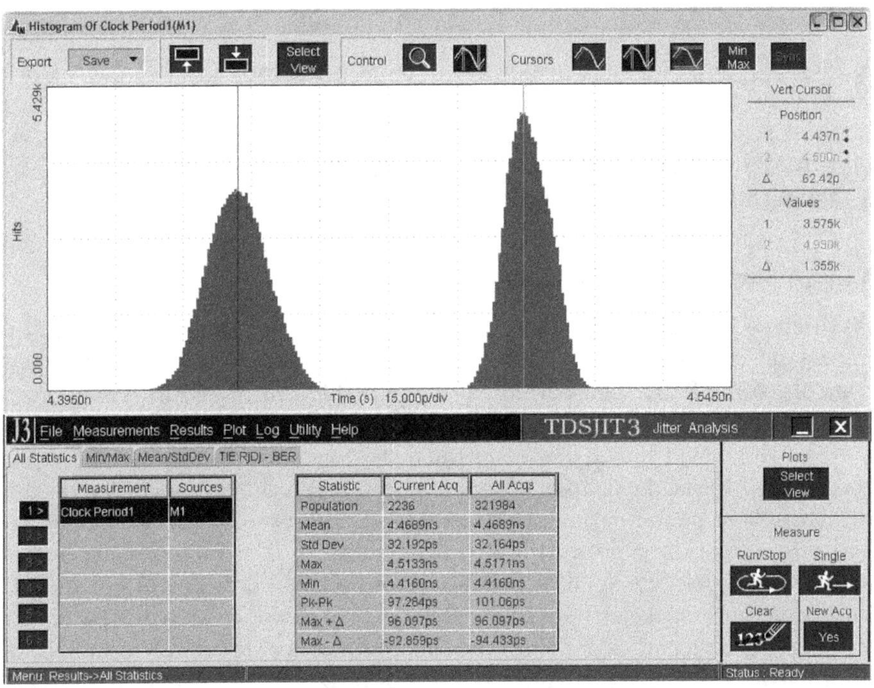

Fig. 4.77. Flying-adder output at FREQ = 8.9375 after $M = 8$, effective fraction = 0.5.

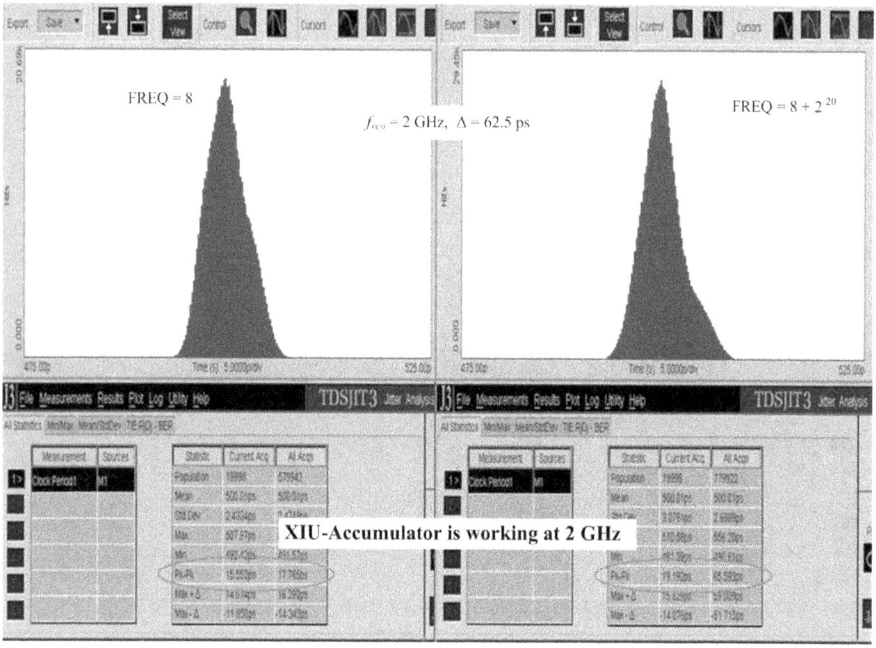

Fig. 4.78. XIU-Accumulator output at $\Delta = 62.5$ ps: FREQ = 8 (left) and FREQ = 8 + 2^{-20} (right).

in FREQ[19:0]). The peak-peak jitter of 66 ps is the indication of the "error" accumulation from the XIU-accumulator (T_A and T_B both exist). Figure 4.79 is the measurement result when $f_{VCO} = 2.5$ GHz ($\Delta = 50$ ps). The standard deviation of 2 ps and peak-peak value of 58 ps are caused by the $(2^{20} - 1)$:1 T_A to T_B ratio. This ratio is realized by the XIU-accumulator at 2.5 GHz.

4.21.10 Input Mismatch Observation

As discussed in Section 4.18, the mismatch of the flying-adder's inputs has an impact on its output jitter. This impact depends heavily on the frequency control word FREQ. The periodicity Ω embedded in the address sequence, which is controlled by FREQ, directly corresponds to the number of unique periods. The bigger the Ω, the larger the peak-peak jitter will be and the more complicated shape the distribution would bear. Figure 4.80 shows the measurement result of peak-peak jitter versus FREQ. The two curves correspond to two sets of measurements. In both cases, the VCO is at 2 GHz. Clearly, the jitter number has the smallest values when FREQ = 8 or 16 ($\Omega = 1$ and all mismatches are canceled). FREQ = 9, 11, 13, 15 are the cases with $\Omega = 8$, and they have a larger jitter. FREQ = 10 and 14 have $\Omega = 4$. Their jitter numbers are relatively smaller compared to the cases of $\Omega = 8$. FREQ = 12 is an interesting case. Since $\Omega = 2$ when FREQ = 12, the initial values on MUXs'

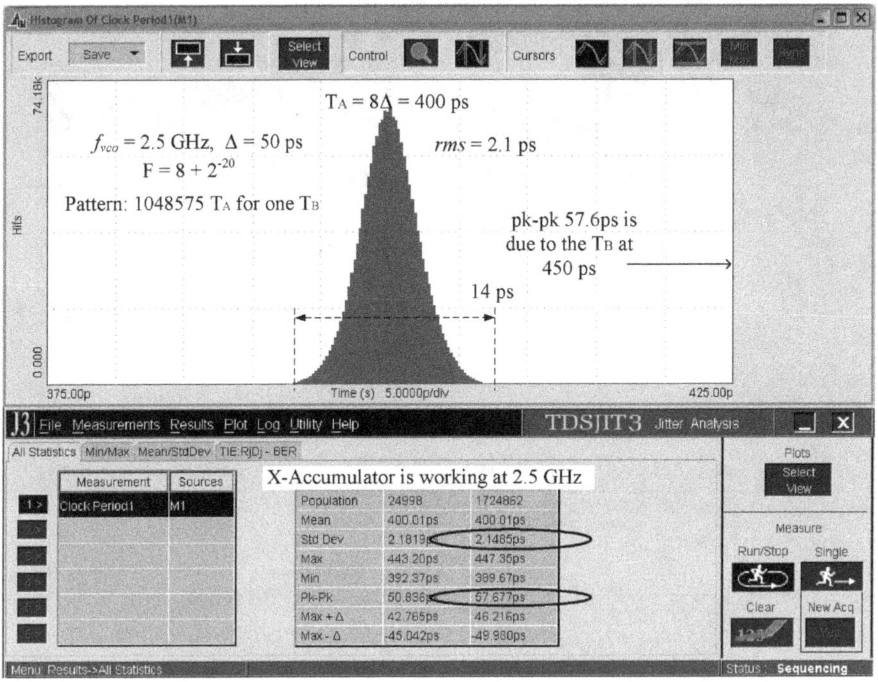

Fig. 4.79. XIU-Accumulator output at Δ = 50 ps, FREQ = 8 + 2^{-20}.

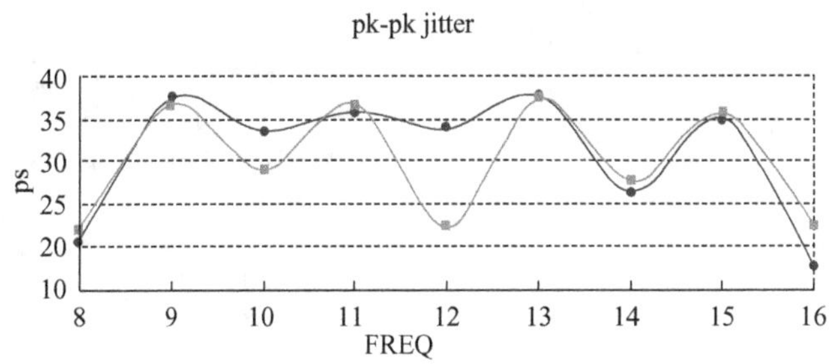

Fig. 4.80. The peak-peak jitter vs. FREQ.

addresses can play bigger role in jitter generation. The FREQ = 12 case on the two curves are obtained from different initial values. Consequently, the jitter numbers show large discrepancy (Xiu 2011).

Figure 4.81 is the period jitter measurement for FREQ = 9 (Ω = 8). There are at least four distinguishable periods visible in the plot. In the period spectrum plot, spurs spaced at $f_s/8$ are clearly seen. They are caused by the input mismatch as explained in Section 4.18.5. The spurs spaced at 125 MHz are PLL reference induced. Figure 4.82 is the case of FREQ = 12. Two distinguishable periods are easily visible. The period spectrum plot shows only one stem at

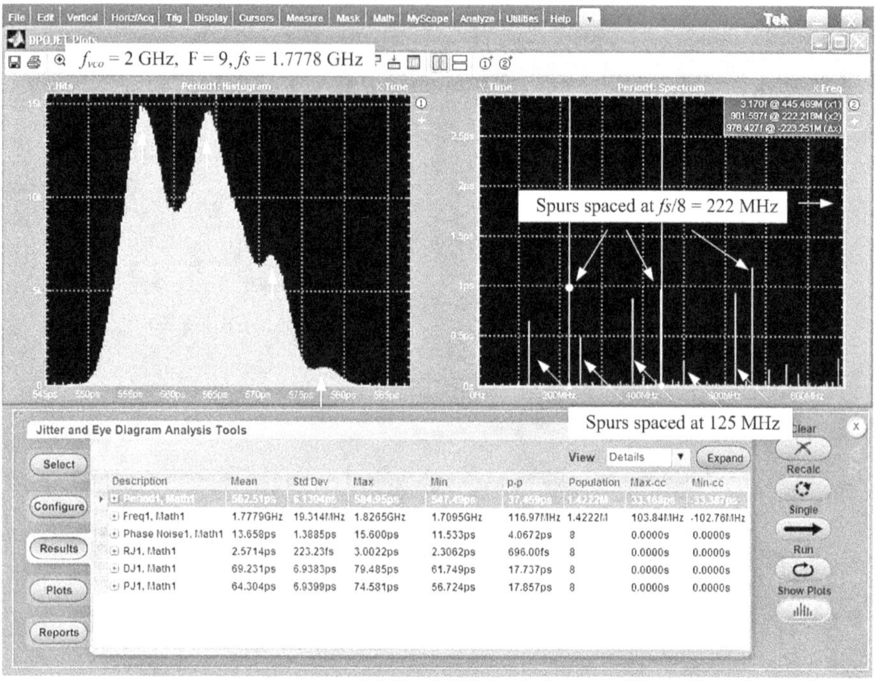

Fig. 4.81. The period jitter measurement at FREQ = 9.

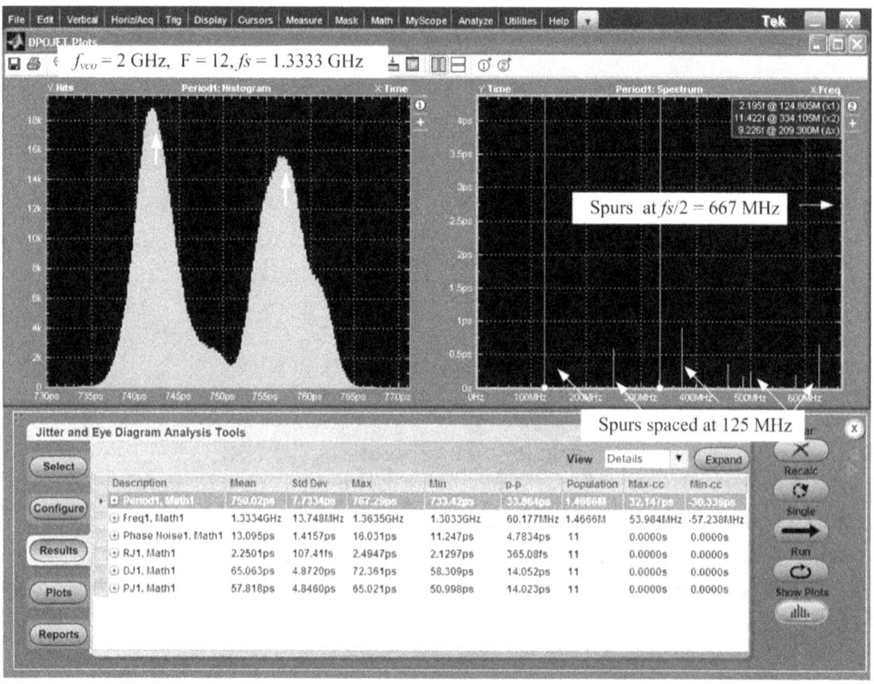

Fig. 4.82. The period jitter measurement at FREQ = 12.

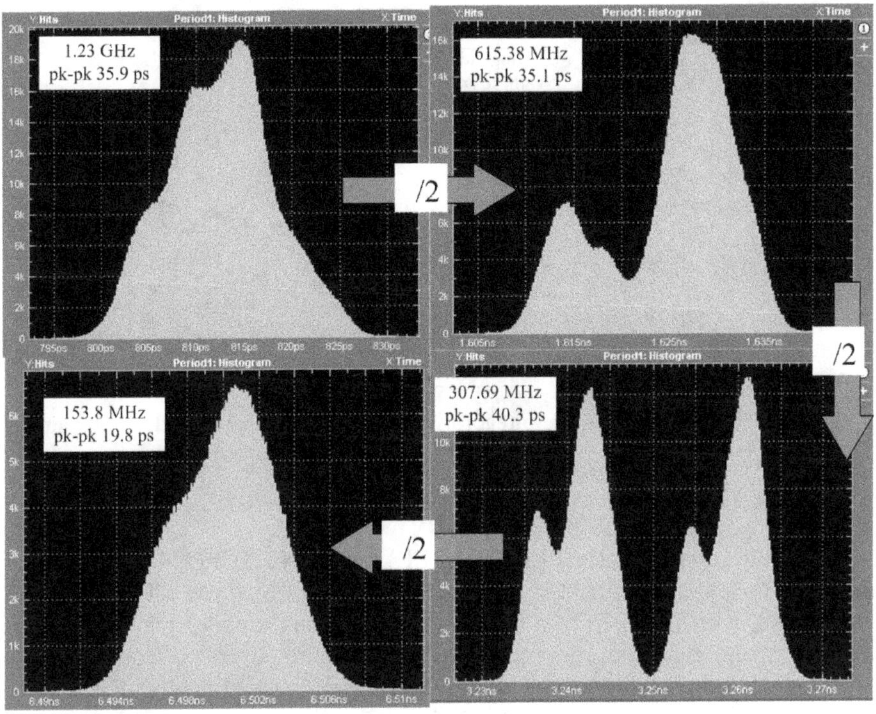

Fig. 4.83. The impact of the divider on the mismatch-induced jitter.

$f_o/2 = 667$ MHz, which agrees with the theoretical analysis. All the rest spurs are related to the 125-MHz reference. Some of them are aliased back from $f_o/2$.

A programmable divider is designed within the FAPLL. Figure 4.83 shows its impact on layout mismatch. In this test, the VCO is still at 2 GHz; the FREQ takes the value of 13. Hence, the FAPLL output is 1.23 GHz. The four measurements in this figure correspond to the post divider ratio of M = 1, 2, 4, and 8, respectively. In the undivided signal (1.23 GHz), there are eight address sequences with eight mismatch-induced periods. In the $M = 4$ case (307 MHz), there are two clearly distinguishable periods since the divider has reduced the address sequence from eight to two. In the $M = 8$ case (158 MHz), all the mismatches are canceled (reset) since $\Omega = 8$ for FREQ = 13. This is supported by the fact that the peak-peak jitter has reduced from 35.9 ps to 19.8 ps. Furthermore, from the jitter numbers in all four measurements, it is seen that the divider itself does not add much jitter (it is a driven system, not cumulative).

4.21.11 The Flying-Adder Fractional Divider Used Inside PLL

Section 4.8 introduces the concept of the flying-adder fractional divider. Section 4.9.2 proposes a structure of inserting this divider inside the PLL loop to improve the synthesizer's resolution.

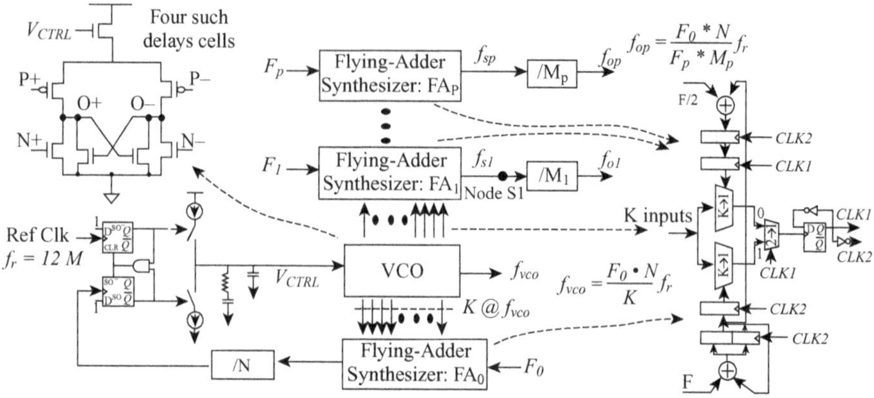

Fig. 4.84. The structure of the integer-flying-adder PLL in an IFAPLL test chip.

An integer-flying-adder PLL (IFAPLL), as shown in Fig. 4.84, has been designed and manufactured in a 55-nm process to test these ideas. It is predicted that, with the help of a flying-adder fractional divider inside the PLL loop, the frequency resolution at VCO output can be improved from f_r to f_r/K. With an additional flying-adder synthesizer outside the PLL loop, the resolution at node S_1 can be further improved to $(f_r/K)/2$. In other words, subinteger resolution can be achieved. In this IFAPLL test chip, the input frequency used is f_r = 12 MHz. The VCO circuit topology is shown in the top left side of the figure. It has four differential delay stages with eight outputs (K = 8). Its frequency range is designed from about 700 MHz to 2.5 GHz.

Figure 4.85 shows the measured clock spectrums at the VCO output when N = 128 and F_0 takes 4, 9, 11, and 12, respectively. The spurious tone in these spectrums is the 12-MHz PLL reference. These plots confirm the equation $f_{vco} = (F_0 \cdot N/K)f_r$. Figure 4.86 shows some interesting cases. In the top two plots, the F_0 takes values of 10 and 10 + 4/128, respectively. Since the fraction 4/128 is recoverable by N = 128, we expect no flying-adder-associated TAF spurs at the VCO output. This is confirmed by the spectrum in the top right plot. In the bottom two plots, the F_0 takes values of 8 and 8 + 1/512. Since the fraction 1/512 is not recoverable by N = 128, there are residual spurs at the spectrum. In all the cases, including the non-PDFR compatible fractional case, the frequencies measured agree with that predicated from $f_{vco} = (F_0 \cdot N/K)f_r$.

Figure 4.87 shows two phase noise plots when N = 64 and F_0 takes 8 + 4/64 and 8 + 5/64. The in-band noise is at the level of around 100 dBc/Hz. This noise is PLL/VCO induced. It is independent of the flying-adder synthesizer. It can also be seen that the resolution is f_r/K = 12/8 = 1.5 MHz as expected, since the two PDFR-compatible fractions are 1/64 apart.

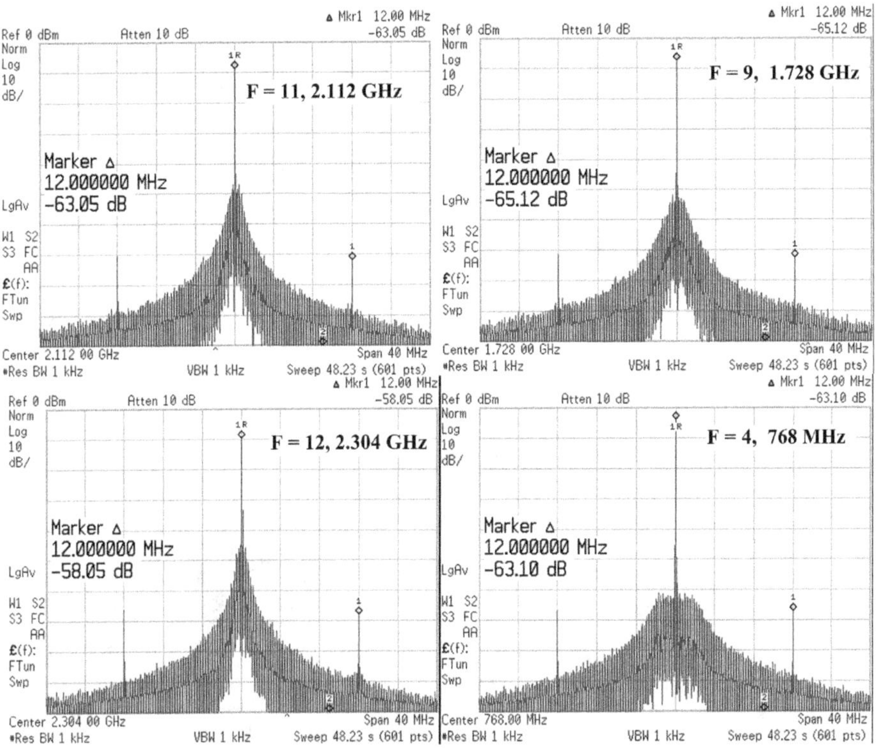

Fig. 4.85. The measured clock spectrums at the VCO output when $N = 128$ and F_0 takes 4, 9, 11, and 12.

4.21.12 The Integer-Flying-Adder PLL

When the flying-adder circuit is used both inside and outside the PLL, the system is termed the integer-flying-adder PLL (IFAPLL). As discussed in Section 4.9.3, the frequency resolution at node S_1 (Fig. 4.84) is (f_r/F_1), instead of (f_r/K). When F_1 takes its maximum value of $2K$, the resolution can reach $(f_r/K)/2$. Figure 4.88 is the experimental data that support this prediction. The two clock spectrums are obtained under the setting of $N = 128$, $F_1 = 16$, and F_0 taking $14 + 4/128$ and $14 + 5/128$. Since $F_1 = 16$ and the two PDFR-compatible fractions for F_0 are 1/128 apart, we reach the resolution of $(f_r/K)/2 = 750$ KHz.

One of the main purposes of IFAPLL is to produce a TAF-spurs-free clock for supporting applications where spectra purity is of high concern, such as when driving SoC on-chip ADC/DAC, when providing frequency reference to other frequency generator, etc. For example, Table 4.10 lists all the frequencies required for an audio application. These frequencies are used to drive an audio ADC. Therefore, spectra purity is a primary design concern. TAF spurs are not appropriate for this application. In the past, from a fixed

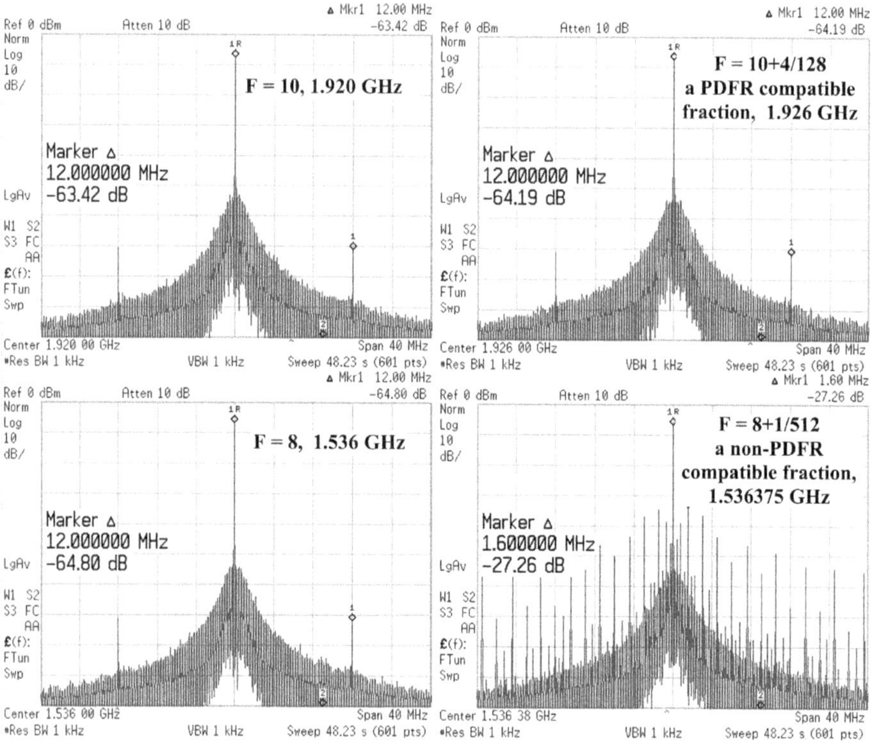

Fig. 4.86. The measured clock spectrums at the VCO output when N = 128 and F_0 takes 10, 10+4/128, 8, and 8+1/512.

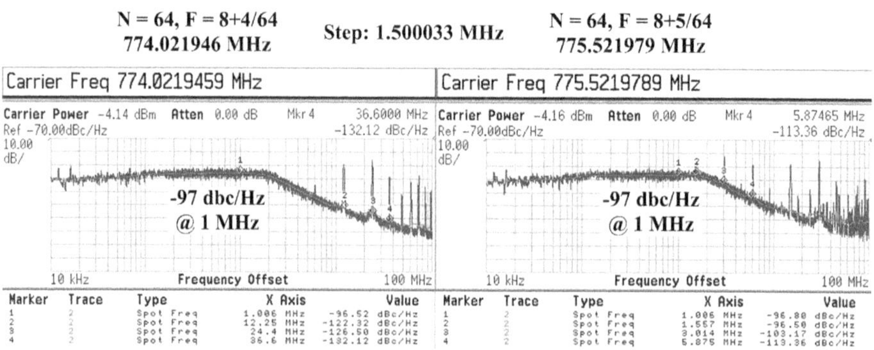

Fig. 4.87. The phase noise plots at VCO output when N = 64 and F_0 takes 8+4/64 and 8+5/64.

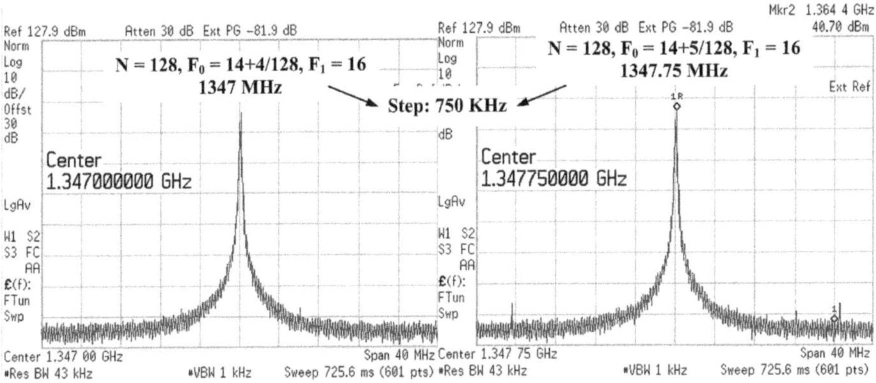

Fig. 4.88. The clock spectrum at node S_1 when $N = 128$, $F_1 = 16$, and F_0 takes 14+4/128 and 14+5/128.

TABLE 4.10. Audio Frequencies in MHz[a]

	128	256	384	512	768	1024
16	2.0480	4.0960	6.1440	8.1920	12.2880	16.3840
22.05	2.8224	5.6448	8.4672	11.2896	16.9344	22.5792
24	3.0720	6.1440	9.2160	12.2880	18.4320	24.5760
32	4.0960	8.1920	12.2880	16.3840	24.5760	32.7680
44.1	5.6448	11.2896	16.9344	22.5792	33.8688	45.1854
48	6.1440	12.2880	18.4320	24.5760	36.8640	49.1250
64	8.1920	16.3840	24.5760	32.7680	49.1520	65.5360
88.2	11.2896	22.5792	33.8688	45.1584	67.7376	90.3168
96	12.2880	24.5760	36.8640	49.1520	73.7280	98.3040
128	16.3840	32.7680	49.1520	65.5360	98.3040	131.0720
176.4	22.5792	45.1584	67.7376	90.3168	135.4752	180.6336
192	24.5760	49.1520	73.7280	98.3040	147.4560	196.6080

[a]The first column is the sampling frequency in KHz, and the first row is the oversample rate.

frequency reference (such as a 12-MHz crystal), it is difficult to generate all these frequencies from one PLL. Usually, the goal can only be accomplished by cascading two or more integer-N PLLs together. Using IFALL, this goal can be achieved with the powerful equation of $f_{op} = ([F_0 \cdot N]/[F_p \cdot M_p])f_r$ (refer to Figs. 4.84 and 4.34). By examining Table 4.10, it is clear that the six frequencies in the low right corner (red) are the key. All the rest of the frequencies can be generated from them through frequency division. Figures 4.89 and 4.90 show the measured results using IFAPLL to generate four frequencies. The fixed reference is 12 MHz. All the settings (N, F_0, F_1, M) are listed in the figures.

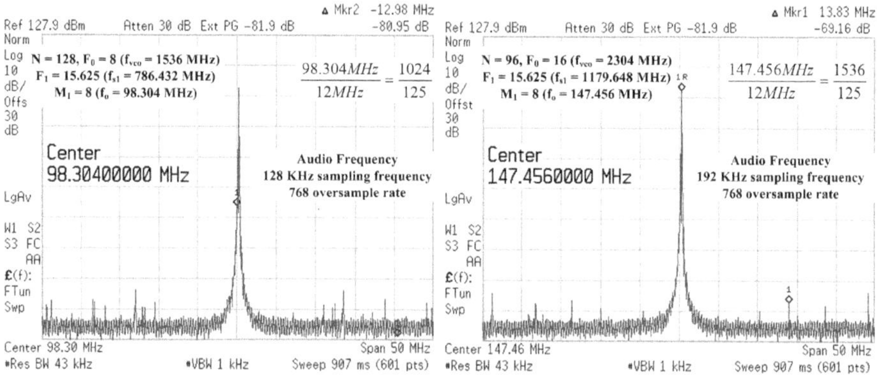

Fig. 4.89. Using IFAPLL to generate two audio frequencies that are free of TAF-spurs: 98.304 and 147.456 MHz.

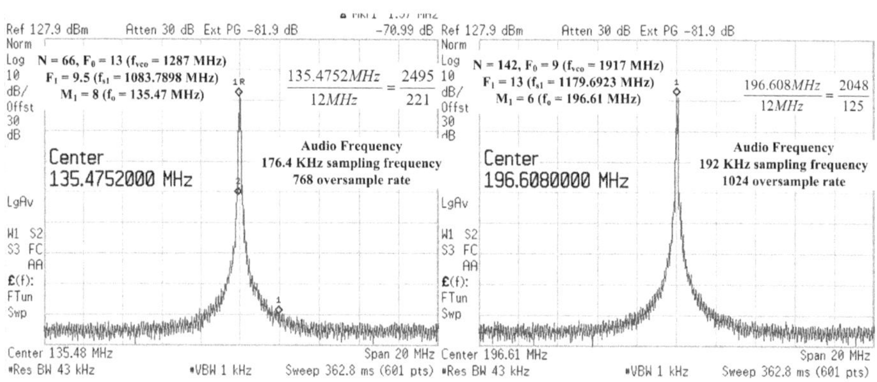

Fig. 4.90. Using IFAPLL to generate two audio frequencies that are free of TAF-spurs: 135.4752 and 196.608 MHz.

4.22 TIME-AVERAGE-FREQUENCY AND *SETUP* CONSTRAINT: REVISIT

In Chapter 3, Section 3.3, time-average-frequency was defined as $T_{TAF} = (1 - r)\cdot T_A + r\cdot T_B$, where r is the weight factor, which is the same r in our later flying-adder implementation of FREQ = $I + r$. After the experimental demonstration of time-average-frequency in Section 4.21.7, it is worth revisiting the relationship between T_{TAF} and the timing closure *setup* constraint. This discussion is extremely important to SoC integration (timing closure). When a conventional frequency is used in driving a digital circuit of designed speed $f = 1/T$, the constraint used for the setup is T. The constraint becomes T_A when time-average-frequency is used. It is the left branch in time-average-frequency

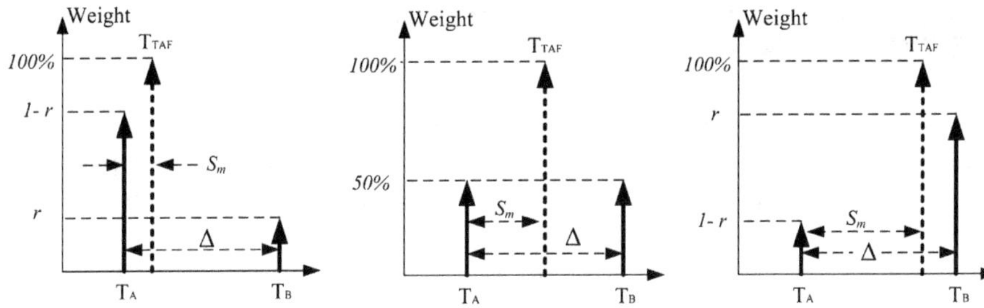

Fig. 4.91. Time-average-frequency and setup constraint margin, r small (left), $r = 0.5$ (middle), r large (right). T_{TAF} is the conventional setup constraint. T_A is the setup constraint when a TAF clock is used.

period distribution, as discussed in Chapter 3. Therefore, the extra margin S_m required can be calculated below:

$$S_m = T_{TAF} - T_A = (1-r)\cdot T_A + r\cdot T_B - T_A = r\cdot(T_B - T_A) \tag{4.19}$$

In flying-adder implementation, $T_B - T_A = \Delta$. Therefore, Eq. 4.19 can be derived as:

$$S_m = r\cdot\Delta \tag{4.20}$$

As is clearly shown, the margin linearly depends on the size of the fraction after Δ is fixed. This can be understood intuitively. When r is small, the location of T_{TAF} is close to that of T_A. In this case, using T_{TAF} or T_A as the setup constraint does not make too much difference. On the other hand, when r is large, the T_{TAF} is closer to T_B. This is graphically illustrated in Fig. 4.91. In the experimental measurements of Figs. 4.71 and 4.72, the location of T_{TAF} is marked in the plots. It is a virtual parameter that cannot be displayed graphically by equipment. However, the numerical value of T_{TAF} can be obtained from the mean value of measured period distribution. Although invisible in a graphic plot, T_{TAF} has real power. It is a breakthrough, which eases the difficult task of frequency generation (and thus clock implementation). It enables innovation on the system level. This point will become clear in the presentations of Chapter 6.

In SoC timing closure practice, as long as the designer uses T_A as the setup constraint (speed up the circuit by $r\cdot\Delta$ ps), the circuit will behave the same as one driven by a conventional frequency-based clock (refer to Chapter 3). Actually, ***the hard evidence is the flying-adder circuit itself***. Due to self-clocking, the accumulator- and adder-register pair inside the flying-adder are driven by the TAF clock. The successful generation of the desired output frequency is a direct indication that time-average-frequency can safely drive a digital circuit.

4.23 SENSE THE FREQUENCY DIFFERENCE: THE TIME-AVERAGE-FREQUENCY WAY

In a conventional frequency-based clock, two signals' waveforms are distinguishably dissimilar in every cycle if their frequencies are not equal. This is due to the definition of $T = 1/f$. In a clock based on time-average-frequency, the situation is slightly different since two types of cycles are involved. By the definition of $T_{TAF} = (1\text{-}r)\cdot T_A + r\cdot T_B$, where $T_A = I\cdot\Delta$ and $T_B = (I + 1)\cdot\Delta$, it is understandable that it could take multiple cycles for the frequency difference to be sensed. If the frequency change is large enough so that the size of T_A (and T_B) needs to be adjusted (the FREQ's integer part I changes), the waveform will show the difference immediately. However, when the frequency difference is small and only the fractional part r is adjusted, it will take longer to sense the difference in waveform. This is because the sizes of the cycles T_A and T_B are not updated; only the number of occurrences is changed. If r is expressed as $r = p/q$, where the greatest common divisor of p and q is 1, the needed minimum number of cycles to see the effect of the new frequency is q.

In flying-adder implementation, the frequency control word F contains both integer I and fraction r: $F = I + r$. Adjusting fraction r can fine-tune the output frequency between $f_B = 1/T_B$ and $f_A = 1/T_A$, where $T_A = I{*}\Delta$, $T_B = (I + 1){*}\Delta$ and Δ is a constant. On the other hand, adjusting integer I can change a frequency in bigger steps since the size of T_A and T_B now are changed. In this mode, the frequency can be adjusted anywhere from $f_{ref}/2$ to f_h, where f_{ref} is the frequency of the multiphases inputs and f_h is the highest possible output frequency (constrained by process speed). Figure 4.92 graphically illustrates this point. The plots in Figs. 4.66–4.69 are the evidence of large frequency change where I is changed. The right plot in Fig. 4.64, which demonstrates small frequency variation, is caused by r adjustment. The cases presented in Table 4.9 are also controlled by r. For example, FREQ takes 14 + 986893/1048576 = 14.941174507 (61683 T_A and 986893 T_B for every 1048576 cycles) and 14 + 986894/1048576 = 14.941175461 (61682 T_A and 986894 T_B) in the first two cases in that table. These settings result in 535.433141 MHz and 535.433107 MHz at the

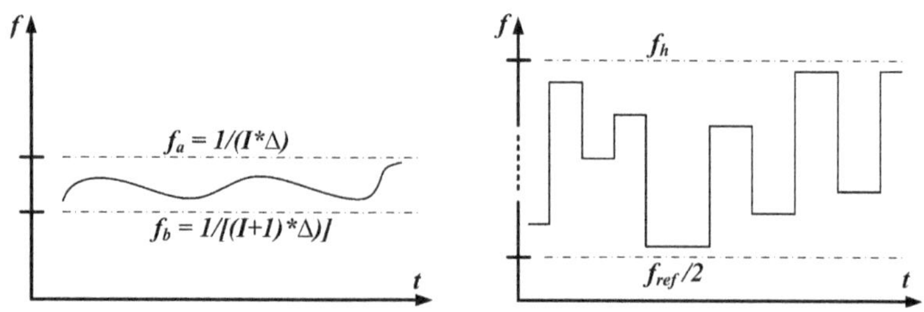

Fig. 4.92. Frequency change in a flying-adder TAF-based clock. In a small frequency variation, only r is changed (left). In a large frequency swing, both I and r are changed (right).

flying-adder synthesizer's output (before the M divider), respectively. The frequency difference is roughly 34 Hz. It takes at minimum 1048576 cycles, or about 1.958 μs, to observe the effect.

With FADPS, the most prominent feature is that frequency can be changed in seamless fashion both in fine-tune and large step mode. No glitch is produced at its output (see Section 4.21.6, Figs. 4.66–4.68), no control is needed except the control word update, and no extra supporting logic is required. From outside the FAPLL, it is just a black box with a frequency-adjustment knob. This is especially convenient for it to be integrated at the chip level (minimum or no system level assistance required).

4.24 FLYING-ADDER AND DIRECT DIGITAL SYNTHESIS (DDS): THE DIFFERENCE

Flying-adder architecture and direct digital synthesis (DDS) are both open-loop methods. In contrast to the indirect approach of PLL, they both directly construct the output waveform. However, they are significantly different. The fundamental difference is the frequency concept. Time-average-frequency is deliberately used in the flying-adder. Architecturally, there is also an important difference. As depicted in Fig. 4.93, the DDS uses an external fixed frequency reference clock to drive its internal circuits. Hence, the base time unit is that clock's period. Everything is recorded or triggered using this clock. The continuous time domain behavior can be solely marked by the clock since clock cycle is linearly proportional to the continuous time t. This is not the case for the flying-adder. As shown, the flying-adder uses its own output to drive its internal circuits. Since its output frequency (period) can vary depending on the control word (when the fractional part is nonzero), the discrete time (indexed by clock cycle) is not linearly proportional to the continuous time (marked by t). This makes flying-adder architecture a completely new challenge in term of theoretical analysis. This will become apparent in Section 5.4.

Circuit-wise, the use of self-clocking provides an efficient way of utilizing the K inputs. However, compared to DDS, the flying-adder's significant improvement in efficiency (when used as a clock generator) comes from the

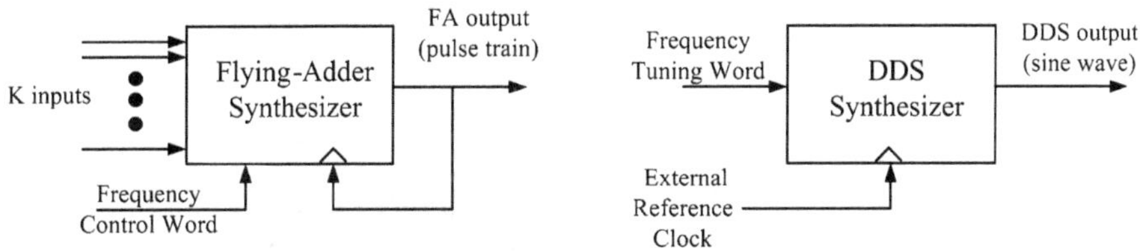

Fig. 4.93. The difference between flying-adder architecture and direct digital synthesis architecture.

fact that it directly generates a clock pulse. In DDS, the output is a sinusoidal wave that has to be converted to a pulse by some means. This requires additional resources. In a flying-adder, the output is a pulse train, which can be used directly to drive a circuit. As a matter of fact, the accumulator registers inside the synthesizer are driven by this clock. Moreover, operationally, a flying-adder can output a frequency that is higher than its input (refer to Fig. 4.63). In contrast, the highest achievable output frequency from DDS is half of the reference frequency (limited by Nqyuist criteria).

4.25 FLYING-ADDER FOR PHASE (DELAY) SYNTHESIS

Phase (delay) synthesis is just as important as frequency synthesis. In VLSI chips, all signals can be classified into three categories: data, clock, and power. "Data" are the information. "Clock" is used to control the flow of information. "Power" is used to deliver the energy needed. When clock is concerned, its frequency determines the rate of the flow; its phase selects the appropriate time for sending or receiving the data. In implementation, phase synthesis can be differentiated into two scenarios: one-clock and two-clocks. In the one-clock scenario, as shown in the left drawing of Fig. 4.94, the clock phase movement is relative to itself (current edge vs. previous edge). In this case, the clock phase can be considered as being compared to a virtual reference (e.g., the data that are controlled by another unreachable clock). In the two-clock scenario, two clock signals are presented explicitly as shown in the right drawing of Fig. 4.94. The phase relationship is between the two same-frequency clocks. Both scenarios are commonly seen in data communication. In the one-clock case, the clock associated with the launching side is invisible; only data are observable. In the two-clock cases, both the source clock and the destination clock are visible, but only the destination clock is controllable. In the two-clock case, a DLL is often used to generate the phase movement.

The flying-adder synthesizer is naturally suitable for phase synthesis since its operation is based on the counting of a base unit Δ. For one-clock phase

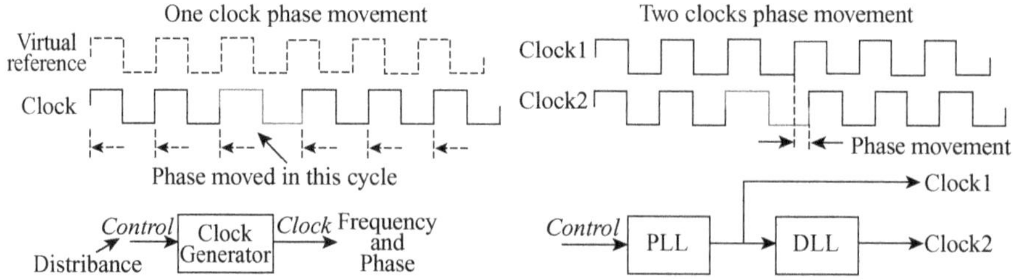

Fig. 4.94. Two scenarios of phase (delay) synthesis: one-clock (left) and two-clock (right).

synthesis, we can simply change the FREQ = $I + r$ (increase or decrease I by 1) to move its phase one Δ forward or backward. Although it is unintentional, the fractional-overflow-induced $T_A \rightarrow T_B$ can be regarded as one Δ phase forward movement. This fact indicates that the circuit itself is fully compatible for one-clock phase synthesis. Figure 4.95 is an example of a transistor-level simulation that illustrates this feature. An asynchronous signal *trigger* is used to increase the FREQ by 1 for less than one cycle of time (from 8 to 9 then back to 8). This change is captured by the flying-adder synthesizer using its own output clock. As shown, the first cycle captures the FREQ, and the second cycle immediately takes the new value. In the third cycle, the result is seen (recall that the two-path flying-adder synthesizer has a latency of two cycles [Fig. 4.25]). Relative to its previous edge, the clock phase is moved one Δ forward (period is changed from 1 ns to 1.25 ns, Δ = 125 ps).

The flying-adder synthesizer can conveniently produce controlled phase movement between two clock signals as well. The circuit structure is shown in the left drawing of Fig. 4.96. Two almost identical blocks, *Frequency Generation* and *Phase Generation*, are used for generating the *CLK1* and its phase-controlled version *CLK1_P*. The two blocks take the same *K* inputs. The input FREQ controls the output frequency. PHASE determines the *CLK1_P*'s delay amount (relative to *CLK1*). The two address signals, *SEL_LOW_P* and *SEL_UP_P*, are made in such a way that *SEL_LOW_P* = *SEL_LOW* + *PHASE* and *SEL_UP_P* = *SEL_UP* + *PHASE*. In other words, the phase generation block is PHASE Δs behind the frequency generation block.

In principle, this scheme is very straightforward. However, there is a key issue of synchronizing the two blocks. As shown, the two flip-flops, DFF_F and DFF_P, are independently configured as toggle flip-flops. Their initial values are uncontrolled and unrelated to each other. Consequently, the *CLK1* and *CLK1_P* phases have an uncertainty of π. This problem is elegantly solved by the flying-adder's self-clocking feature. As shown in the right side of the figure, the SEL_LOW selects an input, Φ_F, out of the *K* inputs. This Φ_F takes effect at the rising edge of the CLK2 (remember that *SEL_LOW* is controlled by CLK2; see Fig. 4.17). Therefore, Φ_F has a deterministic relationship with the edges of *CLK1* and *CLK2*. Since Φ_P, which is controlled by SEL_LOW_P (through PHASE), bears a known and fixed time relationship with Φ_F, we conclude that *CLK1* and *CLK1_P* have a fixed and desired time delay.

Figure 4.97 is the SPICE simulation result for illustrating the result of this circuit. The input references have eight signals ($K = 8$) of frequency 1 GHz. This results in Δ = 125 ps. The settings are FREQ = 8 and PHASE = 6. The *CLK1* (and *CLK1_P*) frequency is calculated as 1 GHz (8Δ = 1 ns). The delay between *CLK1* and *CLK1_P* is 6Δ = 750 ps. In Fig. 4.97, the top signal is CLK1, which shows 1 GHz. The second signal, *CLK1_P*, and third signal, *CLK1_PP*, are produced by the phase generation block. The difference between them is the flip-flop's initial value. One starts with "1" and the other begins with "0." As expected, after the initial setup, the circuit produces the correct 750-ps

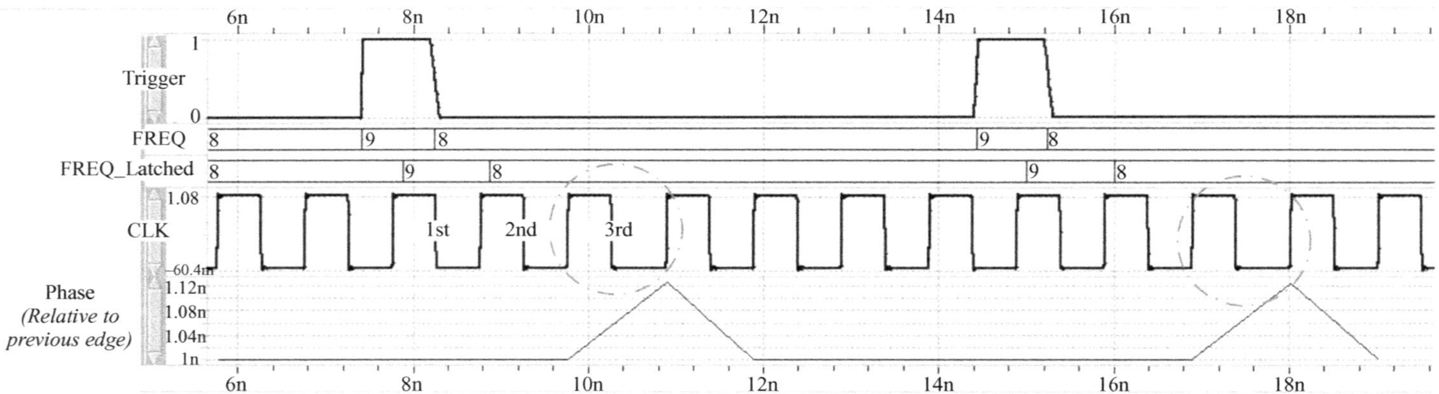

Fig. 4.95. Simulation demonstration of one-clock phase synthesis.

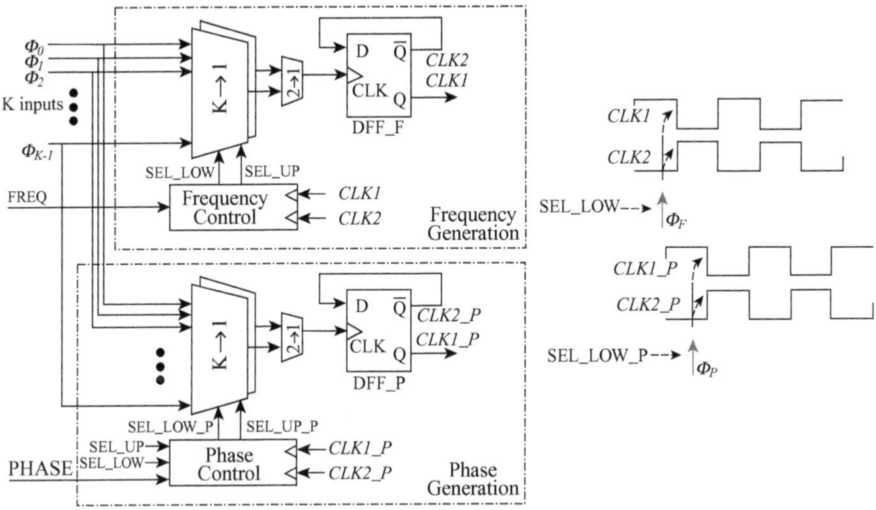

Fig. 4.96. A flying-adder structure for generating delay between two clocks.

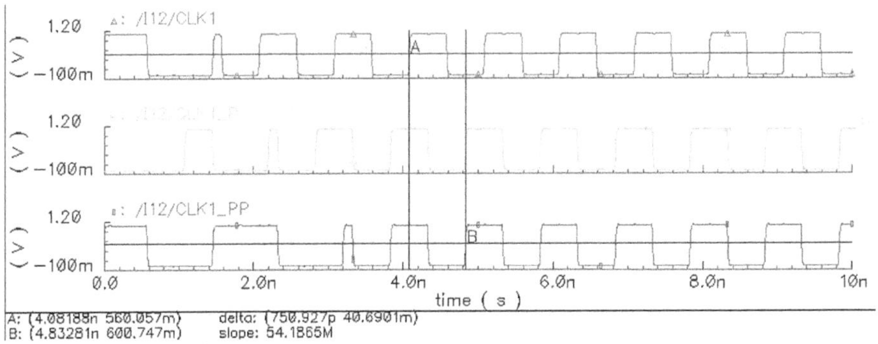

Fig. 4.97. A flying-adder delay synthesis simulation: different DFF initial values produce the same delay: FREQ = 8, PHASE = 6.

delay for both cases. Figure 4.98 is another SPICE simulation: FREQ = 7 and PHASE = 3. The expected frequency is $7\Delta = 875$ ps (1.14 GHz). The phase delay between the two signals is $3\Delta = 375$ ps. This simulation confirms these predications with additional detail of address values SEL_LOW and SEL_LOW_P shown.

For this flying-adder delay synthesis circuit, the phase adjustment range that can be directly produced is 0 to K Δs. Since the output frequency can go as low as $2K$ Δs (two paths are used), the phase in the range of $K + 1$ to $2K$ Δs can be achieved by inverting the *CLK1_P*. This circuit will be used in flying-adder DLL (FADLL) in Chapter 6.

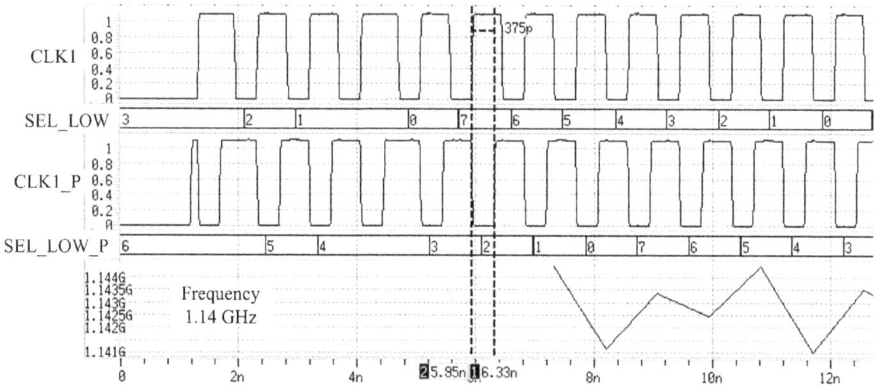

Fig. 4.98. A flying-adder delay synthesis simulation: FREQ = 7, PHASE = 3.

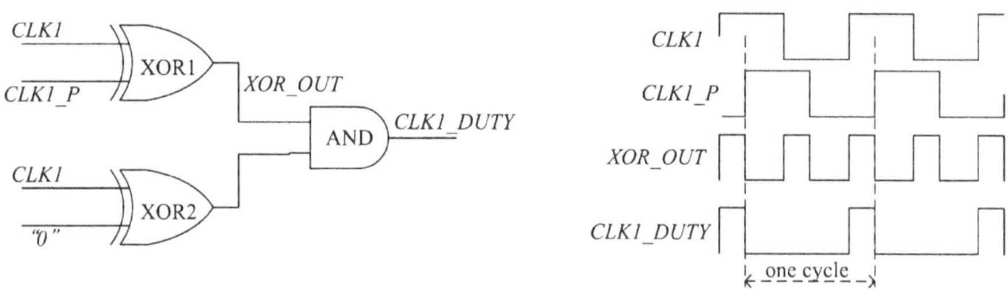

Fig. 4.99. A flying-adder duty cycle synthesis: the circuit (left) and the waveforms (right).

4.26 FLYING-ADDER FOR DUTY CYCLE CONTROL

Clock duty cycle synthesis is a technique that has some important applications. From the discussion in Section 4.3, it is apparent that the flying-adder synthesizer can adjust its output duty cycle readily. This is simply because both the "high" and "low" portions of the output are constructed by counting the base unit Δ. However, since the flying-adder circuit can produce an output whose period is longer than the input reference's period of $K \cdot \Delta$, care must be taken to make the duty cycle programmable in the flying-adder's full frequency range.

The flying-adder phase synthesis circuit presented in Fig. 4.96 can be used to help achieve this goal. This idea is depicted in Fig. 4.99. As shown, the delayed version *CLK1_P* (refer to Fig. 4.96) is XOR'd with the *CLK1* through the gate XOR1. This results in two pulses in one cycle of time. The waveform traces of four key signals—*CLK1, CLK1_P, XOR_OUT*, and *CLK1_DUTY*—are depicted in the right-hand drawing Fig. 4.99. *XOR_OUT* is furthered selected by *CLK1* with the help of gate AND. The gate XOR2 is used for the delay match. The final output *CLK1_DUTY* is the output with the desired duty cycle. The length of the "high" portion is exactly equal to the PHASE

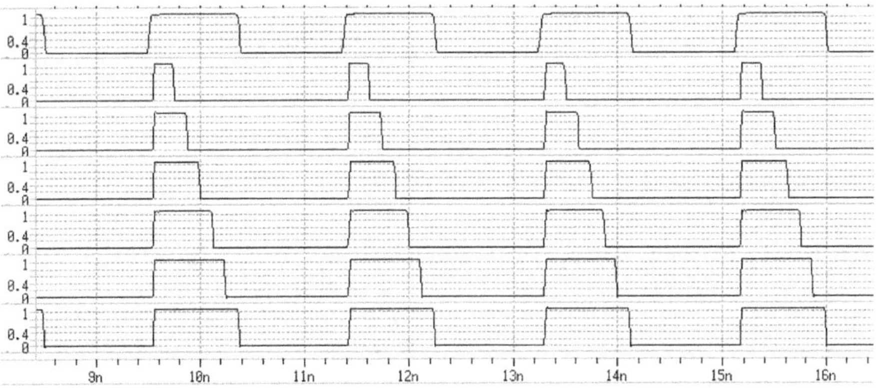

Fig. 4.100. Transistor-level simulation of flying-adder duty cycle synthesis: the nominal clock (top) and various duty cycle clocks (from second to seventh traces).

setting (in unit of Δ). Figure 4.100 shows the transistor-level simulation result from this duty cycle synthesis circuit. There are eight inputs to the synthesizer ($K = 8$) with a frequency of 1 GHz ($\Delta = 125$ ps). The FREQ setting is 15 (533 MHz). The nominal clock output is shown at the top. Its period is 1.875 ns with a "high" of 7Δ (875 ps) and "low" of 8Δ (1 ns). From the second to the seventh trace, the waveforms correspond to PHASE settings of 1 to 6, respectively. As can be seen, the output duty cycle is controlled as expected. The length of the "high" portion equals the PHASE in all six cases.

4.27 FLYING-ADDER SYNTHESIZER IN REDUCING THE NUMBER OF PLLs IN SoC

One of the clear advantages of using flying-adder synthesizer in a SoC environment is to reduce the number of on-chip PLLs. This idea is first presented in Xiu (2007) and it has been used successfully in many commercial products (http://focus.ti.com/lit/ug/sprugx9/sprugx9.pdf; http://focus.ti.com/lit/ug/sprugx7/sprugx7.pdf; http://focus.ti.com/lit/ug/sprugx8/sprugx8.pdf). This structure is depicted in Fig. 4.43. As shown, in this particular FAPLL configuration, there are four flying-adder synthesizers attached to the same VCO, resulting in four independent clock outputs. Depending on the load it drives, each synthesizer can decide whether to use a fraction or not in its control word. In addition, the output from the VCO can also be used to drive the load. Since this one block can replace four or five conventional PLLs, the benefit of power, area, and pin-count saving can be appreciated without doubt. However, attaching more synthesizers (more than four or five) to the same VCO is, in general, not recommended since the interference among the frequency sources will degrade the clock signals' quality. Furthermore, many clock sources clustered in a small physical proximity can cause clock distribution problems in the SoC environment. Therefore, this issue has to be studied carefully case by case. Simply attaching many synthesizers to one PLL/VCO without considering other effects is not encouraged.

BIBLIOGRAPHY

"AM389x Sitara ARM Microprocessors Technical Reference Manual," http://focus.ti.com/lit/ug/sprugx7/sprugx7.pdf, Texas Instruments Inc., 2011.

Alioto, M. and G. Palumbo. 2002. "Analysis and Comparison on Full Adder Block in Submicron Technology," *IEEE Trans. on VLSI Syst.*, vol. 10, no. 6, pp. 806–823.

Bui, H. T., Y. Wang, and Y. Jiang. 2002. "Design and Analysis of Low-Power 10-Transistor Full Adders Using Novel XOR-XNOR Gates," *IEEE Trans. on Circuit Syst. II*, vol. 49, no. 1, pp. 25–30.

Chau, Y. A. and C. F. Chen. 2008. "High-performance Glitch-free Digital Frequency Synthesizer," *Electron. Lett.*, vol. 44, pp. 1063–1064.

Chau, Y. A., Y. Y. Yang, and J. F. Chen. *"All-Digital Frequency Synthesizer with Dual Resolutions,"* *ISPACS'06*, pp.630–633, 2006.

Chien, J. C. and L. H. Lu. 2007. "A 32-GHz Rotary Traveling-Wave Voltage Controlled Oscillator in 0.18-um CMOS," *IEEE Microw. Wireless Component Lett.*, vol. 17, pp. 724–727.

Gharaee, H. and E. Tathesari. "*A New High Resolution Frequency and Phase Synthesis Method based on 'Flying-Adder' Architecture*," *ICSE'06*, pp.520–523, 2006.

Goel, S., A. Kumar, and M. A. Bayoumi. 2006. "Design of Robust, Energy-Efficient Full Adders for Deep-Submicrometer Design Using Hybrid-CMOS Logic Style," *IEEE Trans. on VLSI Syst.*, vol. 14, no. 12, pp. 1309–1321.

Hajimiri, A. and T. H. Lee. 1998. "A General Theory of Phase Noise in Electrical Oscillator," *IEEE J. Solid-State Circuits*, vol. 33, pp. 179–194.

Leung, G. C. and H. C. Loung. 2004. "A 1-V 5.2 GHz CMOS Synthesizer for WLAN Application," *IEEE J. Solid-State Circuits*, vol. 39, pp. 1873–1882.

Lin, J. F., Y. T. Hwang, M. H. Sheu, and C. C. Ho. 2007. "A Novel High-Speed and Energy Efficient 10-Transistor Full Adder Design," *IEEE Trans. on Circuit Sys. I*, vol. 54, no. 5, pp. 1050–1059.

Mair, H. and L. Xiu. 2000. "An Architecture of High-Performance Frequency and Phase Synthesis," *IEEE J. Solid-State Circuits*, vol. 35, pp. 835–846.

Mair, H., L. Xiu, and S. A. Fahrenbruch. 2001. "Precision Frequency and Phase Synthesis," patent US6329850, Dec.

Nagendra, C., M. J. Irwin, and R. M. Owens. 1996. "Area-Time-Power Tradeoffs in Parallel Adders," *IEEE Trans. on Circuit Syst. II*, vol. 43, no. 10, pp. 689–702.

Shams, A. M., T. K. Darwish, and M. A. Bayoumi. 2002. "Performance Analysis of Low-Power 1-Bit CMOS Full Adder Cells," *IEEE Trans. on VLSI Syst.*, vol. 10, no. 1, pp. 20–29.

Sung, G. N., S. C. Liao, J. M. Huang, Y. C. Lu, and C. C. Wang. 2010. "All Digital Frequency Synthesizer Using Flying-Adder," *IEEE Trans. on Circuit Syst. II*, vol. 57, pp. 597–601.

Takinami, K., et al. "A Rotary-Traveling-Wave-Oscillator based All-Digital PLL with a 32-Phase Embedded Phase-to-Digital Converter in 65 nm CMOS," *ISSCC Dig. Tech. Papers*, pp.100–102, Feb., 2011.

"TMS320C6A8x Integra DSP+ARM Processors Technical Reference Manual," http://focus.ti.com/lit/ug/sprugx9/sprugx9.pdf, Texas Instruments Inc., 2011.

"TMS320DM816x DaVinci Digital Media Processors Technical Reference *Manual*," http://focus.ti.com/lit/ug/sprugx8/sprugx8.pdf, Texas Instruments Inc., 2011.

Wood, J., T. C. Edwards, and S. Lipa. 2001. "Rotary Traveling-Wave Oscillator Arrays: A New Clock Technology," *IEEE J. Solid-State Circuits*, vol. 36, pp. 1654–1665.

Xiu, L. 2006. "Method and Apparatus for Reducing Jitter in Output Signals from a Frequency Synthesizer Using a Control Word Having a Fractional Bit," US patent pending, serial no. 11/489982.

Xiu, L. 2007. "A Flying-Adder Based On-chip Frequency Generator for Complex SoC," *IEEE Trans. on Circuit Syst. II*, vol. 54, pp. 1067–1071.

Xiu, L. 2009. "A Fast and Power-Area Efficient Accumulator for Flying-Adder Frequency Synthesizer," *IEEE Trans. on Circuit Syst. I*, vol. 56, pp. 2439–2448.

Xiu, L. 2010. "Adder Circuit and XIU-Accumulator Circuit Using the Same," US patent pending, serial no. 12/892516.

Xiu, L. and Z. You. 2002. "A Flying-Adder Architecture of Frequency and Phase Synthesis with Scalability," *IEEE Trans. on VLSI*, vol. 10, pp. 637–649.

Xiu, L. and Z. You. 2003. "A New Frequency Synthesis Method Based on 'Flying-Adder' Architecture," *IEEE Trans. on Circuit Syst. II*, vol. 50, pp. 130–134.

Xiu, L. and Z. You. 2005a."Scalable High Speed Precision Frequency and Phase Synthesis," patent US6940937, Dec.

Xiu, L. and Z. You. 2005b. "A 'Flying-Adder' Frequency Synthesis Architecture of Reducing VCO Stages," *IEEE Trans. on VLSI*, vol. 13, pp. 201–210.

Xiu, L. and Z. You. 2008. "Precision Frequency and Phase Synthesis with Fewer Voltage-Controlled Oscillator Stages", US patent 7372340, May.

Xiu, L., K. H. Lin, and M. Lin. 2011. "The Impact of Input Mismatch on Flying-Adder Direct Period Synthesizer Output Jitter," *IEEE Trans. on Circuit Syst. I*, available at http://ieeexplore.ieee.org/search/searchresult.jsp?newsearch=true&queryText=The+Impact+of+Input+Mismatch+on+Flying-Adder+Direct+Period&x=11&y=25.

CHAPTER 5

DIGITAL-TO-FREQUENCY CONVERTER

5.1 TWO WAYS OF REPRESENTING INFORMATION

Information in the real world presents itself through the mediums of pressure, temperature, weight, light, etc. To describe information, as discussed in Section 3.1, two scales of measurement are needed: level and time. Level represents the strength of the signal (the information); time records the moment at which that particular strength happens. In the field of VLSI circuit design, all chips are designed for one purpose: to process information. Information is collected, transformed, manipulated, and delivered by various chips. In the collection and delivery stages, information is exchanged between two different worlds: the human-central real world and the circuit-central electronic world. In the boundary between the two, the sensor and the actuator are the bridges. Various kinds of sensors convert the information from its original native form into voltage and/or current. In the majority of cases, the transformation is carried out between the strength of the information and the level of the voltage/current, which is subsequently processed by the VLSI chip. After processing, the information is converted back to some native form by the actuator and returned to the human world for us to use.

VLSI chips are made of transistors. Transistors can be used in two ways to manipulate information: *amplification* and *on–off switch*. In the amplification method, level (magnitude) is used to represent information. In the on–off switch approach, the *rate-of-switching* is the information. The two approaches

Nanometer Frequency Synthesis Beyond the Phase-Locked Loop, First Edition. Liming Xiu.

Published 2012 by John Wiley & Sons, Inc.

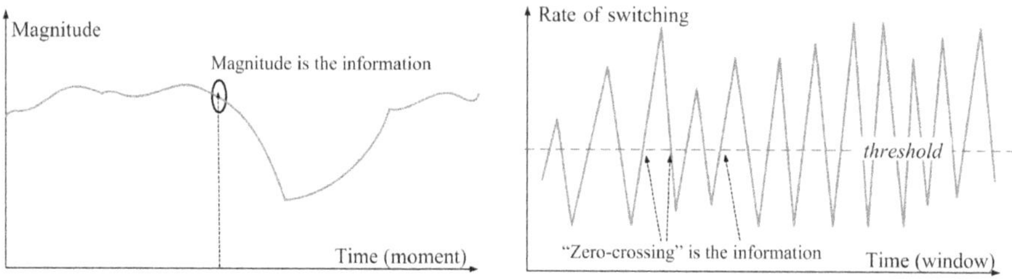

Fig. 5.1. Two ways of representing information: level (magnitude) and rate-of-switching.

are illustrated in Fig. 5.1. Up until now, level/magnitude has been the dominant method for representing information inside the chip because magnitude is directly proportional to the number of electrons flowing inside a device. In this method, a specially created signal—clock—is used to function as a measure of time. In the rate-of-switching approach, magnitude is no longer significant. Instead, how many times the signal crosses a threshold within a time window is what we care about. It is the information. In this approach, the clock is also used to mark the time aspect of the information.

In the rate-of-switching approach, the distinguished characteristic is that there are two parameters that are related to time. One is the signal itself (i.e., the information, the rate-of-switching). The other is the scale of the measurement (the clock). Compared to the magnitude approach, where the clock is used to record "the moment," the purpose of the clock signal in the rate-of-switching approach is to define a "time window." Although the same clock signal is used, its usages are different in these two approaches. Mathematically, the rate-of-switching and the clock signal have the same dimensions. Therefore, instead of absolute value, their ratio is used for calculation. This characteristic can potentially provide an advantage for some applications. In both the magnitude and rate-of-switching methods, the higher the clock rate, the better it can be used to describe the information, with finer detail.

5.2 THE CONVERTERS FOR TRANSFORMING INFORMATION

The sensor and the actuator are used to bridge the human and electronic worlds. After information is received into the chips, information processing is mainly carried out in a binary fashion for efficiency. In other words, a second conversion stage is needed to convert the magnitude (multiple-level analog domain) to a two-level system (digital domain). This is achieved through an analog-to-digital converter (ADC). At the other end of the spectrum, a digital-to-analog converter (DAC) is needed when information is ready to be sent out to outside world though the actuator. It converts the two-level signal to a multiple-level one (magnitude). These two converters can be characterized as a *level⇔ digital* converter.

As semiconductor process continually advances, the power supply voltage keeps dropping at each node. Unfortunately, the noise level does not go down proportionally. As a result, the useful room for processing signal is reduced quickly in this *level⇔ digital* approach. On the other hand, the transistor switches faster as its size becomes smaller. At each process advancement, we reach finer resolution in time. Therefore, the natural choice for future information processing is *time⇔ digital.* In this new approach, the floor keeps going down (we reach finer resolution every time), and the ceiling is infinite (there is no end in time). Hence, accuracy improves with each process's advancement. "It is time to use time."* Naturally, we need another set of converters, digital-to-frequency converter (DFC) and frequency-to-digital converter (FDC), for future chip design (frequency here corresponds to rate-of-switching). Figure 5.2 illustrates this point.

Figure 5.3 shows the two information processing approaches based on these two sets of converters. The ADC and DAC are the components essential for *level⇔ digital.* FDC and DFC are the needed members for *time⇔ digital.* The FDC is a matured component. It can be realized either by a time-to-digital converter (TDC) or simply by a digital counter. However, the DFC is non-trivial. The difficulty lies in the fact that generating ample frequencies in a cost-effective way is not easy. Moreover, switching between frequencies at a

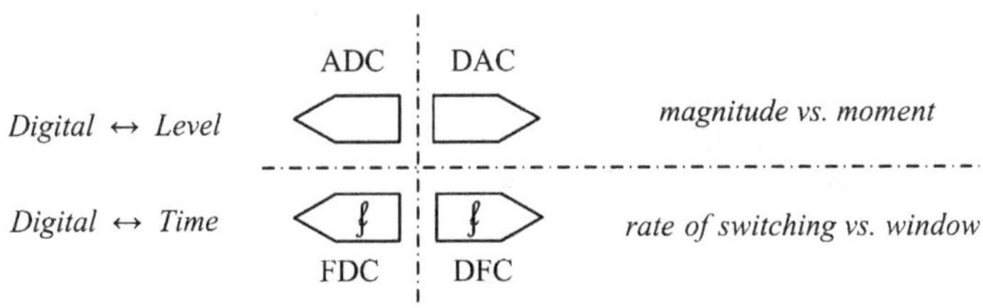

Fig. 5.2. Two sets of converters.

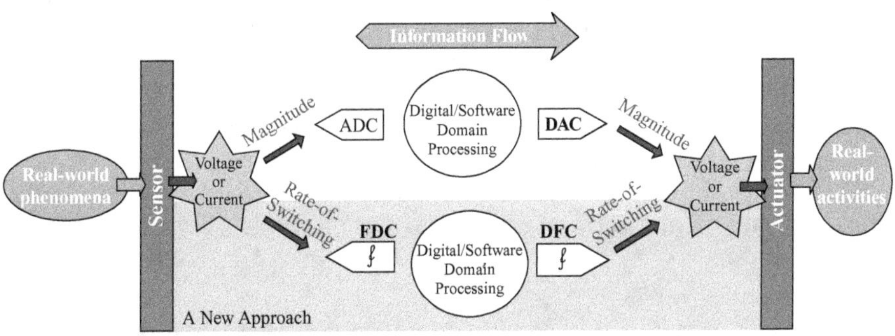

Fig. 5.3. The two information processing flows enabled by the two sets of converters.

* To author's best knowledge, this phrase is first used by R.B. Staszewski in the 2010 IEEE CAS summer school.

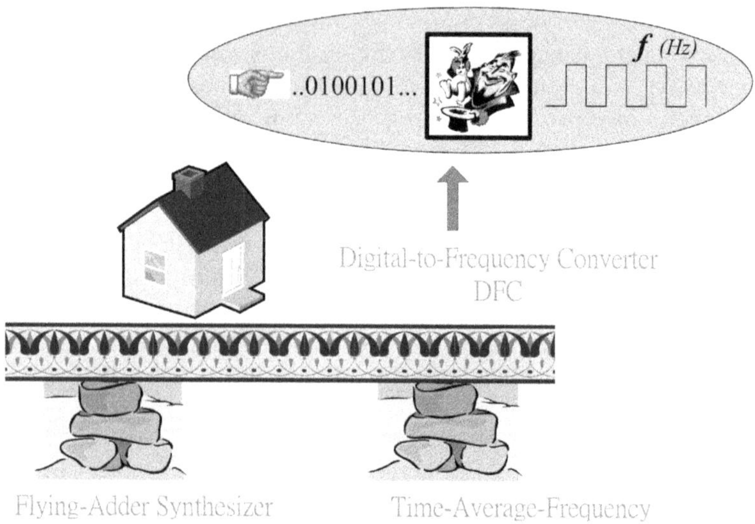

Fig. 5.4. The two cornerstones of the digital-to-frequency converter.

fast pace is hard to achieve. These two problems have delayed the emergence of DFCs until the time-average-frequency's arrival.

5.3 THE TWO CORNERSTONES OF THE DIGITAL-TO-FREQUENCY CONVERTER*

There are two cornerstones that establish the foundation for the emerging DFC, as is illustrated graphically in Fig. 5.4. One is on the theoretical side; the other is on the practical side. Time-average-frequency, introduced in Chapter 3, is the theoretical foundation of this device. The flying-adder direct period synthesizer is the workhorse of this device. Flying-adder architecture has being existed as a circuit technique for more than a decade. Its arbitrary frequency generation† and instantaneous switching‡ have prepared the DFC from the implementation side. The arrival of time-average-frequency in 2008 formally established this critical component of future.

The term of the DFC is created to mimic its counterpart, the DAC. In the case of a DAC, theoretically, an arbitrary voltage level (in a certain range) can

* The DFC used in this book is specifically used to describe the time-average-frequency based timing device.

† Arbitrary frequency generation: any frequency can be generated as long as enough fractional bits are used.

‡ Instantaneous switching: output frequency can be changed in next two cycles after the command is received.

be produced as long as enough resources are used. In practice, fine voltage resolution can be achieved with reasonable resources. Moreover, the switching between voltage levels can be accomplished within, at most, a few clock cycles. A similar level of qualification is required for the DFC in these two regards. Direct analog synthesis can switch its output frequency at a fast pace, but the frequency step is coarse, and the associated cost is very high. Direct digital synthesis can achieve fine resolution, but cost has prevented it from being a major player in on-chip frequency synthesis. The indirect-PLL-based technique inherently has the issue of speed impotence since it uses feedback to direct its output toward the input.

Time-average-frequency is a fundamental breakthrough in clock technology. Unlike all previous techniques, it uses mathematically rigorous counting and an open-loop-circuit structure to truly solve the arbitrary frequency generation problem: $T_{TAF} = (1 - r)\cdot T_A + r\cdot T_B$ (refer to Figs. 4.64, 4.71–4.74, Table 4.9, etc., for hard evidence). Based on the carefully constructed cycles T_A and T_B, the weight r can be precisely set to achieve any arbitrarily desired frequency. On the speed side, the flying-adder's two-cycles switching capability ensures that the DFC can respond quickly like DAC (refer to Figs. 4.66–4.69 for hard evidence). It is only by these two unique features that this new component qualifies as a converter.

Figure 5.5 graphically illustrates the output of the DFC. Unlike the DAC, where the output is a fixed voltage level when the input is fixed, the output from the DFC is a pulse train, which is made of one or two types of pulses (T_A, T_B). These two types of pulses are base units whose sizes are predetermined. In other words, the instantaneous periods* of T_A and T_B are known by their creator. The intended DFC output is the average frequency (the rate-of-switching, the number of zero-crossing). When r is non-zero, its output period alternates between the two values, T_A and T_B. Its DC value (the average) is determined by input FREQ = $I + r$. I is the integer part that determines the sizes of T_A and T_B; r is the fractional part that assigns weights between T_A and T_B.

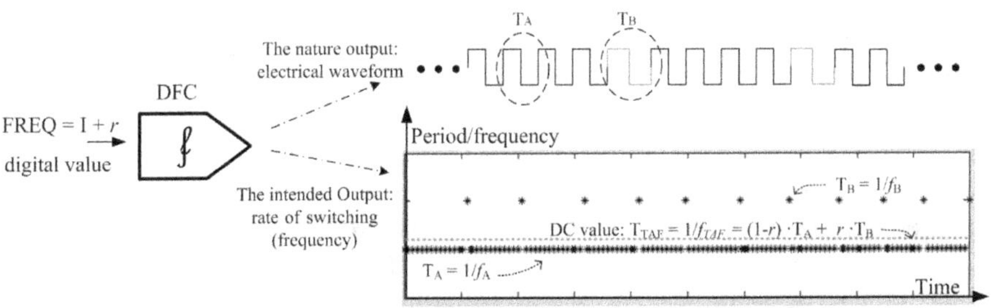

Fig. 5.5. The illustration of the DFC output.

* By definition, frequency is a concept of long term (refer to Section 3.2). The concept of instantaneous frequency is not valid. In Fig. 5.5, instantaneous frequency f_A is defined as the inverse of instantaneous period T_A, which can be measured.

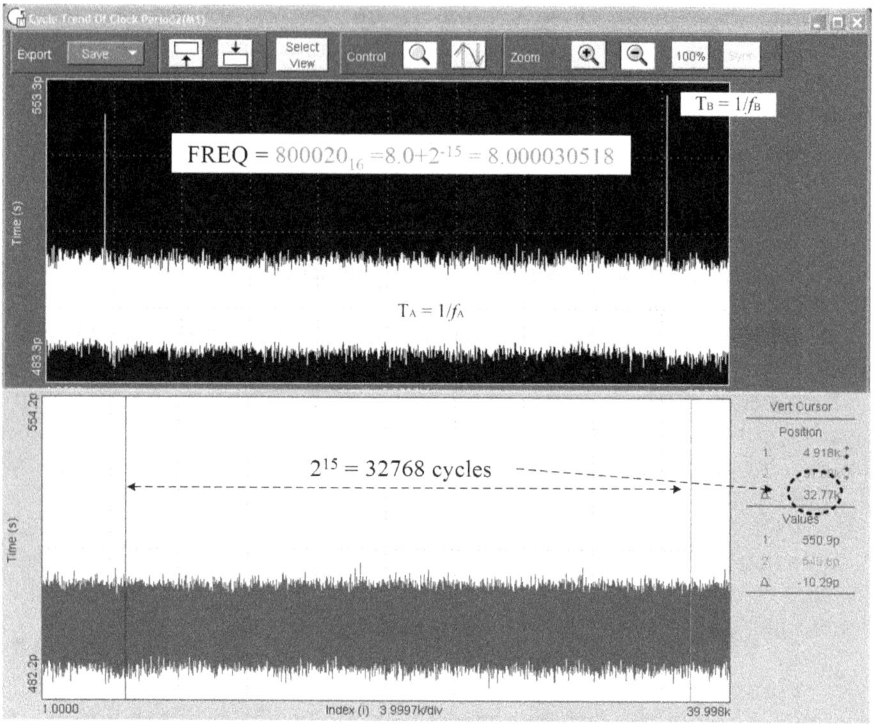

Fig. 5.6. The DFC output from a real chip measurement.

Figure 5.6 is a DFC output from a real chip measurement. In this case, $\Delta = 62.5$ ps and FREQ = $8 + 2^{-15}$. Therefore, $T_A = 8 \cdot \Delta = 500$ ps and $T_B = 9 \cdot \Delta = 562.5$ ps. For every $1/r = 32768$ cycles, there are 32767 T_A and one T_B. The DC value (time-average-frequency) is 500.0019 ps (1.99999 GHz).

5.4 THE THEORETICAL FOUNDATION OF FLYING-ADDER DIGITAL-TO-FREQUENCY CONVERTER*

Direct digital synthesis (DDS) has been studied by researchers for several decades. A large amount of the literature has been published in understanding the theoretical aspect of this technique (*A Technical Tutorial* 1999; Blythe 1985; Curticapean and Niittylahti 2003; Goldberg 1999; Jenq 1988a, 1988b; Ken et al. 2003; Kroupa 1993, 1998; Mehrgardt 1983; Morgan and Aridgides 1985;

* In the following sections, the term "flying-adder DFC" is referred to as "flying-adder direct period synthesizer plus time-average-frequency theory." This term is more than just the flying-adder synthesizer, which only represents the circuit itself. In other words, **flying-adder DFC = flying-adder synthesizer + time-average-frequency theory**.

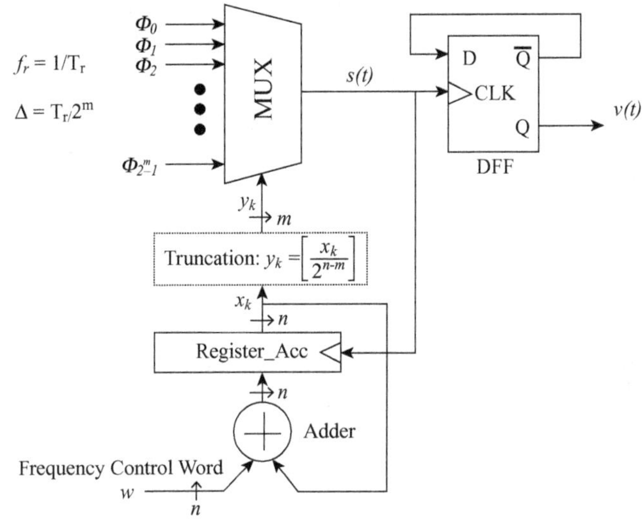

Fig. 5.7. The flying-adder DFC mathematical model.

Nicholas and Samueli 1987). As discussed in Section 4.24, the characteristic of the flying-adder synthesizer is significantly different from that of DDS. Unlike DDS, where everything is driven by a fixed rate clock, a flying-adder circuit is driven by its own clock whose instant period varies from time to time. For this reason, a new theory is needed to fully understand its behavior.

5.4.1 Flying-Adder DFC Mathematical Model* and Its State Variables

Figure 5.7 is the mathematical model that we will use to develop the flying-adder DFC's signal properties. This model resembles the flying-adder circuit in Fig. 4.4. In flying-adder operations, there are two variables that are used to control the flow of time: discrete time k and continuous time t. Discrete time k is indexed by counting the rising (or falling) edges of signal s. Continuous time t is the continuous time recorded in units of Δ. As discussed before, unlike the case in DDS, where discrete time (the clock cycle) is linearly proportional to continuous time, k is not linearly related to t due to the feedback clocking in the flying-adder circuit. They must be treated separately.

From an analysis point of view, the flying-adder DFC can be treated in state space. Its current state is defined by state variables. For this purpose, the notations below are introduced first.

* The flying-adder mathematical model is first created by Paul P. Sotiriadis in 2010. The work in this Section 5.4.1 is largely based on his three papers (Sotiriadis 2010a, 2010b, 2010c).

- n: the size of the flying-adder DFC's register and accumulator (in bits).
- m: the number of bits that control the MUX. $1 \leqq m \leqq n$. There are 2^m inputs for the flying-adder DFC.
- Φ_i: a family of 2^m periodic, square-wave, 50% duty cycle signals, $i = 0, 1, \ldots, 2^m - 1$. They have same frequency but relative phase offsets with steps of $2\pi/2^m$ *rad* between any two adjacent signals.
- $f_r = 1/T_r$: the frequency and period of the 2^m input signals, Φ_i.
- Δ: the time delay between any two temporally adjacent input signals, $\Delta = T_r/2^m$.
- w: frequency control word; n-bits long. $w \in \{0, 1, \ldots, 2^n - 1\}$
- x_k: the state of the flying-adder DFC. The sequence $\{x_k\}$, $k = 0, 1, 2, \ldots$, indicates the state that the flying-adder circuit currently is in. $x_k \in \{0, 1, \ldots, 2^n - 1\}$.
- y_k: the truncated state of the flying-adder DFC. The sequence $\{y_k\}$, $k = 0, 1, 2, \ldots$, is formed by the m MSB of x_k. $y_k \in \{0, 1, \ldots, 2^m - 1\}$ This variable is used to select one input from the 2^m input signals, Φ_i.
- t: continuous time in units of Δ.
- k: discrete time index $k = 0, 1, 2, \ldots$. This is the discrete time reference that is defined as the number of rising edges that occur in signal s within the continuous time interval $(0^-, t^+)$.
- k^{th} discrete time interval: the continuous time interval between the kth and $k + 1^{\text{th}}$ rising edge of s.
- $s(t)$: signal out of MUX. It coincides with input phase $\Phi_{yk}(t)$ during the discrete time interval k.
- $v(t)$: signal at the flying-adder DFC's output. It results from counting the modulo 2 rising edge of $s(t)$.

5.4.2 Flying-Adder DFC as a Finite State Machine (FSM)

The flying-adder DFC is controlled by the accumulator. The fixed-point accumulator is a finite state machine (FSM) and must produce a periodic output sequence (Golomb 1982). Using notations introduced in Section 5.4.1, the state equation that describes the flying-adder DFC's operation can be expressed in Eq. 5.1. This equation solely determines the flying-adder DFC's next state x_{k+1}.

$$x_{k+1} = (x_k + w) \bmod 2^n \tag{5.1}$$

Without loss of generality, the initial state can be assumed zero, $x_0 = 0$. Therefore, we have

$$x_k = (kw) \bmod 2^n \tag{5.2}$$

The truncated state y_k is formed by taking the m MSB of the n-bits state x_k. Thus, it can be expressed in Eq. 5.3, where [] is the operator of "taking the integer part."

$$y_k = \left[\frac{x_k}{2^{n-m}}\right] = \left[\frac{kw}{2^{n-m}}\right] \bmod 2^m \tag{5.3)*$$

The instantaneous period of output signals s (and v) is determined by the advancement of the truncated state y_k when k advances (refer to Fig. 5.7). Thus, an important variable d_k is defined in Eq. 5.4, which is used to describe the "distance" between y_{k-1} and y_k. Using Eq. 5.3, Eq. 5.5 can be derived.

$$d_k = (y_k - y_{k-1}) \bmod 2^m \tag{5.4}$$

$$d_k = \left(\left[\frac{kw}{2^{n-m}}\right] - \left[\frac{(k-1)w}{2^{n-m}}\right]\right) \bmod 2^m \tag{5.5)†$$

To fully describe the flying-adder DFC operation, a modified difference sequence is defined in Eq. 5.6.

$$\delta_k = \begin{cases} d_k, & \text{when} \quad d_k \neq 0 \\ 2^m, & \text{when} \quad d_k = 0 \end{cases} \tag{5.6}$$

5.4.3 The Periodicity in Discrete Time Domain

The state variables x_k, y_k, d_k, and δ_k are used to describe the flying-adder DFC operation in discrete time domain (k is used to represent the elapsed time). When frequency control word w is given, the periodicity embedded in these states is the key to understanding the DFC's characteristics. Several important properties associated with those state variables are listed below. The formal proof can be found in Sotiriadis (2010a).

The state variables x_k and y_k both have a period of K‡, which is defined in Eq. 5.7, where operator *gcd* stands for *greatest common divisor*. It is valid under all initial conditions.

$$K = \frac{2^n}{2^r}, \quad \text{where} \quad 2^r = \gcd(w, 2^n) \tag{5.7}$$

Within the period K, the state variable x_k **only** takes values in the set X. All the values in this set are taken exactly once. The truncated state y_k **only** takes values in the set Y. All the values in this set are taken at least once.

* See Fact1 in the appendix of Sotiriadis 2010a.

† See Fact2 in the appendix of Sotiriadis 2010a.

‡ This K is not the K used in Section 4.1, Fig. 4.1 (the number of flying-adder input).

$$X = \{k2^r \mid k = 0, 1, 2, \dots, 2^{n-r} - 1\} \tag{5.8}$$

$$Y = \{[(k2^r)/2^{n-m}] \mid k = 0, 1, 2, \dots, 2^{n-r} - 1\} \tag{5.9}$$

The state variables d_k and δ_k both have a period of L as defined in Eq. 5.10.

$$L = \frac{2^{n-m}}{\gcd(w, 2^{n-m})} \tag{5.10}$$

To assist with the understanding of these important periodicities, three numerical examples are provided below. The program used to generate those examples is presented in Appendix 5.A. Experimenting with it is strongly recommended. In Example 1, the DFC structure is $n = 8$, $m = 4$, and the frequency control word is $w = 48$. Under this condition, from Eq. 5.7 and Eq. 5.10, the periodicities for state x_k and y_k, d_k and δ_k are $K = 16$ and $L = 1$, respectively. From Eq. 5.8, the valid values for x_k is $X = \{0, 16, 32, 48, 64, 80, 96, 112, 128, 144, 160, 176, 192, 208, 224, 240\}$. As is clear from Table 5.1, all these values have been taken by x_k and each one is only taken once within one $K = 16$ period. Similarly, from Eq. 5.9, the valid values for y_k is $Y = \{0, 1, 2, 3, 4, 5, 6, 7, 8, 9, 10, 11, 12, 13, 14, 15\}$. Example 2 uses the same DFC configuration (same n and m) but with a different w of 72. In this case, $K = 32$. Thus, there are 32 valid values for x_k. However, according to Eq. 5.9, there should be only 16 valid values for y_k since $r = 3$ now. Table 5.2 shows that this indeed is true. Each of the 16 values has been used twice within the $K = 32$ period. Within this period, the y_k sequence is random. However, for the next $K = 32$ period, it will repeat (not shown in this table). This observation aligns with the statement in Eq. 5.7 that K is the period for both x_k and y_k.

Example 3 is a case that mimics many real flying-adder implementations: $n = 32$ and $m = 4$. In other words, there are 32 bits in the accumulator. The number of inputs is $2^m = 16$ (8 inputs with two paths). Under the condition of $w = 1{,}140{,}850{,}688$, L is calculated as 4 from Eq. 5.10. Consequently, the distance state variables d_k and δ_k repeat every four k's advancements. This results in time-average-frequency as will be explained in next section.

5.4.4 The Periodicity in Continuous Time Domain

The flying-adder DFC output is the continuous time signal $s(t)$ (or $v[t]$). As stated earlier, the flying-adder's continuous time signal is not linearly proportional to its discrete time counterpart due to the feedback clocking. The bridge between the two is the modified difference variable δ_k. For signal $s(t)$, its fundamental continuous time period can be found as in Eq. 5.11. The fundamental continuous time period for signal $v(t)$ is expressed as Eq. 5.12 since $v(t)$ is formed by counting two rising edges of $s(t)$. $L = 1$ is a special case because δ_k is a constant under this condition (see Eqs. 5.10 and 5.5). Refer back to Tables

TABLE 5.1. Example 1: $n = 8$, $m = 4$, $w = 48 \rightarrow (K = 16, L = 1, r = 4$, $FREQ[7:0] = 6.0^a = 0110.0000_2)$

k	x_k	y_k	Note	δ_k	T_s^b	Note	T_v^b	Note
0	0^c	0	A value from X and Y	3	3Δ	δ_k constant since $L = 1$		
1	48	3	A value from X and Y	3	3Δ	δ_k constant since $L = 1$	6Δ	$L = 1$, 50% duty
2	96	6	A value from X and Y	3	3Δ	δ_k constant since $L = 1$		
3	144	9	A value from X and Y	3	3Δ	δ_k constant since $L = 1$	6Δ	$L = 1$, 50% duty
4	192	12	A value from X and Y	3	3Δ	δ_k constant since $L = 1$		
5	240	15	A value from X and Y	3	3Δ	δ_k constant since $L = 1$	6Δ	$L = 1$, 50% duty
6	32	2	A value from X and Y	3	3Δ	δ_k constant since $L = 1$		
7	80	5	A value from X and Y	3	3Δ	δ_k constant since $L = 1$	6Δ	$L = 1$, 50% duty
8	128	8	A value from X and Y	3	3Δ	δ_k constant since $L = 1$		
9	176	11	A value from X and Y	3	3Δ	δ_k constant since $L = 1$	6Δ	$L = 1$, 50% duty
10	224	14	A value from X and Y	3	3Δ	δ_k constant since $L = 1$		
11	16	1	A value from X and Y	3	3Δ	δ_k constant since $L = 1$	6Δ	$L = 1$, 50% duty
12	64	4	A value from X and Y	3	3Δ	δ_k constant since $L = 1$		
13	112	7	A value from X and Y	3	3Δ	δ_k constant since $L = 1$	6Δ	$L = 1$, 50% duty
14	160	10	A value from X and Y	3	3Δ	δ_k constant since $L = 1$		
15	208	13	A value from X and Y	3	3Δ	δ_k constant since $L = 1$	6Δ	$L = 1$, 50% duty
16	0	0	Repeat since $K = 16$	3	3Δ	δ_k constant since $L = 1$		
17	48	3	Repeat since $K = 16$	3	3Δ	δ_k constant since $L = 1$	6Δ	$L = 1$, 50% duty
18	96	6	Repeat since $K = 16$	3	3Δ	δ_k constant since $L = 1$		
19	144	9	Repeat since $K = 16$	3	3Δ	δ_k constant since $L = 1$	6Δ	$L = 1$, 50% duty
20	192	12	Repeat since $K = 16$	3	3Δ	δ_k constant since $L = 1$		
21	240	15	Repeat since $K = 16$	3	3Δ	δ_k constant since $L = 1$	6Δ	$L = 1$, 50% duty

(*Continued*)

TABLE 5.1. (*Continued*)

k	x_k	y_k	Note	δ_k	T_s^b	Note	T_v^b	Note
22	32	2	Repeat since $K = 16$	3	3Δ	δ_k constant since $L = 1$		
23	80	5	Repeat since $K = 16$	3	3Δ	δ_k constant since $L = 1$	6Δ	$L = 1$, 50% duty
24	128	8	Repeat since $K = 16$	3	3Δ	δ_k constant since $L = 1$		
25	176	11	Repeat since $K = 16$	3	3Δ	δ_k constant since $L = 1$	6Δ	$L = 1$, 50% duty
26	224	14	Repeat since $K = 16$	3	3Δ	δ_k constant since $L = 1$		
27	16	1	Repeat since $K = 16$	3	3Δ	δ_k constant since $L = 1$	6Δ	$L = 1$, 50% duty
28	64	4	Repeat since $K = 16$	3	3Δ	δ_k constant since $L = 1$		
29	112	7	Repeat since $K = 16$	3	3Δ	δ_k constant since $L = 1$	6Δ	$L = 1$, 50% duty
30	160	10	Repeat since $K = 16$	3	3Δ	δ_k constant since $L = 1$		
31	208	13	Repeat since $K = 16$	3	3Δ	δ_k constant since $L = 1$	6Δ	$L = 1$, 50% duty

[a] Refer to Section 5.4.11.
[b] Refer to Section 5.4.4.
[c] The initial state can start from any value that is less than $2^n - 1 = 255$.

TABLE 5.2. Example 2: $n = 8$, $m = 4$, $w = 72 \rightarrow (K = 32, L = 2, r = 3, FREQ = 9.0 = 1001.0000_2)$

k	x_k	y_k	Note	δ_k	T_s	Note	T_v	Note
0	72	4	32 values for X, 16 values for Y	4		δ_k has two values since $L = 2$		
1	144	9	32 values for X, 16 values for Y	5	9Δ	δ_k has two values since $L = 2$	9Δ	$L = 2$, non-50% duty
2	216	13	32 values for X, 16 values for Y	4		δ_k has two values since $L = 2$		
3	32	2	32 values for X, 16 values for Y	5	9Δ	δ_k has two values since $L = 2$	9Δ	$L = 2$, non-50% duty
4	104	6	32 values for X, 16 values for Y	4		δ_k has two values since $L = 2$		

TABLE 5.2. (*Continued*)

k	x_k	y_k	Note	δ_k	T_s	Note	T_v	Note
5	176	11	32 values for X, 16 values for Y	5	9Δ	δ_k has two values since $L = 2$	9Δ	$L = 2$, non-50% duty
6	248	15	32 values for X, 16 values for Y	4		δ_k has two values since $L = 2$		
7	64	4	32 values for X, 16 values for Y	5	9Δ	δ_k has two values since $L = 2$	9Δ	$L = 2$, non-50% duty
8	136	8	32 values for X, 16 values for Y	4		δ_k has two values since $L = 2$		
9	208	13	32 values for X, 16 values for Y	5	9Δ	δ_k has two values since $L = 2$	9Δ	$L = 2$, non-50% duty
10	24	1	32 values for X, 16 values for Y	4		δ_k has two values since $L = 2$		
11	96	6	32 values for X, 16 values for Y	5	9Δ	δ_k has two values since $L = 2$	9Δ	$L = 2$, non-50% duty
12	168	10	32 values for X, 16 values for Y	4		δ_k has two values since $L = 2$		
13	240	15	32 values for X, 16 values for Y	5	9Δ	δ_k has two values since $L = 2$	9Δ	$L = 2$, non-50% duty
14	56	3	32 values for X, 16 values for Y	4		δ_k has two values since $L = 2$		
15	128	8	32 values for X, 16 values for Y	5	9Δ	δ_k has two values since $L = 2$	9Δ	$L = 2$, non-50% duty
16	200	12	32 values for X, 16 values for Y	4		δ_k has two values since $L = 2$		
17	16	1	32 values for X, 16 values for Y	5	9Δ	δ_k has two values since $L = 2$	9Δ	$L = 2$, non-50% duty
18	88	5	32 values for X, 16 values for Y	4		δ_k has two values since $L = 2$		
19	160	10	32 values for X, 16 values for Y	5	9Δ	δ_k has two values since $L = 2$	9Δ	$L = 2$, non-50% duty

(*Continued*)

TABLE 5.2. (*Continued*)

k	x_k	y_k	Note	δ_k	T_s	Note	T_v	Note
20	232	14	32 values for X, 16 values for Y	4		δ_k has two values since $L = 2$		
21	48	3	32 values for X, 16 values for Y	5	9Δ	δ_k has two values since $L = 2$	9Δ	$L = 2$, non-50% duty
22	120	7	32 values for X, 16 values for Y	4		δ_k has two values since $L = 2$		
23	192	12	32 values for X, 16 values for Y	5	9Δ	δ_k has two values since $L = 2$	9Δ	$L = 2$, non-50% duty
24	8	0	32 values for X, 16 values for Y	4		δ_k has two values since $L = 2$		
25	80	5	32 values for X, 16 values for Y	5	9Δ	δ_k has two values since $L = 2$	9Δ	$L = 2$, non-50% duty
26	152	9	32 values for X, 16 values for Y	4		δ_k has two values since $L = 2$		
27	224	14	32 values for X, 16 values for Y	5	9Δ	δ_k has two values since $L = 2$	9Δ	$L = 2$, non-50% duty
28	40	2	32 values for X, 16 values for Y	4		δ_k has two values since $L = 2$		
29	112	7	32 values for X, 16 values for Y	5	9Δ	δ_k has two values since $L = 2$	9Δ	$L = 2$, non-50% duty
30	184	11	32 values for X, 16 values for Y	4		δ_k has two values since $L = 2$		
31	0	0	32 values for X, 16 values for Y	5	9Δ	δ_k has two values since $L = 2$	9Δ	$L = 2$, non-50% duty

5.1–5.3; T_s and T_v are listed in the tables for reference. The internal state x_k and y_k also have their corresponding fundamental continuous time period. They can be derived in Eq. 5.13. In all these equations, we only consider the cases of $2^{n-m} \leqq w < 2^n$. For $w < 2^{n-m}$, the DFC output is very irregular. It can hardly be used as a clock (but could be useful for pattern generation in other applications, such as communication).

TABLE 5.3. Example 3: $n = 32$, $m = 4$, $w = 1140850688 \rightarrow (K = 64, L = 4, r = 26,$ $FREQ = 8.5 = 8.8000000_{16})$

k	x_k in units of 2^{26}	y_k	δ_k	T_s	Note	T_v	T_{TAF}	Note
0	17	4	4		δ_k has two values since $L = 4$			
1	34	8	4		δ_k has two values since $L = 4$		8.5Δ	Time-average-frequency
2	51	12	4		δ_k has two values since $L = 4$			
3	4	1	5	17Δ	δ_k has two values since $L = 4$	17Δ	8.5Δ	Time-average-frequency
4	21	5	4		δ_k has two values since $L = 4$			
5	38	9	4		δ_k has two values since $L = 4$		8.5Δ	Time-average-frequency
6	55	13	4		δ_k has two values since $L = 4$			
7	8	2	5	17Δ	δ_k has two values since $L = 4$	17Δ	8.5Δ	Time-average-frequency
8	25	6	4		δ_k has two values since $L = 4$			
9	42	10	4		δ_k has two values since $L = 4$		8.5Δ	Time-average-frequency
10	59	14	4		δ_k has two values since $L = 4$			
11	12	3	5	17Δ	δ_k has two values since $L = 4$	17Δ	8.5Δ	Time-average-frequency
12	29	7	4		δ_k has two values since $L = 4$			
13	46	11	4		δ_k has two values since $L = 4$		8.5Δ	Time-average-frequency
14	63	15	4		δ_k has two values since $L = 4$			
15	16	4	5	17Δ	δ_k has two values since $L = 4$	17Δ	8.5Δ	Time-average-frequency
16	33	8	4		δ_k has two values since $L = 4$			
17	50	12	4		δ_k has two values since $L = 4$		8.5Δ	Time-average-frequency
18	3	0	4		δ_k has two values since $L = 4$			
19	20	5	5	17Δ	δ_k has two values since $L = 4$	17Δ	8.5Δ	Time-average-frequency
20	37	9	4		δ_k has two values since $L = 4$			

(*Continued*)

TABLE 5.3. (*Continued*)

k	x_k in units of 2^{26}	y_k	δ_k	T_s	Note	T_v	T_{TAF}	Note
21	54	13	4		δ_k has two values since $L = 4$		8.5Δ	Time-average-frequency
22	7	1	4		δ_k has two values since $L = 4$			
23	24	6	5	17Δ	δ_k has two values since $L = 4$	17Δ	8.5Δ	Time-average-frequency
24	41	10	4		δ_k has two values since $L = 4$			
25	58	14	4		δ_k has two values since $L = 4$		8.5Δ	Time-average-frequency
26	11	2	4		δ_k has two values since $L = 4$			
27	28	7	5	17Δ	δ_k has two values since $L = 4$	17Δ	8.5Δ	Time-average-frequency
28	45	11	4		δ_k has two values since $L = 4$			
29	62	15	4		δ_k has two values since $L = 4$		8.5Δ	Time-average-frequency
30	15	3	4		δ_k has two values since $L = 4$			
31	32	8	5	17Δ	δ_k has two values since $L = 4$	17Δ	8.5Δ	Time-average-frequency
32	49	12	4		δ_k has two values since $L = 4$			
33	2	0	4		δ_k has two values since $L = 4$		8.5Δ	Time-average-frequency
34	19	4	4		δ_k has two values since $L = 4$			
35	36	9	5	17Δ	δ_k has two values since $L = 4$	17Δ	8.5Δ	Time-average-frequency
36	53	13	4		δ_k has two values since $L = 4$			
37	6	1	4		δ_k has two values since $L = 4$		8.5Δ	Time-average-frequency
38	23	5	4		δ_k has two values since $L = 4$			
39	40	10	5	17Δ	δ_k has two values since $L = 4$	17Δ	8.5Δ	Time-average-frequency
40	57	14	4		δ_k has two values since $L = 4$			
41	10	2	4		δ_k has two values since $L = 4$		8.5Δ	Time-average-frequency
42	27	6	4		δ_k has two values since $L = 4$			

TABLE 5.3. (*Continued*)

k	x_k in units of 2^{26}	y_k	δ_k	T_s	Note	T_v	T_{TAF}	Note
43	44	11	5	17Δ	δ_k has two values since $L = 4$	17Δ	8.5Δ	Time-average-frequency
44	61	15	4		δ_k has two values since $L = 4$			
45	14	3	4		δ_k has two values since $L = 4$		8.5Δ	Time-average-frequency
46	31	7	4		δ_k has two values since $L = 4$			
47	48	12	5	17Δ	δ_k has two values since $L = 4$	17Δ	8.5Δ	Time-average-frequency
48	1	0	4		δ_k has two values since $L = 4$			
49	18	4	4		δ_k has two values since $L = 4$		8.5Δ	Time-average-frequency
50	35	8	4		δ_k has two values since $L = 4$			
51	52	13	5	17Δ	δ_k has two values since $L = 4$	17Δ	8.5Δ	Time-average-frequency
52	5	1	4		δ_k has two values since $L = 4$			
53	22	5	4		δ_k has two values since $L = 4$		8.5Δ	Time-average-frequency
54	39	9	4		δ_k has two values since $L = 4$			
55	56	14	5	17Δ	δ_k has two values since $L = 4$	17Δ	8.5Δ	Time-average-frequency
56	9	2	4		δ_k has two values since $L = 4$			
57	26	6	4		δ_k has two values since $L = 4$		8.5Δ	Time-average-frequency
58	43	10	4		δ_k has two values since $L = 4$			
59	60	15	5	17Δ	δ_k has two values since $L = 4$	17Δ	8.5Δ	Time-average-frequency
60	13	3	4		δ_k has two values since $L = 4$			
61	30	7	4		δ_k has two values since $L = 4$		8.5Δ	Time-average-frequency
62	47	11	4		δ_k has two values since $L = 4$			
63	0	0	5	17Δ	δ_k has two values since $L = 4$	17Δ	8.5Δ	Time-average-frequency

$$T_s = \Delta \sum_{j=1}^{L} \delta_k = \frac{w}{\gcd(w, 2^{n-m})} \Delta \quad for \quad 2^{n-m} \le w < 2^n \tag{5.11}*$$

$$T_v = \begin{cases} 2T_s, & if\ L = 1 \\ T_s, & if\ L > 1 \end{cases} \tag{5.12}†$$

$$T_{xy} = \frac{K}{L} T_s = \frac{w}{\gcd(w, 2^n)} \Delta \quad for \quad 2^{n-m} \le w < 2^n \tag{5.13}‡$$

5.4.5 The Time-Average-Frequency

Previous sections have developed the base for us to work on the most important parameter of flying-adder DFC: time-average-frequency. Before we proceed to it, Table 5.4 summarizes the state variables, periodicities, and relationships around the flying-adder DFC's operation.

In Section 3.3, time-average-frequency is defined as the number of cycles that exist within a given time frame of a minimum repeatable waveform. In flying-adder DFC output $v(t)$, it can be conveniently calculated by counting the cycles within the time frame of T_v (the fundamental period). Again, we only consider the meaningful case of $2^{n-m} \leqq w < 2^n$. When $L = 1$, from the previous discussion, we know that there is one $v(t)$ cycle for every $2T_s$ of time (refer to Table 5.1). When $L > 1$, there are exactly $L/2$ $v(t)$ cycles for every T_s of time (refer to Tables 5.2 and 5.3). Therefore, under all the cases, there are L cycles for every $2T_s$ of time. Using Eq. 5.12, this is equivalent to the statement that there are $L/2$[§] cycles for every T_v of time. Hence, the time-average-frequency T_{TAF} can be calculated as $2T_v/L$.

$$T_{TAF} = \frac{2T_v}{L} \quad \text{or} \quad f_{TAF} = \frac{2^{n-1}}{w} f_r \tag{5.14}¶$$

The fundamental frequency of signal $v(t)$, f_{FD}, is defined from its fundamental continuous time period T_v

$$f_{FD} = 1/T_v \tag{5.15}$$

Based on above, we reach an important fact: time-average-frequency is the qth harmonic of signal $v(t)$. This is equivalent of saying that there are q cycles of signal v for every time frame of T_v.

* See Sotiriadis 2010a, Section IV for proof.

† T_v is the T_{FD} in Section 3.5.

‡ See Sotiriadis 2010a, Section IV for proof.

§ L is an even number when $L > 1$. See Sotiriadis 2010a, Corollary 3.2.

¶ See Sotiriadis 2010a, Section IV for details.

TABLE 5.4. The Important State Variables, Periodicities, and Relationships

Variable	Value	Note
k	$0, 1, 2, \ldots$	Index for discrete time.
x_k	$\{k2^r \mid k=0, 1, 2, \ldots, 2^{n-r}-1\}$	State variable for describing the flying-adder DFC's state.
y_k	$Y=\{[(k2^r)/2^{n-m}] \mid k=0, 1, 2, \ldots, 2^{n-r}-1\}$	Truncated state variable for controlling the MUX.
K	2^{n-r}, $r=\log_2(gcd(w, 2^n))$	The periodicity in x_k and y_k.
δ_k	$([kw/2^{n-m}]-[(k-1)w/2^{n-m}])$ mod 2^m or 2^m if above is 0	The modified difference state. It is the bridge between discrete time domain and continuous time domain.
L	$2^{n-m}/gcd(w, 2^{n-m})$	The periodicity in δ_k.
t	In units of Δ	The continuous time.
T_s	$(w/gcd(w, 2^{n-m}))\Delta^a$	Fundamental continuous time period of signal s.
T_v	T_s if $L>1$, $2T_s$ if $L=1$	Fundamental continuous time period of signal v.
T_{xy}	$(w/gcd(w, 2^n))2^m\Delta^a$	Fundamental continuous time period of x_k and y_k.
T_{TAF}	T_v/q	Time-average-frequency on signal v.

[a]Under the condition of $2^{n-m} \leqq w < 2^n$.

$$f_{TAF} = q \cdot f_{FD} \quad q = \begin{cases} 1, & if\ L=1 \\ L/2, & if\ L>1 \end{cases} \tag{5.16}$$

In the frequency domain, it is expected that there are spectral components of $i{\cdot}f_{FD}$, i = 0, 1, 2, . . . , outputted from the DFCs. The closest of them to f_{TAF} are $f_{TAF} \pm f_{FD}$. From Eq. 5.17, this distance is inversely proportional to q (or L).

$$\left| \frac{f_{TAF} - (f_{TAF} \pm f_{FD})}{f_{TAF}} \right| = \frac{1}{q} \tag{5.17}$$

5.4.6 Pulse and Cycle in Time-Average-Frequency Signal

Definitions:

- 1-interval (t_a, t_b): a continuous time interval of $v(t) = 1$ for all $t \in (t_a, t_b)$ and $v(t_a^-) = v(t_b^+) = 0$. A 1-interval is also called *pulse*.
- 0-interval (t_a, t_b): a continuous time interval of $v(t) = 0$ for all $t \in (t_a, t_b)$ and $v(t_a^-) = v(t_b^+) = 1$.
- Cycle: a pair of 0- and 1- consecutive intervals in either 0-to-1 or 1-to-0 order.

In flying-adder DFC output $s(t)$, the length of the kth 0- or 1-intervals is $\delta_k\Delta$. When $L = 1$, δ_k is constant that results in a 50% duty cycle square waveform

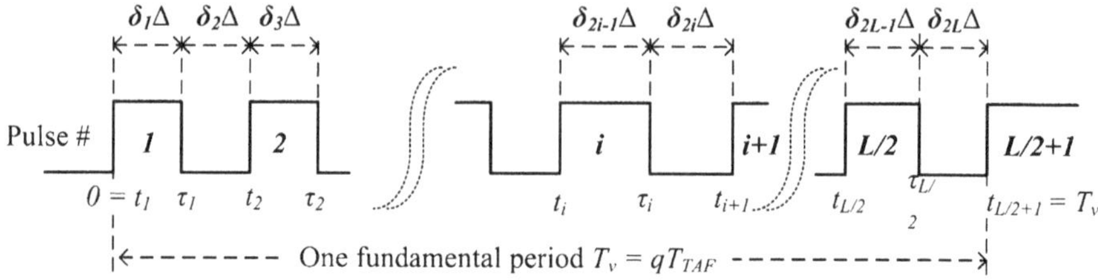

Fig. 5.8. The flying-adder DFC output $v(t)$ when $L > 1$.

in $v(t)$. Therefore, in the following time-average-frequency discussion, we assume $L > 1$. Figure 5.8 shows $v(t)$ for one fundamental period T_v. There are $L/2$ 1-0 cycles, and t_i and τ_i are used to mark the rising and falling edges, respectively. For the ith 1-interval (pulse), its length is $\tau_i - t_i = \delta_{2i-1}\Delta$. For the ith cycle, its length is $t_{i+1} - t_i = (\delta_{2i-1} + \delta_{2i})\Delta$. It is proven* that when $L > 1$, the length of a 0- and 1-interval in signal $v(t)$ can only take value from the two values in Eq. 5.18. And moreover, within every fundamental period T_v, there are at least two intervals. One of their lengths takes the low value and the other takes the high value. The length of cycle in $v(t)$ can only take one of the two values from Eq. 5.19. Refer to Tables 5.1–5.3 for examples.

$$\left\lfloor \frac{w}{2^{n-m}} \right\rfloor \Delta \quad or \quad \left(\left\lfloor \frac{w}{2^{n-m}} \right\rfloor + 1 \right) \Delta \tag{5.18}$$

$$\left\lfloor \frac{2w}{2^{n-m}} \right\rfloor \Delta \quad or \quad \left(\left\lfloor \frac{2w}{2^{n-m}} \right\rfloor + 1 \right) \Delta \tag{5.19}$$

5.4.7 Timing Irregularity in the Time-Average-Frequency Signal

In this section, we study the time domain difference between the time-average-frequency signal and an ideal square waveform of $f_{TAF} = 1/T_{TAF}$ with a 50% duty cycle. Since the time-average-frequency waveform repeats itself in T_v, the comparison is done in one T_v time frame. Referring to Fig. 5.8, based on the definition of δ_k and the condition that $t_1 = 0$, we have the following:

$$t_i = \Delta \sum_{k=1}^{2i-2} \delta_k \quad and \quad \tau_i = t_i + \Delta\delta_{2i-1} \tag{5.20}$$

They can be further expressed as Eq. 5.21.†

$$t_i = \left\lfloor \frac{(2i-2)w}{2^{n-m}} \right\rfloor \Delta \quad and \quad \tau_i = \left\lfloor \frac{(2i-1)w}{2^{n-m}} \right\rfloor \Delta \tag{5.21}$$

* See Sotiriadis 2010b, Section III for proof.

† See Sotiriadis 2010b, Section III for proof.

From Eqs. 5.10–5.12, the length of T_v can be derived as Eq. 5.22 (we only consider the $L > 1$ case since $L = 1$ is trivial).

$$T_v = \frac{wL}{2^{n-m}}\Delta \tag{5.22}$$

The ideal 50% duty cycle square waveform is formed by dividing the T_v into $L/2$ equal-length cycles (since the fundamental period T_v has $L/2$ cycles). If we further assume that this ideal waveform's first rising edge coincides with that of the time-average-frequency signal $v(t)$, the timing of its rising and falling edges can be derived in Eq. 5.23.

$$\tilde{t}_i = \frac{(2i-2)w}{2^{n-m}}\Delta \quad \text{and} \quad \tilde{\tau}_i = \frac{(2i-1)w}{2^{n-m}}\Delta \tag{5.23}$$

From Eq. 5.21 and Eq. 5.23, we have Eq. 5.24, which implies the bounds on the time-average-frequency signal $v(t)$'s rising and falling edges (Eq. 5.25).

$$\frac{t_i}{\Delta} = \left\lfloor \frac{\tilde{t}_i}{\Delta} \right\rfloor \quad \text{and} \quad \frac{\tau_i}{\Delta} = \left\lfloor \frac{\tilde{\tau}_i}{\Delta} \right\rfloor \tag{5.24}$$

$$\tilde{t}_i - \Delta < t_i \le \tilde{t}_i \quad \text{and} \quad \tilde{\tau}_i - \Delta < \tau_i \le \tilde{\tau}_i \tag{5.25}$$

The relationship in Eq. 5.25 is the upper and lower limit of the timing irregularity associated with the time-average-frequency signal $v(t)$. It shows that the ith rising and falling edges of $v(t)$ appear before or simultaneously to the corresponding ones of the ith pulse of the ideal waveform. The offset is less than one Δ. Figure 5.9 shows two examples to illustrate this point. In the top plot, the DFC configuration is $n = 4$, $m = 2$ and the frequency control word w is 15 (FREQ = 7.5). In this case, $K = 16$ and $L = 4$. One T_v is 15Δs. The plot covers

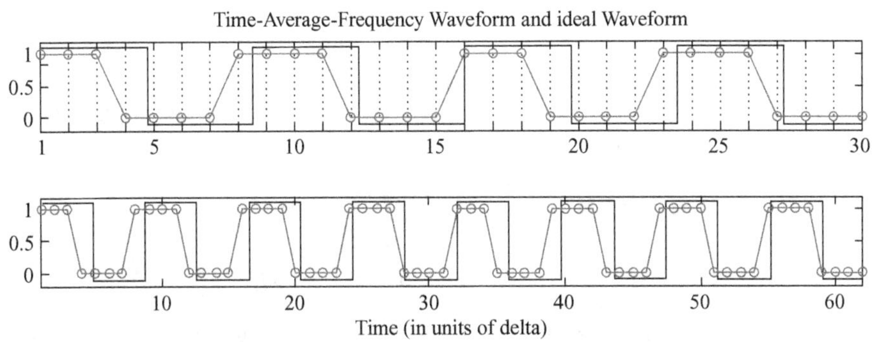

Fig. 5.9. The time irregularity of $v(t)$. Top: $n = 4$, $m = 2$, $w = 15$ (FREQ = 7.5). Bottom: $n = 6$, $m = 2$, $w = 62$ (FREQ = 7.75).

two T_vs. The trace of red color is the ideal one with a 50% duty cycle. The blue one is the time-average-frequency signal $v(t)$. The waveform moves forward in units of Δ as indicated by the dots on the trace. As expected, $v(t)$ is made of two type of cycles: T_A and T_B, with $T_A = 7\Delta$ and $T_B = 8\Delta$. They appear alternatively. We can see that the edges of the ideal waveform are always behind or coincide (at the T_v boundary) with $v(t)$. The bottom plot is under the condition of $n = 6$, $m = 2$, and $w = 62$ (FREQ = 7.75). Here, $K = 32$, $L = 8$, and $T_v = 31\Delta$s. Similarly, two T_vs of the waveform are plotted. There are one T_A and three T_B within one T_v. Again, we can see that the ideal waveform is always behind. The time offsets between the signals' edges are all less than one Δ.

5.4.8 The Sample and Hold Method for Modeling DFC Output

After the investigation of time domain behavior, one remaining task is to derive the exact frequency spectrum of the time-average-frequency signal. Attempts to perform Fourier analysis on $v(t)$ directly have been carried out in the past (Xiu 2008a 2008b; Xiu et al. 2010). However, a closed form expression cannot be reached due to the complexity of the signals $s(t)$ and $v(t)$. An elegant method is proposed in Sotiriadis (2010c), which relates the $v(t)$ to the ideal 50% duty cycle waveform of the fundamental frequency f_{FD}. By performing Fourier analysis on the transformed signal, the exact spectrum can be derived. In this section, the method of relating $v(t)$ to the ideal waveform is described. This ideal waveform is formally expressed in Eq. 5.26, where f_a is the time-average-frequency f_{TAF} in Eqs. 5.14 and 5.16; *sgn* is the *sign* function.

$$\gamma(t) = \frac{1}{2} + \frac{1}{2}\operatorname{sgn}[\sin(2\pi f_a t)] \tag{5.26}$$

From the bounds of Eq. 5.25 and the waveforms in Fig. 5.9, it is intuitive that $v(t)$ might be represented as a sampled version of the ideal waveform $\gamma(t)$. However, since the edges of $v(t)$ and $\gamma(t)$ coincide at T_v boundaries, an extra step is needed to avoid mathematical complication. In other words, we need to time shift $\gamma(t)$ by a small amount ε so that the sampling operation can be safely carried out. For signal $\gamma(t)$ and $v(t)$, the timing of their rising and falling edges is expressed in Eqs. 5.23 and 5.21, respectively. Therefore, we have following relationship between the rising edges of the two signals where α and β are a unique pair of integers:

$$\begin{gathered}\frac{\tilde{t}_i - t_i}{\Delta} = \frac{(i-1)w}{2^{n-m-1}} - \left\lfloor \frac{(i-1)w}{2^{n-m-1}} \right\rfloor = \frac{\beta}{2^{n-m-1}} \\ \text{where} \quad (i-1)w = \alpha 2^{n-m-1} + \beta \quad \text{and} \quad 0 \le \beta < 2^{n-m-1}\end{gathered} \tag{5.27}$$

From Eq. 5.27, the inequality in Eq. 5.28 can be easily derived.

$$0 \le \tilde{t}_i - t_i \le \Delta - \Delta/2^{n-m-1} = \Delta - \varepsilon \quad \text{where} \quad \varepsilon = \Delta/2^{n-m-1} \tag{5.28}$$

$$t_i + \varepsilon \le \tilde{t}_i + \varepsilon \le t_i + \Delta - \varepsilon \tag{5.29}$$

Equation 5.29 is equivalent to Eq. 5.28. By definition, ε is a positive real number that can be used to shift the $\gamma(t)$ to $\hat{\gamma}(t)$.

$$\hat{\gamma}(t) = \gamma(t - \varepsilon) \tag{5.30}$$

Two more functions are needed. The first one is the impulse sequence $\theta(t)$ as expressed in Eq. 5.31. The second one is the pulse function $p(t)$, which is defined in the right drawing in Fig. 5.10.

$$\theta(t) = \sum_{j=-\infty}^{\infty} \delta(t - j\Delta) \tag{5.31}$$

With the help of these two functions, the time-average-frequency signal $v(t)$ can be created by sampling $\hat{\gamma}(t)$ with $\theta(t)$ and holding it with $p(t)$. This is graphically illustrated in the left drawing of Fig. 5.10. The bound in Eq. 5.29 ensures that this sampling operation is valid. This sample-and-hold operation can be mathematically expressed in Eq. 5.32, where * stands for convolution. The circuit interpretation is illustrated in Fig. 5.11, where the impulse is replaced by a virtual clock signal $h(t)$.

$$v(t - \Delta) = (\hat{\gamma}(t)\theta(t)) * p(t - \Delta) \tag{5.32}$$

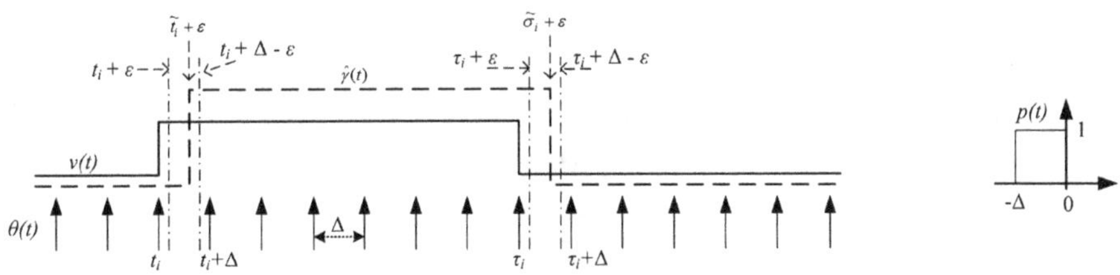

Fig. 5.10. The DFC output $v(t)$, the time-shifted ideal waveform $\hat{\gamma}(t)$, and the impulse sequence $\theta(t)$ (left). The pulse function used for hold $p(t)$ (right).

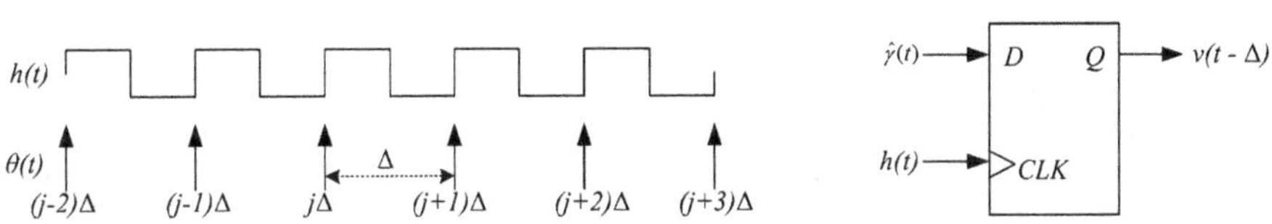

Fig. 5.11. The circuit interpretation of sample and hold method for the time-average-frequency signal.

5.4.9 Frequency Spectrum of DFC Output

Mathematically, Eq. 5.32 is equivalent to Eq. 5.33. Note that $p(t)$ is a non-casual function (it is defined in this way since $\hat{\gamma}(t)$ is behind $v(t)$; refer to Fig. 5.10). This is not a problem because we are not concerned with implementation but rather with mathematical expression. Equation 5.34 is the definition of $v(t)$'s Fourier transform. It can be derived through Eq. 5.35, where $\hat{\Gamma}$ and Θ are the Fourier transforms of $\gamma(t)$ and $\theta(t)$, respectively.

$$v(t) = (\hat{\gamma}(t)\theta(t)) * p(t) \tag{5.33}$$

$$V(f) = \int_{-\infty}^{\infty} v(t)e^{-2\pi i f t}\,dt \tag{5.34}$$

$$V(f) = \left(\hat{\Gamma} * \Theta\right)(f) \cdot P(f) \tag{5.35}$$

The $\gamma(t)$ of Eq. 5.26 can be transformed into Eq. 5.36 using Eq. 5.14 and the definition of $f_r = 1/(2^m\Delta)$. Its Fourier transform is Eq. 5.37. The Fourier transform of the time-shifted version $\hat{\gamma}(t)$ can be derived in Eq. 5.38. The Fourier transform of the impulse sequence $\theta(t)$ is Eq. 5.39. From Eqs. 5.38 and 5.39, Eq. 5.40 is derived. The Fourier transform of the pulse function is Eq. 5.41. Based on the above, the flying-adder DFC output $v(t)$'s Fourier transform is finally arrived as Eq. 5.42.

$$\gamma(t) = \frac{1}{2} + \frac{1}{\pi}\sum_{l}\frac{1}{l}\sin\left(\frac{2^{n-m}\pi l}{w\Delta}t\right), \quad l:\text{odd} \tag{5.36}$$

$$\Gamma(f) = \frac{\delta(f)}{2} + \frac{1}{\pi i}\sum_{l}\frac{1}{l}\delta\left(f - \frac{2^{n-m-1}l}{w\Delta}\right), \quad l:\text{odd} \tag{5.37}$$

$$\hat{\Gamma}(f) = e^{-\frac{\pi i \Delta f}{2^{n-m}}}\Gamma(f) = \frac{\delta(f)}{2} + \frac{1}{\pi i}\sum_{l}\frac{e^{-\frac{\pi i l}{2w}}}{l}\delta\left(f - \frac{2^{n-m-1}l}{w\Delta}\right), \quad l:\text{odd} \tag{5.38}$$

$$\Theta(f) = \frac{1}{\Delta}\sum_{r=-\infty}^{\infty}\delta\left(f - \frac{r}{\Delta}\right) \tag{5.39}$$

$$\left(\hat{\Gamma} * \Theta\right)(f) = \frac{1}{2\Delta}\sum_{r=-\infty}^{\infty}\delta\left(f - \frac{r}{\Delta}\right) + \frac{1}{\pi\Delta i}\sum_{r}\sum_{l}\frac{e^{-\frac{\pi i l}{2w}}}{l}\delta\left(f - \frac{rw + 2^{n-m-1}l}{w\Delta}\right), \quad l:\text{odd} \tag{5.40}$$

$$P(f) = \int_{-\Delta}^{0} e^{-2\pi i f t}\,dt = \Delta e^{\pi i f \Delta}\,\text{sinc}(f\Delta) \quad \text{where} \quad \text{sinc}(x) = \begin{cases} 1 & x = 0 \\ \sin(\pi x)/(\pi x) & x \neq 0 \end{cases} \tag{5.41}$$

$$V(f) = \frac{\delta(f)}{2} + \frac{1}{\pi i}\sum_{r=-\infty}^{\infty}\sum_{l:odd}(-1)^r\frac{e^{\frac{\pi i l}{2w}(2^{n-m}-1)}}{l}\,\text{sinc}\left(\frac{rw + 2^{n-m-1}l}{w}\right)\cdot\delta\left(f - \frac{rw + 2^{n-m-1}l}{w\Delta}\right) \tag{5.42}$$

Using the inverse Fourier transform of Eq. 5.43, the time domain waveform $v(t)$ can be derived as Eq. 5.44.

$$v(t)=\int_{-\infty}^{\infty} V(f)e^{2\pi ift}df \tag{5.43}$$

$$\begin{aligned} v(t)=\frac{1}{2}+\frac{1}{\pi}\sum_{r=-\infty}^{\infty}\sum_{l:odd}\frac{(-1)^r}{l}\operatorname{sin}c\left(\frac{rw+2^{n-m-1}l}{w}\right) \\ \cdot\sin\left(\frac{(2^{n-m}-1)l\pi}{2w}+\frac{2(rw+2^{n-m-1}l)\pi}{w\Delta}t\right) \end{aligned} \tag{5.44}$$

5.4.10 Amplitude of the Time-Average-Frequency

In $v(t)$ of Eq. 5.44, the frequency term is controlled by two integers r and l (refer to Eqs. 5.36 and 5.39 for the introduction of r and l). The term $f_{r,l}$ is related to these two integers by Eq. 5.45. Since l is an odd number, it can be replaced by another integer p as $l = 2p + 1$, where p can be any integer. Therefore, Eq. 5.45 can be transformed into Eq. 5.46 using f_a (f_{TAF}) in Eq. 5.14. It can be proven that the term $r(w/g) + p(2^{n-m}/g)$ can take any integer.* Therefore, Eq. 5.46 is equivalent to Eq. 5.47.

$$f_{r,l}=\frac{rw+2^{n-m-1}l}{w\Delta} \tag{5.45}$$

$$f_{r,l}=f_a\left(1+\frac{g}{2^{n-m-1}}\left(r\frac{w}{g}+p\frac{2^{n-m}}{g}\right)\right),\quad p,r\in Z,\quad g=\gcd(w,2^{n-m}) \tag{5.46†}$$

$$f_{r,l}=f_a\left(1+j\frac{g}{2^{n-m-1}}\right)=f_a\left(1+\left(j\frac{2}{L}\right)\right),\quad j\in Z \tag{5.47}$$

From Eq. 5.15, we know that $v(t)$ can be expanded in terms of fundament frequency f_{FD}. Eq. 5.47 shows that $v(t)$ can also be expanded using average frequency f_a as the base term. When $L = 1$ (δ_k is constant), $v(t)$ is a 50% duty cycle square wave. In this case, the fundamental frequency is the average frequency, and only odd terms exist. When $L = 2$, the fundamental frequency equals the average frequency as well; $v(t)$ is still a square wave but its duty cycle is different from 50% (since δ_k is not a constant). Hence, both even and odd frequency terms are presented. When $L > 2$, average frequency will be higher than fundamental frequency. Their ratio is an integer (refer to Eq. 5.16). Equation 5.47 shows that sub-f_a terms (also called subharmonic terms) will appear. As a matter of fact, all the frequency terms are based on units of $f_{FD} = 2f_a/L$, in steps of f_{FD} (remember that L is an even number).

* See Sotiriadis 2010c, Section III for proof.

† Z is the set of integer numbers.

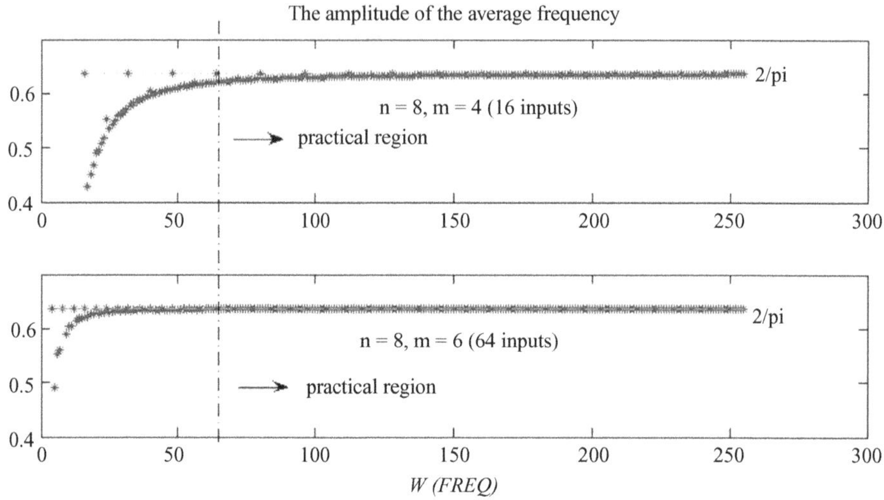

Fig. 5.12. The average frequency's amplitude vs. frequency control word.

In Eq. 5.45, the average frequency f_a (f_{TAF}) is reached when $r = 0$ and $l = 1$ (refer to Eq. 5.14 and $f_r = 1/[2^m\Delta]$). This term is of special interest since it is the frequency of the corresponding ideal 50% duty cycle square wave. We want to know how much of the total energy is allocated to this term. Plugging $r = 0$ and $l = 1$ into Eq. 5.44, the term of $v_{fa}(t)$ can be derived as in Eq. 5.48. Its amplitude is given by Eq. 5.49. Using it, Fig. 5.12 shows two cases that illustrate the trends of amplitude versus frequency control word. In the top plot, the DFC has the configuration of $n = 8$ and $m = 4$. The frequency control word w is varied between 2^{n-m} and $2^n - 1$. The line in the plot is positioned at $2/\pi$, which is the amplitude of an ideal 50% duty cycle square wave. The bottom plot is the case of $n = 8$ and $m = 6$. A vertical line separates the whole area into two parts. On the right side, the DFC output is in the range of $f \leqq 2f_r$, which is usually the practical region for the DFC to be used as a clock generator in real designs.

$$v_{f_a}(t) = \frac{2}{\pi L} \cdot \frac{\sin\left(\frac{2^{n-m}\pi}{2w}\right)}{\sin\left(\frac{g\pi}{2w}\right)} \cdot \sin\left(2\pi f_a\left(t + \frac{L-1}{2L}\Delta\right)\right) \tag{5.48}$$

$$\left\| v_{f_a} \right\| = \frac{2}{\pi L} \cdot \frac{\sin\left(\frac{2^{n-m}\pi}{2w}\right)}{\sin\left(\frac{g\pi}{2w}\right)} \tag{5.49}$$

In this practical region, the amplitude is close to that of an ideal 50% duty cycle square wave ($2/\pi$). This implies the fact that most of its energy is concentrated at the appropriate location: time-average-frequency $f_{TAF} = 1/T_{TAF}$. In terms of **energy distribution, the behavior of the time-average-frequency signal $v(t)$ is close to the ideal square wave**. This is the single most important conclusion obtained from this lengthy analysis.

5.4.11 Relates the Mathematic Model with Real Circuit

The flying-adder DFC's mathematic model is created from Fig. 5.7, which is different from the real working circuit of Fig. 4.17. However, when investigated further, it is found that they are mathematically equivalent. The working circuit has two paths that are responsible for generating the output's ($v[t]$ in the math model) rising and falling edges, respectively. Those two paths can be considered as two identical circuit blocks (the accumulator register in Fig. 5.7) that work in parallel. One is responsible for the 1-interval of the $s(t)$. The other is for its 0-interval. Since the accumulator-register size is twice as larger as that of Fig. 5.7 (the two accumulators/adders work alternatively at half speed), the m is virtually doubled (size increased by 1 bit). Consequently, the frequency control word FREQ = $I + r$ is 1 bit more than that of w. Hence, we have Eq. 5.50. The analysis in this section is fully applicable to the real working circuit.

$$FREQ = 2w, \quad I = \left[\frac{2w}{2^{n-m}}\right], \quad r = \frac{2w}{2^{n-m}} - \left[\frac{2w}{2^{n-m}}\right] \tag{5.50}$$

5.5 CONVERT THE SPURIOUS ENERGY TO NOISE ENERGY

The important conclusion from Section 5.4.10 is that, for most of the practical cases, the majority of DFC output power concentrates at time-average-frequency f_{TAF}. However, there is some portion of energy leaking into other frequency terms (the other harmonics of fundamental frequency f_{FD}). For many applications, such as when DFC output is used to drive ADC and DAC or when the DFC is used for frequency conversion, it is beneficial to reduce the levels of these spurious frequency components ("spurs" hereafter) without significantly impacting the power allocated to the f_{TAF}. From the mathematic analysis of Section 5.4, it is clear that the spurs are caused by the periodicity L. Circuit-wise, it is due to the periodic carry overflow from the fractional accumulation. This fact is illustrated in Fig. 5.13a and 5.13b. The accumulation is controlled by the input FREQ = $I + r$, where I is an integer and r is a fraction. The fraction r can be expressed as $r = a/b$, where a and b are both integers and $a < b$.* This r is the base for fractional accumulation. Whenever it crosses the b boundary, an overflow occurs and the cycle T_A is replaced by T_B. This is depicted in Fig. 5.13b where the x-axis is the clock cycle (the discrete time

* Using the notations in Section 5.4.1, $b = 2^{n-m}$, $a = 2w - (2w/2^{n-m})2^{n-m}$.

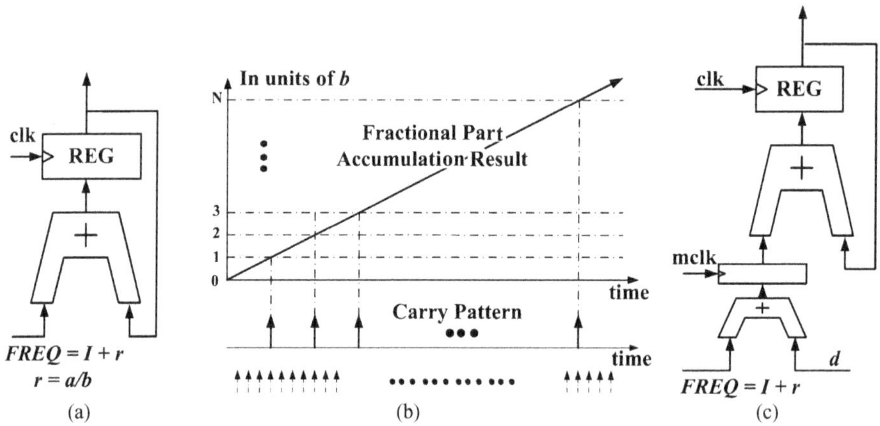

Fig. 5.13. (a) The flying-adder accumulation, (b) carry-overflow vs. clock cycle, (c) adding a disturbance.

index k, not the continuous time t).* Note that the linear increase in accumulation result is in respect to clock cycles whose length-in-time could vary from time to time.

Intuitively, the periodicity embedded in the carry flow can be broken by adding a disturbance d to FREQ as illustrated in Fig. 5.13(c). In such arrangement, the accumulation result $R(t)$ will cross the b boundary in irregular intervals. $R(t)$ can be expressed in Eq. 5.51, where $N(t)^\dagger$ is the time in units of clock cycle. The characteristic of the disturbance can be random or bear a certain pattern. Triangular or sawtooth-shaped patterns are often used because of their implementation convenience. In the case of sawtooth and triangular patterns, d can be expressed as $d_i = g*i + h$, and $R(t)$ is derived as Eq. 5.52. Figure 5.14 shows the case where $r = 0.1$ and a sawtooth disturbance of 255 steps in a full cycle ($mclk = clk$). As can be seen, $R(t)$'s linear dependency on $N(t)$ becomes the order of $O(N^2)$. Consequently, the carry flow shows a more complicated pattern. Its periodicity is broken or prolonged. As a result, we expect that the spurs' levels would be reduced.

$$R(t) = \sum_{i=0}^{N(t)} r + \sum_{i=0}^{N(t)} d_i = N(t)*r + D(t) \tag{5.51}$$

$$\begin{aligned} R(t) &= \sum_{i=0}^{N(t)} r + \sum_{i=0}^{N(t)} d_i = N(t)*r + N(t)*h + g\sum_{i=0}^{N(t)} i \\ &= N(t)*(r+h) + g*N(t)*[N(t)+1]/2 \\ &= \left(\frac{g}{2}\right)*N^2(t) + \left(r+h+\frac{g}{2}\right)*N(t) \end{aligned} \tag{5.52}$$

* The time elapse between the ith and $(i+1)^{\text{th}}$ clock edges is $(\delta_{2i-1} + \delta_{2i})\Delta$.

† N is k counting modulo 2.

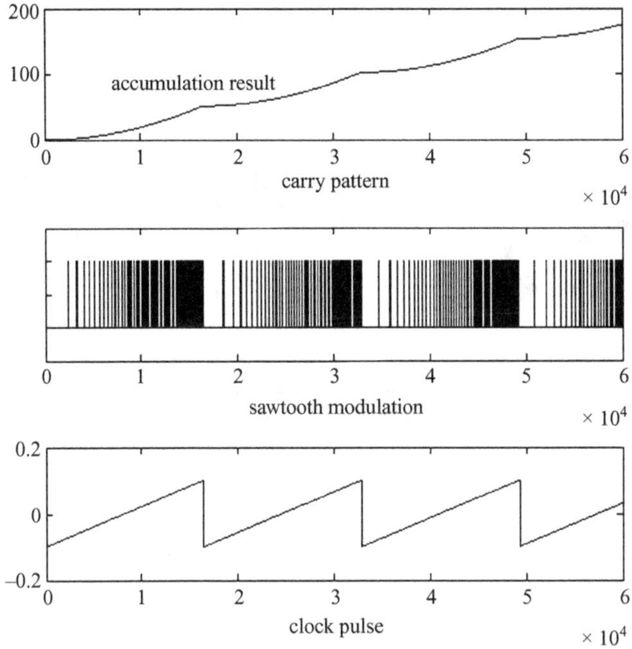

Fig. 5.14. The carry pattern under a sawtooth disturbance.

Figures 5.15–5.17 show a simulation result for the case of FREQ = 10.0625. In this case, the flying-adder synthesizer uses 8 inputs (K* = 8) with a frequency of f_{vco} (f_r) = 2 GHz. This results in the DFC output $f = f_{TAF}$ = (K/FREQ)*f_{vco} = 1.59 GHz. Under this condition, periodicity L is calculated as L = 32, which results in the fundament frequency $f_{FD} = f_{TAF}/(L/2)$ = 99.375 MHz. In Fig. 5.15, the plot at left is the flying-adder DFC output's spectrum. Clearly, around f_{TAF} = 1.59 GHz, there are spurs spaced at 99.375 MHz (f_{TAF} is actually the 16th harmonic of the fundamental frequency). The plot in the middle is the result after a triangular disturbance (magnitude 0.0625 with 32 steps in a full triangular cycle; mclk = 50 MHz). The right plot is from a random disturbance with a magnitude of 0.0625 and mclk = 50 MHz. It can be seen that in the latter two cases, the strong spurs are converted into noise. These simulation results are obtained by using the model in Appendices 5.A–5.D (n = 9, m = 4, w = 161). Experimenting with them is recommended.

In Chapter 3, the time-average-frequency is expressed as $T_{TAF} = (1 - r)T_A + rT_B$, where $r = a/b = p/q$ (p and q have a gcd of 1). From Eq. 5.16, we know that $f_{FD} = f_{TAF}/q$ ($q = L/2$ = 16 in this case). In theory, we can enlarge q to make f_{FD} small. Adding a disturbance to FREQ actually increases the effective q. This can be appreciated by investigating Eq. 5.53. When a disturbance d is

* This K is the number of inputs (see Eq. 4-9, Section 4.8). It is not the periodicity of state variable x_k and y_k.

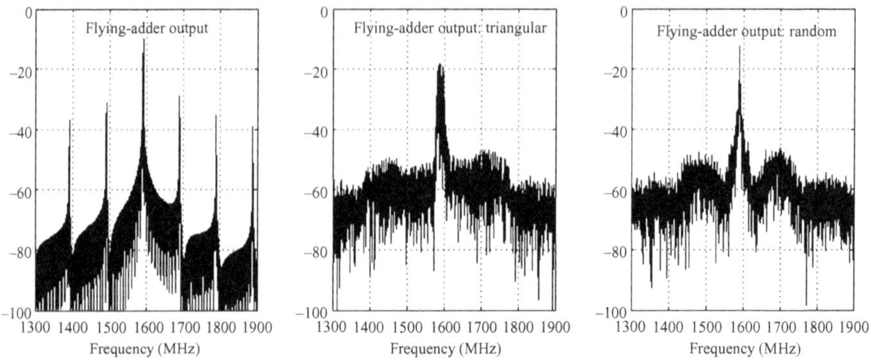

Fig. 5.15. The flying-adder DFC output spectrum at FREQ = 10.0625: original (left), triangular disturbance (center), and random disturbance (right).

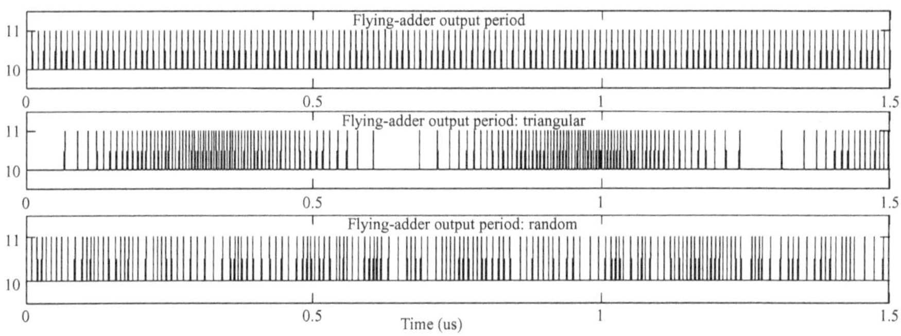

Fig. 5.16. The flying-adder DFC output period vs. time FREQ = 10.0625: original (top), triangular disturbance (middle), and random disturbance (bottom).

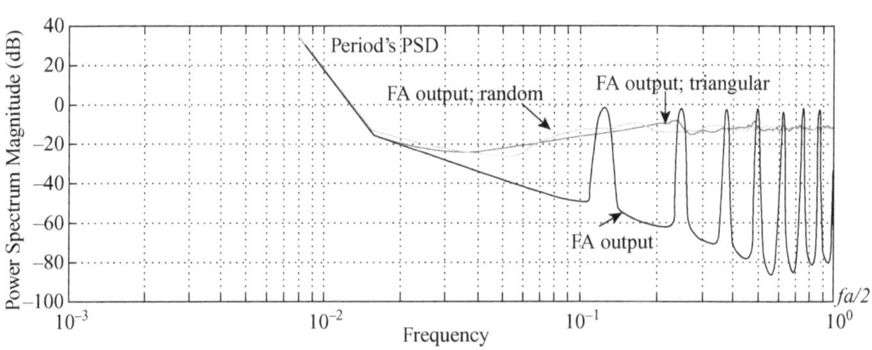

Fig. 5.17. The period's power spectrum density FREQ = 10.0625.

added, instead of a single value, r alternates among several values. It can be proven that $q_{avg} \geqq \max\{q_i\}$. Figure 5.16 demonstrates this point. The flying-adder DFC output period is plotted versus continuous time t. In the top plot, there are 15 T_A and one T_B in every 16 cycles ($r = 0.0625 = 1/16$, $q = 16$). $T_A = 10\Delta$ and $T_B = 11\Delta$ occur in regular patterns. In the middle and bottom plots, the once-every-16 pattern is broken or prolonged, and q_{avg} is enlarged (remember that q is the minimum number of cycles that $v[t]$'s waveform repeats itself). Consequently, the f_{FD} is much smaller now. However, the r_{avg} still equals 0.0625. As a result, in the spectrums of Fig. 5.15, the central frequency is unaltered but spurs are reduced in magnitude. The different spectrums are caused by their different T_A and T_B patterns.

$$\begin{aligned} r = r_{avg} = p_{avg} / q_{avg} &= \left[\sum_{i=1}^{n} r_i(t)\right] / n \\ &= \left[\sum_{i=1}^{n} (p_1 / q_1 + p_2 / q_2 + \ldots + p_n / q_n)\right] / n \end{aligned} \tag{5.53}$$

Figure 5.17 further confirms the point of q enlargement. Power spectrum density has been calculated on the periods of all three cases. The 10^0 in the x-axis stands for half of the sample rate f_a (f_{TAF}). The original eight spurs in this $f_a/2$ range have been converted into noise. The noise level is significantly raised, and is also visible in Fig. 5.15. One important point worth mentioning is that, in all the three cases, the DFC's instantaneous output period is always one of the two values of T_A and T_B. Thus, digital operation is not affected by this mechanism of adding disturbances as long as T_A is used for setup constraint.

Due to the importance of this issue, another example is presented in Figs. 5.18 and 5.19. In this example, FREQ = 10.34. Therefore, $f = f_{TAF} = (K/\text{FREQ})^* f_{vco} = 1.547$ GHz. Since $r = 0.34 = 17/50$, we have $q = 50$. Hence, the fundamental frequency is $f_{TAF}/50 = 30.95$ MHz. These harmonics of the fundamental frequency show their presence as comb-like spurs as illustrated in the left plot of Fig. 5.18. Since $r = 0.34 \approx 1/3$, some periodicity of $f_{TAF}/3 = 516$ MHz is also seen. These two groups of spurs are also clearly visible in Fig. 5.19 of the power spectrum density. The spectrum in the right plot of Fig. 5.18 results after a random disturbance of magnitude 0.3 is added. In Fig. 5.19, the fact that spurs energy is converted into noise energy is again illustrated.

For both approaches of adding random and triangular disturbances, one key requirement is the target frequency preservation. In other words, r_{avg} has to equal to the original r. Another important requirement is the minimum timing irregularity. We only want to use two types of cycles: $T_A = I \cdot \Delta$ and $T_B = (I + 1) \cdot \Delta$. The adding of disturbance d shall not make the DFC use more integers other than I and $I + 1$. In practice, for implementing the random disturbance, one method is to use a temporary storage to reorder the carry sequence and break its inherent periodicity (Xiu et al. 2011). The idea is depicted in Fig. 5.20. In

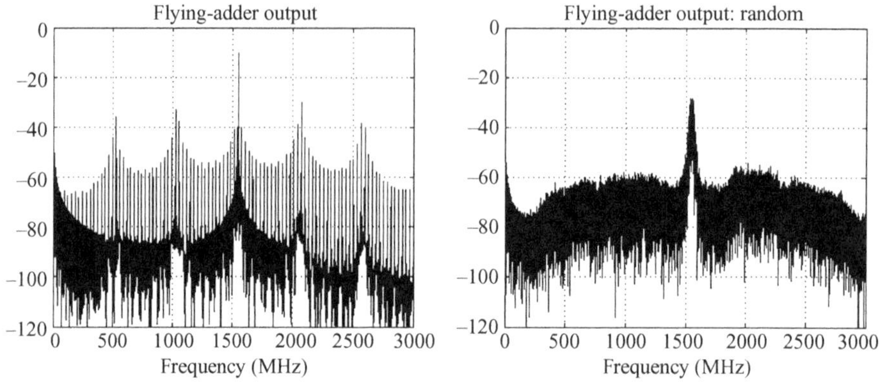

Fig. 5.18. The DFC output spectrum of FREQ = 10.34: original (left) and random disturbance (right).

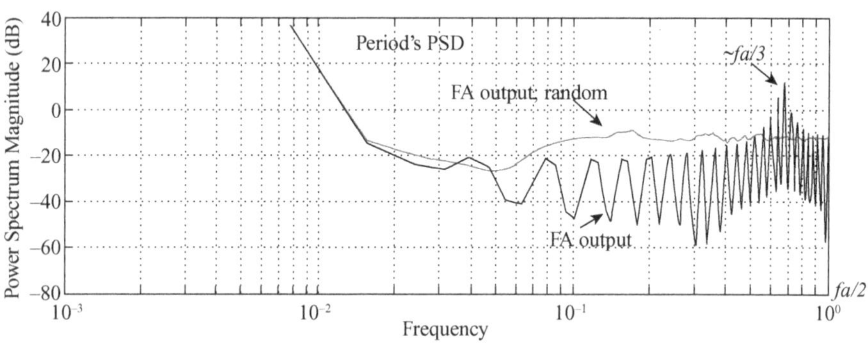

Fig. 5.19. The period's PSD of FREQ = 10.34.

this way, both the minimum timing irregularity and target frequency preservation are automatically fulfilled.

In addition to random and triangular modulation, a high-order sigma-delta modulator can also be used to enlarge the q. The accumulator in Fig. 5.13a is a first-order sigma-delta modulator. Its carry pattern is inherently regular. High-order sigma-delta modulators can introduce some degree of randomness into the carry pattern and thus prolong its fundamental period. One example of a second sigma-delta modulator is presented in Rapinoja et al. (2010).

5.6 MOVE SPURS AROUND

Spurious tones are harmful to many applications. Besides converting them into noise, another method in flying-adder DFC is to move the spurs to other locations so that their negative impact can be minimized. From Section 4.8, Eq. 4.9, we know that the output frequency from DFC is $f_s = (\text{K/FREQ}) \cdot f_{vco} =$

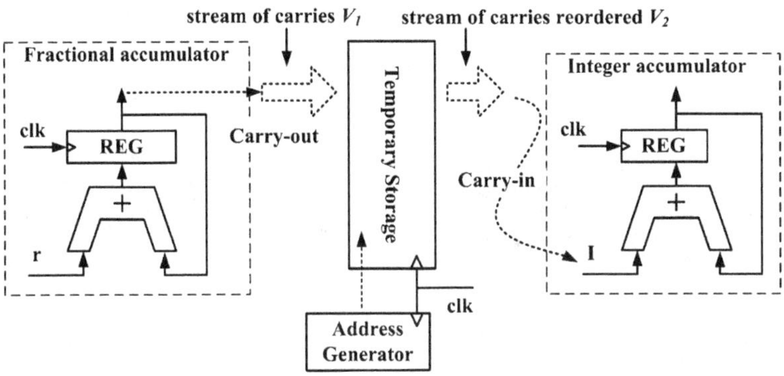

Fig. 5.20. Using temporary storage to reorder the carry sequence and break its inherent periodicity.

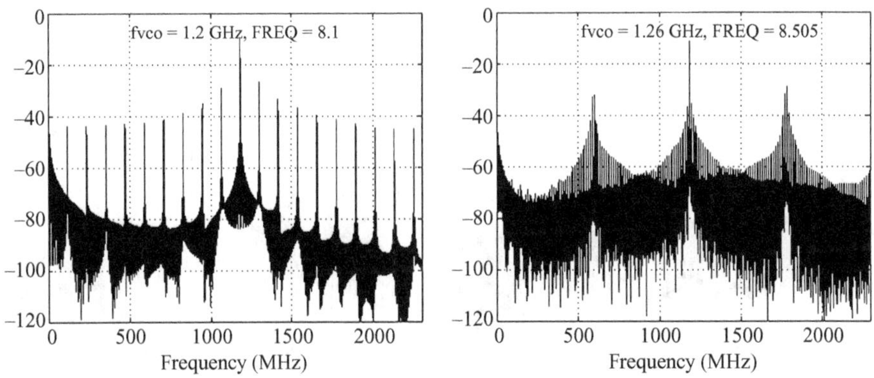

Fig. 5.21. To generate 1.185 GHz: $f_{vco} = 1.2$ GHz, FREQ = 8.1 (left); and $f_{vco} = 1.26$ GHz, FREQ = 8.505 (right).

(K·N/FREQ)·f_r. This can be transformed to Eq. 5.54. Usually, when a design plan is finalized, K and f_r are fixed. For generating an output frequency f_s, we can adjust f_{vco} (the N value) to influence the fraction r and consequently affect the spurs configuration. Assume that there is a flying adder phased-locked loop (FAPLL) with input frequency $f_r = 12$ MHz. The VCO is designed in the range of 960 MHz to 1440 MHz. To generate an output of 1.185 GHz, we can set $N = 100$ ($f_{vco} = 1.2$ GHz) and FREQ = 8.1. The output signal's spectrum is shown in the left plot of Fig. 5.21. Clearly, there are spurs spaced at $f_s/10 =$ 118.5 MHz since $q = 10$ ($r = p/q = 1/10$).

$$FREQ = I + r = \frac{K}{f_s} f_{vco} = \frac{K \cdot N}{f_s} f_r \tag{5.54}$$

The right plot in Fig. 5.21 shows another DFC setting that can also generate 1.185 GHz. In this case, $r = 0.505 = 101/200$. Hence, $q = 200$ and $f_s/200 = 5.9$ MHz.

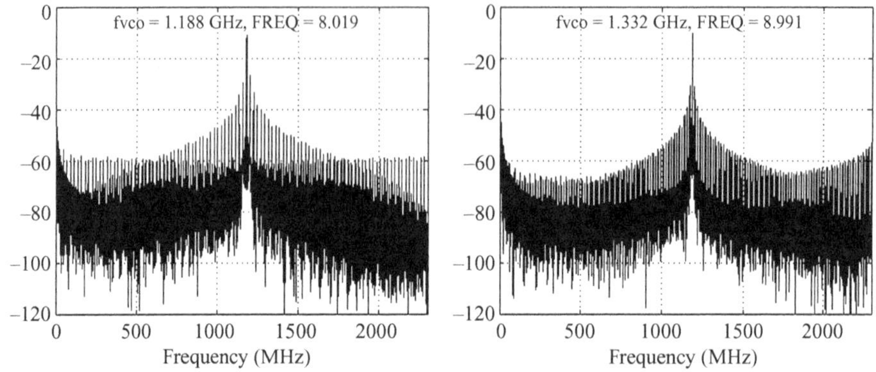

Fig. 5.22. To generate 1.185 GHz: f_{vco} = 1.188 GHz, FREQ = 8.019 (left); and f_{vco} = 1.332 GHz, FREQ = 8.991 (right).

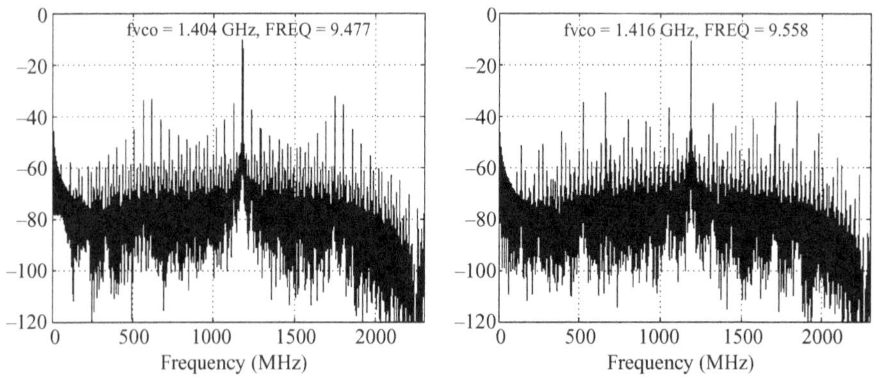

Fig. 5.23. To generate 1.185 GHz: f_{vco} = 1.404 GHz, FREQ = 9.477 (left); and f_{vco} = 1.416 GHz, FREQ = 9.558 (right).

Since $r = 0.505 \approx 1/2$, strong modulation around $f_s/2$ = 593 MHz is also visible. Figures 5.22 and 5.23 show four more cases that can generate the same frequency of 1.185 GHz, each with unique spurs characteristic. In the left plot of Fig. 5.22, $r = 0.019 = 19/1000 \approx 1/50$. The spurs are spaced at about $f_s/50$ = 23 MHz. In the right plot of Fig. 5.22, $r = 0.991 = 991/1000$. Spurs are densely spaced at $f_s/1000$ = 1.185 MHz. They are so densely packed that they can hardly be seen. The visible spurs are the secondary spurs spaced at 11.85 MHz. This is because $991/1000 \approx 990/1000 = 99/100$. The two cases in Fig. 5.23 bear more complicated fractions. The left one has $r = 0.477 = 477/1000$. The fundamental frequency is $f_s/1000$ = 1.185 MHz. There are at least two visible spur patterns. They are caused by $r \approx 500/1000 = 1/2$ (repeatable at $f_s/2$ = 593 MHz) and $r \approx 450/1000 = 9/20$ (spaced at about 59 MHz). The case at the right is $r = 0.558$. There are more than several spurs patterns that coexist. The main one is caused by $r = 558/1000 \approx 5/9$ (spaced at about $f_s/9$ = 129 MHz).

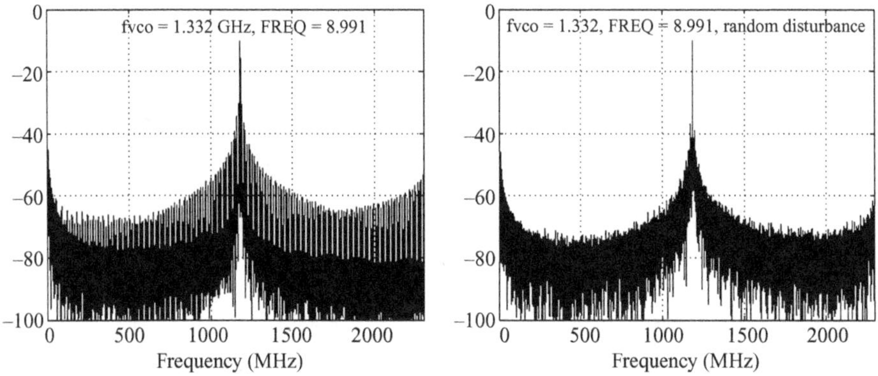

Fig. 5.24. To generate 1.185 GHz: f_{vco} = 1.332 GHz, FREQ = 8.991 (left) and further adding a disturbance (right).

All six DFC settings that generate 1.185 GHz point to the fact that in flying-adder DFC there are many ways of generating a given frequency. In other words, there are many ways that we can make the DFC concentrate its majority energy on the frequency of f_{TAF}. We can adjust the fundamental frequency to change the energy distribution profile but keep the strength of f_{TAF} unaltered. Furthermore, the method of adding disturbance can be combined with this technique to make the result more favorable. Figure 5.24 shows the case of adding a random number of magnitude 0.008 at the rate of 20 MHz to convert the spurs into noise. If the left plot of Fig. 5.21 is compared with the right plot of Fig. 5.24, the improvement is clearly seen. The flying-adder DFC is built with flexibility so that this technique can be applied for almost any frequency asked.

5.7 SPREAD THE ENERGY

In many applications, it is desirable to spread the concentrated clock energy to a broad range. Flying-adder DFC is very flexible in serving this purpose. From time-average-frequency theory, the average frequency f_{TAF} is the qth harmonic of the fundamental frequency f_{FD}. Under the condition of keeping the value of f_{TAF} roughly unaltered, the task of spreading the clock energy can be achieved by making the f_{FD} very small (enlarging q). At the same time, we have to boost the harmonics around the qth and make them roughly the same level as that of the qth. In the process of boosting the harmonics around the qth, the goal of spreading the energy is accomplished. In implementation, this task is often achieved by modulating the output frequency in a certain way, such as letting it follow a certain profile (e.g., a triangular cycle).

Based on the discussion in Section 4.5, we know that the flying-adder DFC's frequency transfer function is $1/x$. And it becomes linear in a small range. This characteristic can be used for modulating its output. The mechanism is

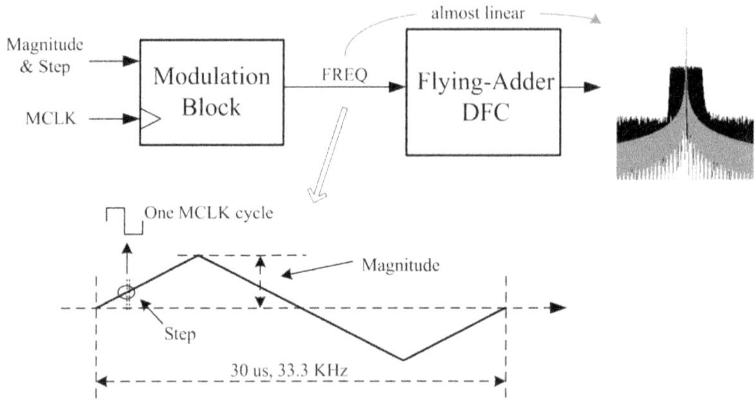

Fig. 5.25. The mechanism of modulating the output to spread the energy.

illustrated in Fig. 5.25. A modulation block is used to create the desired triangular profile in the control word FREQ. Following the conventional approach, 33.3 KHz is used. This is translated into 30 us in a full triangular cycle. The magnitude of the modulation is the control parameter that controls the spread range (modulation range). The step and modulation clock rate is another parameter that impacts the shape of the spread (modulation depth). From Eq. 4.7 of $df_s/f_s = -dF/F$, we can directly apply the same amount on F to achieve a given frequency spread ratio.

In Figures 5.26–5.28, an example is presented to demonstrate this mechanism. A modulation clock of 20 MHz is used to apply the triangular modulation profile on FREQ (50 ns per cycle). This translates into 600 MCLK cycles for one full triangular cycle. For a given output frequency of 941.18 MHz (f_{VCO} = 1 GHz, K = 8, FREQ = 8.5), a 1% spread corresponds to dF = 8.5*0.01 = 0.085. Therefore, the magnitude can be set at 0.085/2 = 0.0425 and the step is calculated as 0.0425/(600/4) = 0.000283. Figure 5.26 shows the resulting FREQ profile and the period distribution. In the left plot, the FREQ varies in a triangular fashion. Its central value is the original 8.5, the maximum and minimum are 8.5 ± 0.0425, respectively. Two full cycles are shown. On the right plot, it is clear that only two types of cycles $T_A = 8\Delta$ and $T_B = 9\Delta$ are used, with equal weights. Figure 5.27 is interesting. It shows the output periods' time trend. The top trace is the original period output. The periods alternate between two values: 8Δ and 9Δ. The modulation profile is hardly visible due to the large scale. The next three traces are the result after the divider of M = 2, 32, and 256, respectively. This operation of dividing is similar to averaging in M cycles (but without multiplying back). The modulation profile gradually becomes visible by this operation.

Figure 5.28 is the output's spectrum. The left plot shows the frequency range from 440 MHz to 1440 MHz. The spectrum trace in red is the original clock output with center frequency 941.18 MHz. The two tones on the sides are

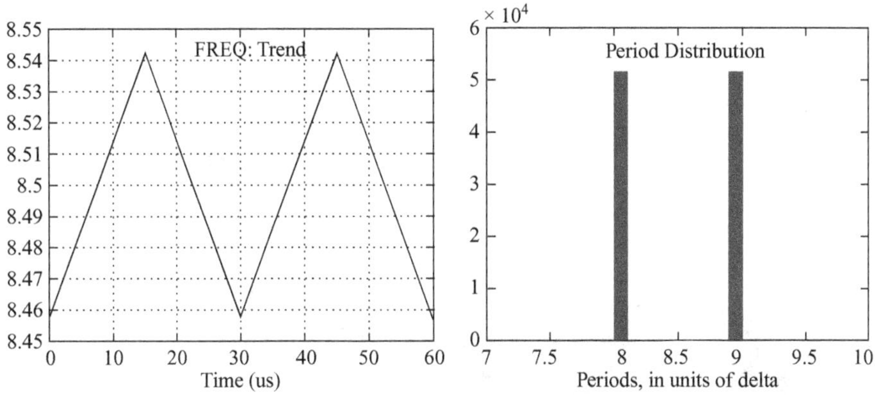

Fig. 5.26. The FREQ profile (left) and period distribution (right).

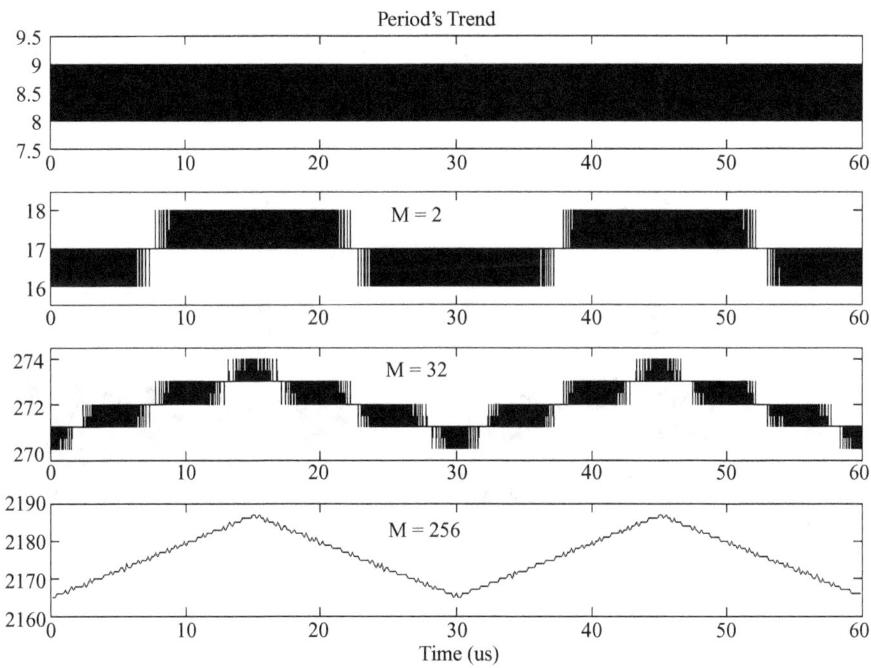

Fig. 5.27. The periods' time trends. From top to bottom, the divide ratio is $M = 1, 2, 32$, and 256.

470.59 MHz and 1411.77 MHz (caused by fraction 0.5). In this spectrum, 470.59 MHz is the fundamental frequency (first harmonic). The output 941.18 MHz is the second harmonic. The spectrum trace in black is the result after the modulation. In the right plot, which is the magnified view around 941 MHz, it can be clearly seen that the original clock peak has been lowed by about 25 db and the spread range is about 9.4 MHz (1% of 941 MHz). In

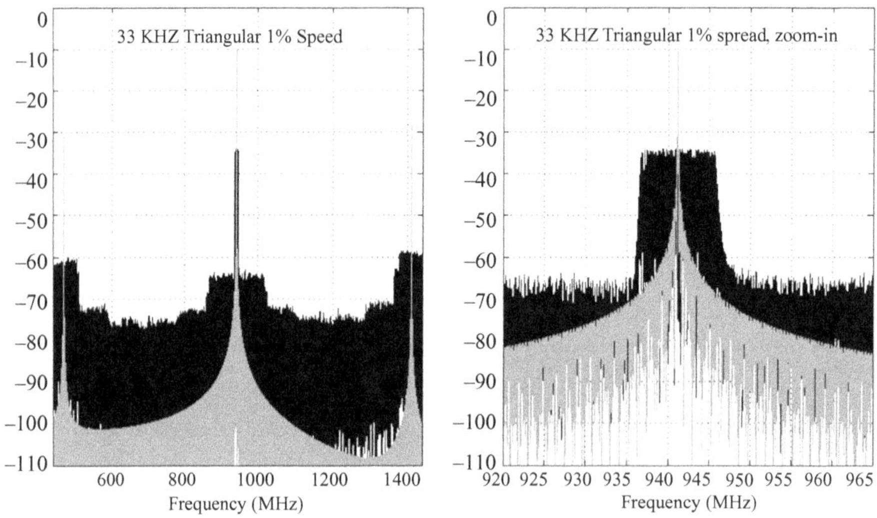

Fig. 5.28. The flying-adder DFC output spectrum: original (light gray), 1% triangular spread (black).

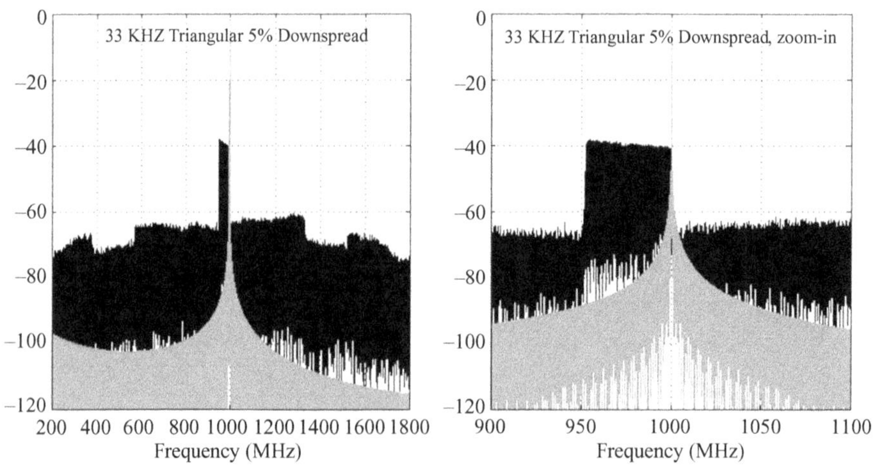

Fig. 5.29. The flying-adder DFC output spectrum: 5% downspread.

the spread spectrum, the fundamental frequency is much smaller now (q is enlarged by perhaps hundreds of times). In the neighborhood of the $q_{\text{new}}^{\text{th}}$ harmonic, there is a group of harmonics that have almost the same level of strength. They form the plateau in that spectrum.

The flying-adder DFC's open loop structure makes it very suitable for precisely controlling the output spectrum. Figure 5.29 is an example of downspread. In this example, the DFC setting is still $f_{vco} = 1$ GHz, K = 8. The unspread clock is designed at 1 GHz (FREQ = 8). Its spectrum is shown as the light gray trace. The modulation block uses a modulation magnitude of 8*0.05/2 = 0.2. The

MCLK is still 20 MHz. Thus, the step is 0.2/(600/4) = 0.001333. Instead of 8, however, this time the central value for spread is FREQ = 8.2 since we want to do downspread. The right plot shows a spread range of ~50 MHz at the lower side, which is the expected 5% of 1 GHz. The peak has been lowed by about 30 db. Figure 5.30 shows the traces of FREQ versus time and period trend versus time. In the top plot of the left side, the period trend and the FREQ trend are plotted together. The angled line is the FREQ trend. The middle and bottom plot are the zoom-in around the areas of 30 μs and 15 μs, respectively. Around the 30 μs area, FREQ uses its lower values (around 8). We can see that, most of the time, the DFC output is 8Δ. On the other hand, around the area of 15 μs, FREQ is close to its high values (around 8.4). Thus, the DFC outputs 9Δ more frequently. In the right plot, the overall period distribution is shown. Clearly, only two types of cycles are used. T_A of 8Δ is used more since, after the spread, the DFC output is closer to 1 GHz (8Δ) than to 889 MHz (9Δ).

5.8 PERFORMANCE MERITS

For the purpose of gauging the flying-adder DFC's performance, several measurement parameters are proposed. They are categorized in time domain, frequency domain, and static.

Time domain:

- **Time Resolution Δ:** See Section 5.4.1 for its definition. This is the base time unit in the flying-adder DFC. Usually, the smaller the Δ is, the better the performance would be.
- **The Sizes of T_A and T_B:** For a given frequency control word $FREQ = I + r$, where I is integer and r is fraction, $T_A = I \cdot \Delta$ and $T_B = (I + 1) \cdot \Delta$. The integer I determines the degree of dissimilarity between T_A and T_B. In general, large I is preferred because it trims the size of dissimilarities.
- **Irregularity ρ:** It is defined in Eq. 5.55. This parameter describes the similarity between the waveforms of the DFC output and that of the corresponding 50% duty cycle ideal square wave. The smaller this number is, the closer the DFC output resembles the ideal one. Using Eq. 5.50, Table 5.5 lists the ρ for some FREQ settings. The irregularity depends on both I and r. Integer part I, which influences the sizes of T_A and T_B, plays a dominant role. Fractional part r, which controls the weights of T_A and T_B, has a secondary role.

$$\rho \equiv \frac{2/\pi - \|v_{f_a}\|}{2/\pi} \cdot 100 = \left(1 - \frac{1}{L} \cdot \frac{\sin\left(\dfrac{2^{n-m}\pi}{2w}\right)}{\sin\left(\dfrac{g\pi}{2w}\right)} \right) \cdot 100 \qquad (5.55)^*$$

* Refer to Eq. 5.49.

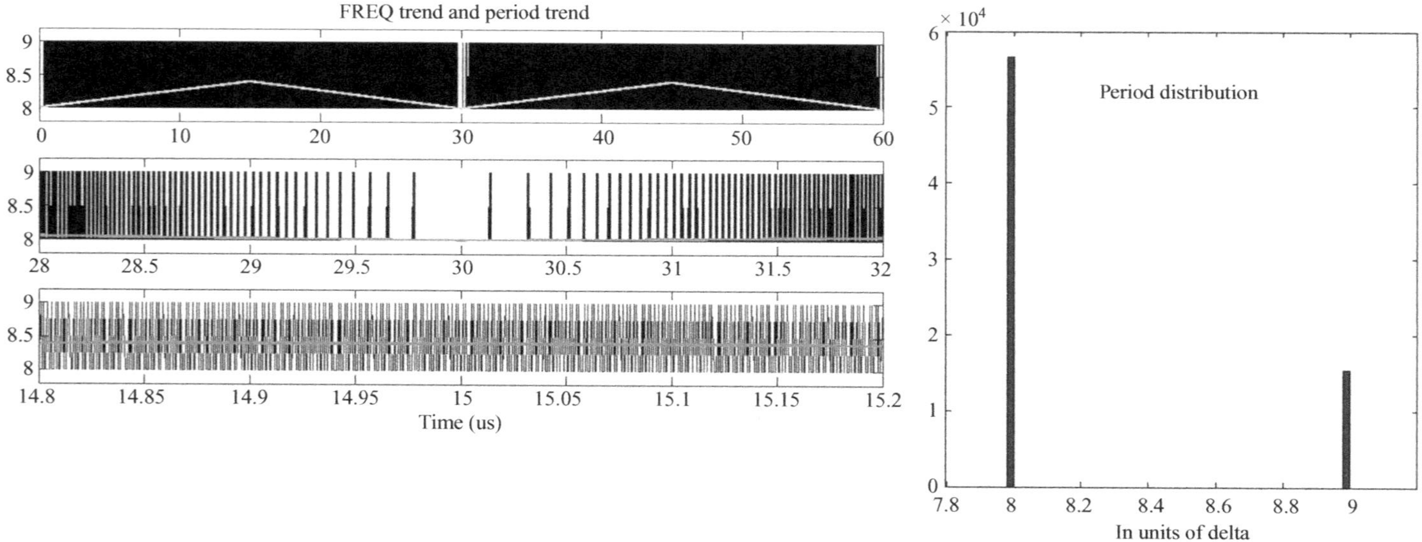

Fig. 5.30. The 5% downspread. The flying-adder DFC's FREQ and period trend (left) and period distribution (right).

TABLE 5.5. Some Numerical Examples of Irregularity

FREQ	8.5	8.25	8.75	8.125	8.875	8.0625	8.9375	8.03125	8.96875	8.5625	$8 + 2^{-10}$	12.5	4.5
ρ	2.122	2.363	2.102	2.464	2.067	2.509	2.045	2.530	2.032	2.226	2.550	0.984	7.458

Frequency domain

- **Fundamental Frequency (f_{FD}):** The fundamental period $T_{FD} = 1/f_{FD}$ is the minimum time window required for the DFC output waveform to repeat itself. The time-average-frequency $f_{TAF} = 1/T_{TAF}$ is the qth harmonic of f_{FD}, where $q = L/2$. q can also be found using $r = p/q$ (the *gcd* of p and q is 1).
- **Spurious-Free Dynamic Range (SFDR):** The ratio of the strength of the qth harmonic (the f_{TAF}) to the next strongest harmonic of the f_{FD}. The harmonics of the f_{TAF} are excluded. In other words, the f_{FD}'s $i{\cdot}q^{th}$ ($i = 2, 3, 4, \ldots$) harmonics shall not be included in the calculation of SFDR.
- **Nearby $\ddot{q}$:** $r = p/q \approx \ddot{p}/\ddot{q}$. $q = L/2$ determines the densest spur space in the DFC spectrum. In most cases, some secondary spur groups are visible when r can be expressed as $r \approx \ddot{p}/\ddot{q}$. For example, in the right plot of Fig. 5.21, $r = 505/1000 \approx 1/2$ ($\ddot{q} = 2$); in the left plot of Fig. 5.23, $r = 477/1000 \approx 1/2$ ($\ddot{q} = 2$) $\approx 12/25$ ($\ddot{q} = 25$); and in the right plot of Fig. 5.23, $r = 558/1000 \approx 1/2$ ($\ddot{q} = 2$) $\approx 5/9$ ($\ddot{q} = 9$). All these secondary spurs groups are visible. More numbers of available $\ddot{q}$ usually corresponds to richer patterns in the spectrum.

Static

- **Frequency Resolution δf:** From Eqs. 4.5 and 4.7, the frequency resolution of Eq. 5.56 can be derived where x is the number of fractional bits used in FREQ and f is the DFC output frequency (also f_{TAF} or f_a).

$$\delta f = -f^2 \cdot \Delta \cdot 2^{-x} \tag{5.56}$$

- **Maximum Output Frequency:** (K/2) f_r in theory. In practice, it is limited by process speed.
- **Minimum Output Frequency:** (1/2) f_r when M = 1 in FAPLL.
- **Period Jitter:** The random noise (e.g., VCO) and deterministic noise (such as the input mismatch) induced uncertainty in length-in-time of T_A. T_B is not considered since it has no impact on setup constraints.
- **Switching Speed:** The time (in clock cycle) required for the DFC output to switch to a new frequency. In flying-adder architecture, the switching speed is two clock cycles. In other words, if the command is received in the current cycle (latched by the DFC output clock), the DFC updates its output two cycles later (refer to Fig. 4.25).

- **Output Period Nonlinearity:** It is used to measure the degree to which the DFC's period transfer function (T_{TAF} vs. FREQ) deviated from the ideal one.
- **Output Frequency Monotonicity:** The DFC's output frequency f_{TAF} is monotonically and inversely proportional to frequency control word FREQ. This is mathematically guaranteed by the architecture. It is not implementation dependent.

BIBLIOGRAPHY

A Technical Tutorial on Digital Signal Synthesis, 1999. Analog Devices.

Blythe, J. H. 1985. "The Spectrum of the Quantized Sinusoid," *GEC J. Res.*, vol. 3, p. 229.

Curticapean, E. and J. Niittylahti. 2003. "Exact Analysis of Spurious Signals in Direct Digital Frequency Synthesizers due to Phase Truncations," *Electron. Lett.*, vol. 39, p. 499.

Goldberg, B. G. 1999. *Digital Frequency Synthesis Demystified*, LLH Technology Publishing.

Golomb, S. W. 1982. *Shift Register Sequences*, Laguna Hill, CA: Aegean park Press.

Jenq, Y. C. 1988a. "Digital Spectra of Nonuniformly Sampled Signals: Fundamentals and High-Speed Waveform Digitizers," *IEEE Trans. Instrum. Meas.*, vol. 37, p. 245.

Jenq, Y. C. 1988b. "Digital Spectra of Nonuniformly Sampled SignalsDigital Look-up Tunable Sinusoidal Oscillator," *IEEE Trans. Instrum. Meas.*, vol. 37, p. 358.

Kroupa, V. F. 1993. "Discrete Spurious Signals and Background Noise in Direct Frequency Synthesizers," *Proc. 47th IEEE Int. Freq. Control Symp.*, pp. 242–250, 1993.

Kroupa, V. F. 1998. *Direct Digital Frequency Synthesis*, IEEE Press.

Mehrgardt, S. 1983. "Noise Spectra of Digital Sine-generators Using the Table-Lookup Method," *IEEE Trans. Acoust.*, vol. ASSP-31, p. 1037.

Morgan, D. R. and A. Aridgides. 1985. "Discrete-Time Distortion Analysis of Quantized Sinusoids," *IEEE Trans. Acoust.*, vol. ASSP-33, p. 323.

Nicholas, H. T. and H. Samueli. 1987. "An Analysis of the Output Spectrum of Direct Digital Frequency Synthesizers in the Presence of Phase-accumulator Truncation," *Proc. 41st Annu. Freq. Control Symp.*, pp. 495–502, 1987.

Rapinoja, T., et al. 2010. "A Digital Frequency Synthesizer for Cognitive Radio Spectrum Sensing Applications," *IEEE Trans. Microw. Theory Tech.*, vol. 58, pp. 1339–1348.

Sotiriadis, P. 2010a. "Theory of Flying-Adder Frequency Synthesizers, Part I: Modeling, Signals' Periods and Output Average Frequency," *IEEE Trans. Circuits Syst. I*, vol. 57, pp. 1935–1948.

Sotiriadis, P. 2010b. "Theory of Flying-Adder Frequency Synthesizers, Part II: Time and Frequency Domain Properties of the Output Signal," *IEEE Trans. Circuits Syst. I*, vol. 57, pp. 1949–1963.

Sotiriadis, P. 2010c. "Exact Spectrum and Time-Domain Output of Flying-Adder Frequency Synthesizers," *IEEE Trans. Ultrason. Ferroelectr. Freq. Control*, vol. 57, pp. 1926–1935.

Xiu, L. 2008a. "The Concept of Time-Average-Frequency and Mathematical Analysis of Flying-Adder Frequency Synthesis Architecture," *IEEE Circuit Syst. Mag.*, 3rd quarter, pp. 27–51, 2008.

Xiu, L. 2008b. "Some Open Issues Associated with the New Type of Component: Digital-to-Frequency Converter," *IEEE Circuit Syst. Mag.*, 3rd quarter, pp. 90–94, 2008.

Xiu, L., C. W. Huang, and P. Gui. 2010. "Analysis of Harmonic Energy Distribution Portfolio for Digital-to-Frequency Converters," *IEEE Trans. Instrum. Meas.*, vol. 59, no. 10, pp. 2770–2778, 2010.

Xiu, L., M. Lin, and H. Jiang. 2011. "A Storage Based Carry Randomization Technique for Spurs Reduction in Flying-Adder Frequency Synthesizer," *IEEE Trans. Circuits Syst. II*, vol. 58, no. 6, pp. 326–330, 2011.

CHAPTER 6

THE NEW FRONTIER IN ELECTRONIC SYSTEM DESIGN

6.1 THE CLOCKING CHALLENGES IN REALITY

6.1.1 The Environment

The task of on-chip clock generation (frequency synthesis) is to generate required frequencies for supporting chip operation. In today's trend of system-on-a-chip (SoC) integration, more and more functions are integrated into one chip. To support this large number of functions, hundreds of frequencies could be required for successful operation. To make the situation even more difficult, all the frequencies are preferably generated from one single reference source (one crystal) for cost considerations. Besides the high quality (low jitter, ample frequencies) requirement, it is also demanded that the clock circuitry should use as few resources (area, power, and pins) as possible. This is especially important for the consumer electronics market, where price is the most effective tool to use in competition.

From a functional perspective, clock circuitry can be responsible for (refer to Fig. 6.1): (I) Driving digital processing units (CPU, DSP, microcontroller, etc.); (II) driving on-chip ADC and DAC; (III) providing frequency reference for on-chip IPs (USB, DDR, LVDS, HDMI, etc.); (IV) local oscillator (LO) for frequency down-conversion or up-conversion; and (V) real-time frequency tracking (synchronization in digital communication). Overall, the digital circuit (type I) accounts for the majority of SoC clock loading. The most important

Nanometer Frequency Synthesis Beyond the Phase-Locked Loop, First Edition. Liming Xiu.

Published 2012 by John Wiley & Sons, Inc.

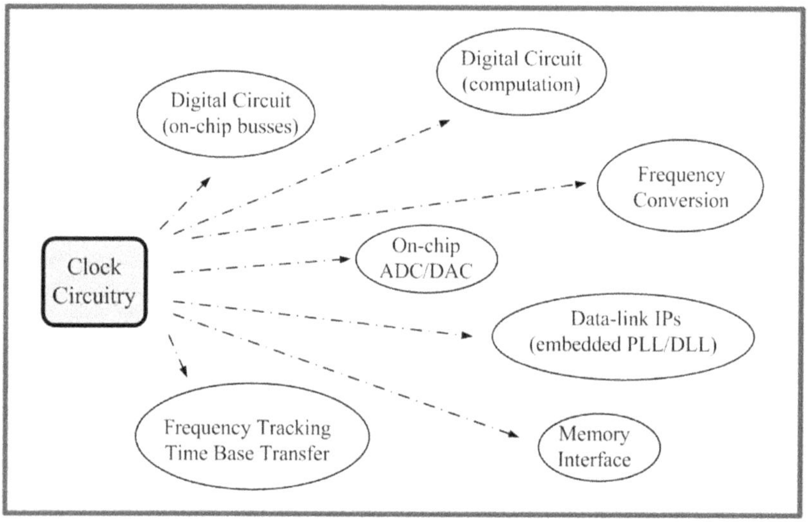

Fig. 6.1. The clock loads in a real-chip environment.

thing to care about in this task is jitter. Driving ADC/DAC, providing reference to IPs, and frequency conversion (types II, II, and IV) require spectral purity in a clock signal. When clock circuitry is used for frequency tracking (type V, also called the time base transfer), the desirable frequency is not predetermined, but only decided in real time from tracking certain targets.

6.1.2 Clock Signal for Computation

The CPU, DSP, and microcontroller perform the computation tasks in a SoC. Operationally, these systems are all built on the synchronous design principle. To drive this type of load, it is desirable that the clock signal bear low jitter. Spectral purity is not of concern (i.e., the spurious tones do not have any impact on this type of operation). Moreover, frequency accuracy is not critically important either. For example, a CPU can run at either 1 GHz or 1.05 GHz without too much user-sensible performance difference. Therefore, for this application, clock jitter is of the highest significance since jitter consumes some of the timing budget from the logic circuit's allowance (setup constraint). It is important to point out that clock jitter has no impact on hold constraint since all clock sinks attached to the particular clock generator see exactly the same amount of clock edge uncertainty. And the fact that the hold check only uses the current clock edge implies that it has nothing to do with clock period/frequency. This is one of the key reasons that the time-average-frequency-based (TAF-based) clock has been used successfully in numerous commercial products. Only clock skew (from clock distribution network) affects the hold-check (refer to Section 1.3).

6.1.3 Clock Signal for Synchronization

In digital systems, operations are carried out either concurrently or in order of precedence. If operations must follow precedence, the role of synchronization is to ensure that operations are performed in the correct order. Synchronization is crucial in digital system and in digital communication interfaces. The problem in synchronization is one of the most frequent causes of unreliable operations of a system. The most important signal in performing synchronization is the clock. The most critical parameters of a clock signal are its frequency and phase. In the abstract level, a clock signal can be described as in Eq. 6.1, where $p(t)$ is a pulse function, f is the nominal frequency, δf is the frequency offset, and $\Phi(t)$ is instantaneous phase whose first derivative has mean value of 0. The average frequency is f_{avg} (Messerschmitt 1990).

$$
\begin{aligned}
&C(t) = p(((f+\delta f)\cdot t + \Phi(t)) \mod 1)\\
&p(t) = \begin{cases} 1, & 0 \le t < 0.5 \\ 0, & 0.5 \le t \le 1 \end{cases}\\
&f_{avg}(t) = f + \delta f + d\Phi(t)/dt = f + \delta f
\end{aligned} \tag{6.1}
$$

Signal can be classified based on this model. In Eq. 6.1, if $f + \delta f$ is a constant, the signal is said to be isochronous, whereas if the frequency is not constant (δf varies with time), this signal is anisochronous. In an anisochronous case, the phase is unbounded. For an isochronous signal, it is generally assumed that $\Phi(t)$ is bounded. For two signals with same nominal frequency, the instantaneous phase difference between them can be expressed as in Eq. 6.2.

$$
\delta\Phi(t) = (\delta f_1 - \delta f_2)\cdot t + (\Phi_1(t) - \Phi_2(t)) \tag{6.2}
$$

For digital communication, based on Eq. 6.2, the type of synchronization can be classified as synchronous and asynchronous. Between the two parties of the communication, if their frequency offsets are the same ($\delta f_1 = \delta f_2$), the two systems are said to be synchronous. Synchronous systems can have fixed and known nonzero phase differences. A common example of a synchronous system is the case where the signals in both parties are generated by the same clock. Any two systems that are not synchronous are asynchronous. Asynchronous systems can be further categorized as mesochronous, plesiochronous, and heterochronous. The signal type and communication type are illustrated in the top drawing of Fig. 6.2.

- **Mesochronous:** Two isochronous signals have the exact same average frequency $f + \delta f$. Their phase difference $\delta\Phi$ is bounded. Example: Two signals are generated from the same clock (even with difference phases), but they suffer from indeterminate interconnect delays relative to each other.

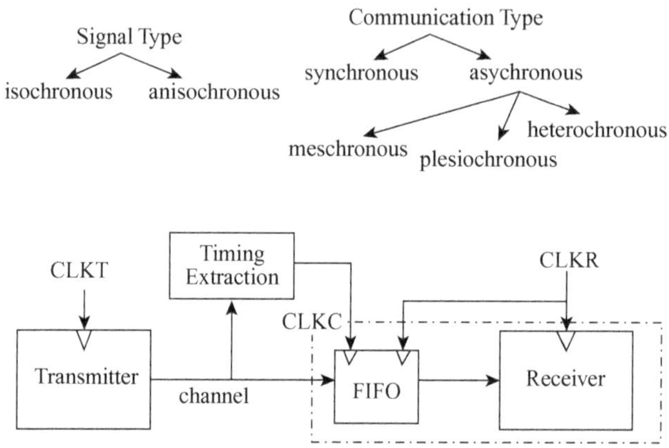

Fig. 6.2. Digital communication: signal type and communication type (top); the scheme of making communication throughput independent of channel's interconnect delay (bottom).

- **Plesiochronous:** Two signals have average frequencies that are nominally the same, but not exactly the same. Their phase differences can be expressed in Eq. 6.2. Example: Two signals are derived from independent oscillators.
- **Heterochronous:** Two signals have nominally different average frequencies.

In digital communication, the interconnect delay can be large. To make throughput independent of the interconnect delay, the synchronization technique depicted in the bottom drawing of Fig. 6.2 is used. All the signals involved are isochronous (meaning that they are all slaved to clocks). But depending on the clock configuration, the communication could be mesochronous (one clock is distributed to all the nodes) or plesiochronous and heterochronous (the transmitter and receiver use independent clocks). As shown, the transmitter and the receiver are driven by their respective clocks: CLKT and CLKR. The block of *timing extraction* is used to extract a third clock (CLKC) from the data in the communication channel. If CLKT and CLKR are originated from a common source, the system is mesochronous. Otherwise, it is plesiochronous or heterochronous. CLKC is synchronous with data in the channel but mesochronous to CLKT since the interconnect delay causes an indeterminate phase. First-in-first-out (FIFO) memory is used for adjusting the phase difference between CLKR and CLKC so that no loss of data can happen (or is below an acceptable level). This generic discussion of digital communication leads to the following two popular application cases.

6.1.3.1 Clock Data Recovery In real systems, the term "communication" describes the data exchange between blocks, modules, or chips. In some cases, the clock signal itself is part of the exchanged information, such as in LVDS, HDMI, or DDR standards (mesochronous systems). In other cases, the clock is not transmitted but has to be extracted from the data (USB, PCI, etc.; plesiochronous or heterochronous systems). This task of extracting the clock from the data is often called clock data recovery (CDR). Frequency accuracy is vital in this case. Sensible long-term frequency mismatch between the transmitter and the receiver would result in data error. General speaking, frequency matching is a difficult task when a clock is not transmitted but has to be extracted from the data.

6.1.3.2 Time Recovery (Real-Time Frequency Tracking) In certain applications, the digital processing unit's time base (frequency of CLKR) is not predetermined (not known at design time). It is implied from a certain source. The source information (CLKT) must be extracted from certain media through the timing extraction block. The desired time base (CLKR) needs to be calculated in real time. And, in most cases, this process has to be carried out continually. Unlike the case of CDR, where clock information is electrically embedded in the voltage level (which is associated with the data stream), the timing information in this case presents itself in other more complex forms, such as a low-frequency analog waveform (horizontal sync, or HS, in an analog video decoder), a digital value (HDMI audio clock recovery), or a digital time stamp (MPEG stream transport). In this type of application, the information associated with the source time base (not the time base itself) is passed between the parties, usually at a much lower rate. The recovered time base has to accurately follow the source time base. Thus, frequency accuracy is of high importance. The key difference between CDR and timing recovery (time base transfer) is that in the case of CDR the frequency information is directly embedded in the data. In a time base transfer, the frequency is indirectly encoded in data; some processing tasks (measure, calculation, transform, etc.) have to be performed to get the desired time base.

6.1.4 IP Reference, Driving ADC/DAC, Frequency Conversion

Unlike previous cases where it has no impact on circuit operation, clock spectral purity is a major factor when clock is used to drive ADC/DAC, or is used as a reference for other clock circuitry (such as PLL/DLL embedded in IPs), or is used in frequency conversion (as a local oscillator (LO) in RF applications). Spurious tones in the clock spectrum will cause undesired spurious content in the ADC/DAC's output and undesired frequencies in the LO output. Spurious tones could leak through the PLL to the output when the clock is used as its reference. Hence, when a clock generator is used for this purpose, care must be taken to address its spectral purity (refer to Section 1.4).

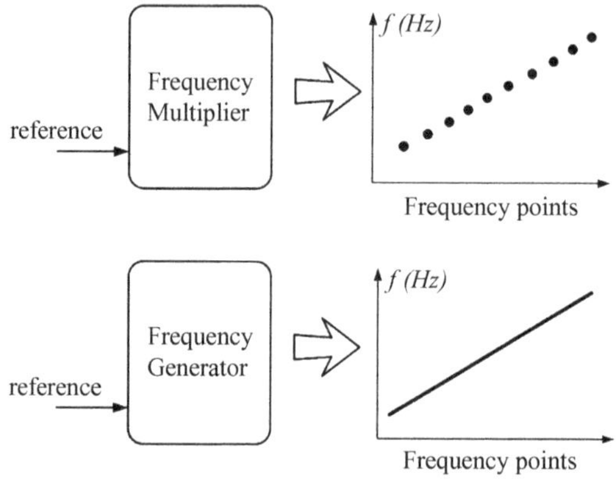

Fig. 6.3. Frequency multiplier vs. frequency generator.

6.1.5 Frequency Multiplier versus Frequency Generator

Driven by the SoC integration trend, the demand for frequency synthesis is constantly escalating. It is believed that we have now reached the stage of needing to distinguish the terms of *frequency multiplier* and *frequency generator*. As illustrated in Fig. 6.3, from a reference frequency, a frequency multiplier simply multiplies it by an integer (or fraction) multiple. The available frequencies from the multiplier are usually limited. Modern SoC development has seen progress leapfrog thanks to the technological advance in processors, memory, and peripherals. Due to the large number of functions on chips, their operation requires many frequencies. To keep up with this pace, the on-chip timing circuitry needs an overhaul. The frequency multiplier is no longer sufficient; we need frequency generators. As depicted in this figure, from a given reference frequency, it is desired that the clock circuitry have the capability of producing ***many*** other frequencies. And moreover, we want the frequency switching to be achieved as fast as possible. Without such a component, the SoC will have to sacrifice a significant amount of budget on silicon real estate, power consumption, and pin-count number just to accommodate the various timing circuitries.

6.2 FLYING-ADDER AND ITS THREE MAJOR APPLICATION AREAS

TAF-based flying-adder direct period synthesis technology can influence modern electronic design from three directions: (1) as a frequency synthesis technique for clock generation, (2) as a message encoder for communication, and (3) as a digital-to-frequency converter (DFC) for rate-of-switching-based information processing.

These three major application areas are graphically depicted in the left part of Fig. 6.4. Chapter 4 established the technical base for the flying-adder as a

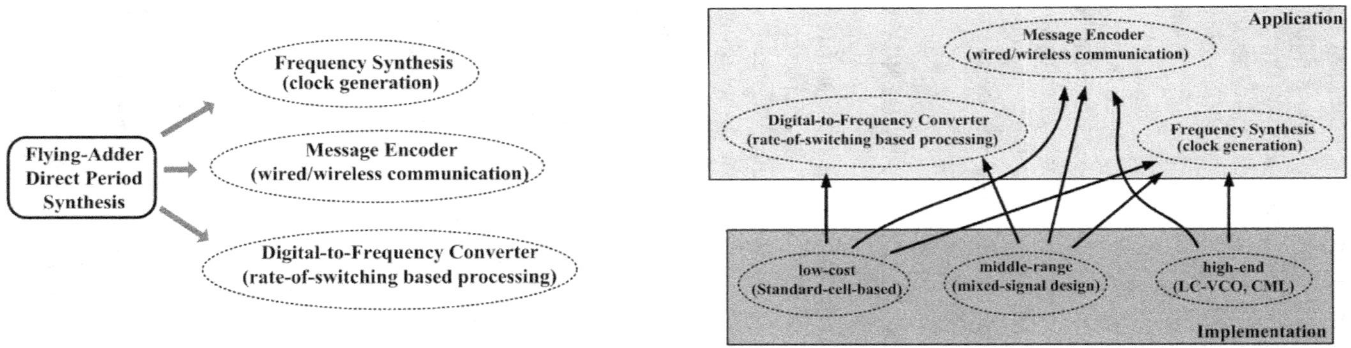

Fig. 6.4. Flying-adder's three major application areas (left); the application areas and the implementation styles (right).

unique circuit technique for frequency synthesis. Chapter 5 presented the theoretical support for the DFC-based information process method. In terms of message encoding (communication), either the time-average-frequency time domain waveform or its frequency domain spectrum can be used as the message. Owing to its flexibility in producing different types of cycles, both the time-average-frequency waveform and the spectrum are rich in content. This is ideal for conveying messages.

For a particular application, one of the three implementation styles (low-cost, middle-range, and high-end; see Section 4.16) can be chosen based on the environment. Together, they create a new frontier in electronic design. This vision is illustrated in Fig. 6.4. Potentially, there are countless possibilities. In this chapter, some exemplary applications will be presented.

6.3 FLYING-ADDER FOR ON-CHIP FREQUENCY GENERATION

Today's SoC is very functionally complicated, and it is very challenging to implement the paper design into silicon. As depicted in Fig. 6.3, the simple frequency multiplier is no longer adequate. We need a frequency generator. For the majority of on-chip processing tasks, operations can be characterized by the scenario of Fig. 1.17, where the clock signal functions as a trigger: The clock controls the time flow, and there is no wall time involvement. Time-average-frequency can play an important role in this environment. Compared to the conventional techniques, the direct benefits of the TAF-based clock generator are: (1) ample frequencies can be generated for building powerful systems, (2) the number of on-chip PLLs can be reduced by attaching multiple flying-adder synthesizers to one integer-N PLL, and (3) frequency switching can be achieved at a fast pace, which enables efficient systems.

Figure 6.5 is a commercial video SoC example that uses a flying-adder PLL as its on-chip clock source. On the left, the overall clock planning is presented. There are several subsystems in this SoC; each has its unique clock requirement. The FAPLL of Fig. 4.27 is used to support all these clock subsystems. For each FAPLL, several synthesizers are attached to the same analog VCO. For example, the ARM/DDR PLL is depicted in the right-hand drawing and has two synthesizers attached.

The system-clock domain presents the simplest design constraint. It only requires 216, 108, 54, 27, and 13.5 MHz. Thus, the VCO is set to run at 864 MHz. Five dividers, along with phase alignment circuit and glitch-free clock switches, are used to produce those frequencies simultaneously. In terms of frequencies required, the audio-clock domain presents the toughest challenge. By using FAPLL, this problem can be solved by just one component. As shown in Fig. 6.5, the input reference of this audio PLL is 86.4 MHz, which can be obtained from the system PLL's VCO frequency of 864 MHz. By setting the FREQ, P, N, and M appropriately, all the required frequencies can be generated under the constraints of (1) the VCO in the optimized range of 700 MHz to 1.2 GHz

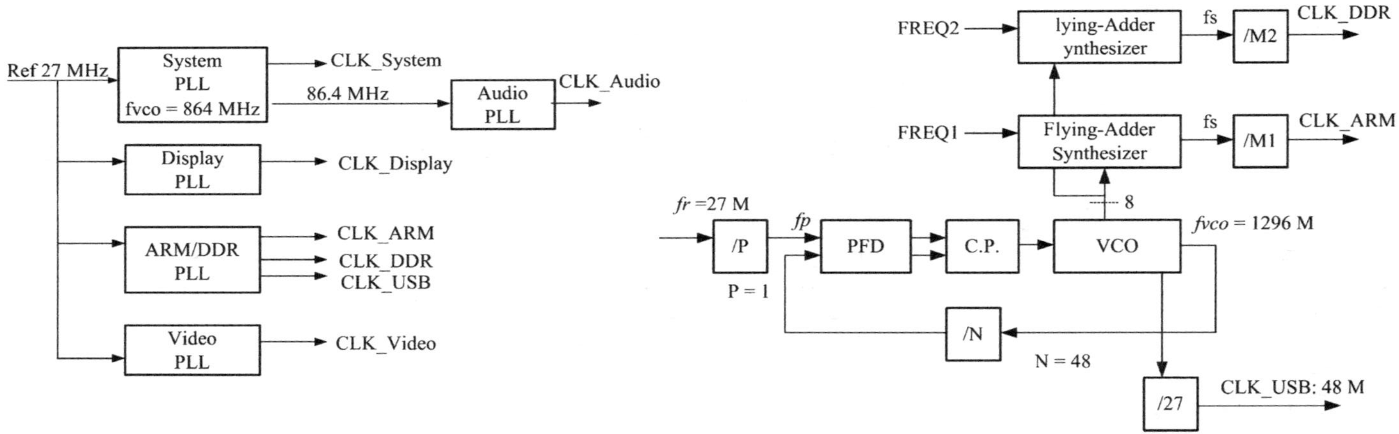

Fig. 6.5. The flying-adder PLL as an on-chip clock generator in a video SoC example.

and (2) f_p in the range of 17.28 to 28.8 MHz. For the example of 45.1584 MHz, one setting could be $P = 5$ ($f_p = 17.28$ MHz), $N = 49$ ($f_{vco} = 846.72$ MHz), FREQ = 15 ($f_s = 451.584$ MHz), and $M = 10$ ($f_o = 45.1584$ MHz). Table 4.10 shows the required frequencies for the audio subsystem. The first column is the sampling frequency in KHz. The first row is the over-sample rate. The numbers presented in the table are the audio clock frequencies in MHz. With the help of Eq. 4.8, all these frequencies can be generated from one FAPLL without using fractions.* If the conventional PLL approach is used, two or three PLLs cascaded in series are required to generate these frequencies.

The ARM/DDR PLL needs to support three independent clocks: *CLK_ARM*, *CLK_DDR*, and *CLK_USB*. From 100 to 250 MHz, several frequencies are required. However, the precise frequency is not important as long as there are enough frequency points available in this range to support the system operation. The more frequencies there are in the range, the more flexibility the system can enjoy in the operation. Moreover, PDFR can be utilized to improve the available frequency points in this range from 21 to 60. The display PLL is used to drive the display engine that supports the display modes of 480i, 480p, 720p, 1080i, 1080p, etc. The major frequencies required are: 13.5, 27, 31.5, 36, 33.75, 40, 50, 49.5, 56.25, 74.25, 75, 78.75, 85.5, 94.5, 108, 148.5, 135, 156, 157.5, and 162 MHz. These frequencies can all be generated by the FAPLL without the use of fractional bits in FREQ. Additional frequencies, such as 25.175, 35.5, 50, 56.25, 65, 68.25,75, 79.5, 101, 102.25, 117.5, and 121.75, are required for some graphic modes. They can be produced with the help of the fractions. The fine resolution of FAPLL has also been used to solve the frame rate synchronization problem. A video PLL is used for an on-chip video decoder that converts an analog NTSC composite signal to a digital component video signal. In this application, the frequencies required are not predetermined. It could be any value in real applications. Due to this special requirement FAPLL is the only choice; there are hardly any other alternatives (in term of performance and cost).

Figure 6.6 shows the SoC's floor plan where the location of the five FAPLLs can be seen. More details on this SoC example can be found in Xiu (2007). Some more recent examples in 40 nm are available (http://focus.ti.com/lit/ug/sprugx9/sprugx9.pdf; http://focus.ti.com/lit/ug/sprugx7/sprugx7.pdf; http://focus.ti.com/lit/ug/sprugx8/sprugx8.pdf).

Figure 6.7 is the block diagram that shows the structure of future FAPLLs for large SoC (second-generation FAPLLs). It includes a flying-adder synthesizer inside the PLL loop (refer to Section 4.9.2, Figs. 4.29 and 4.32). In this exemplary configuration, there are five independent clocks available from this single FAPLL. In this powerful component, the VCO frequency can be controlled by Eq. 6.3.

$$f_{vco} = ([F_0 \cdot N]/[P \cdot K]) \cdot f_r \tag{6.3}$$

* This design was done during a 2003–2004 time frame. Nowadays, these audio frequencies can be more conveniently generated by the IFAPLL of Fig. 4.32. See Figs. 4.89 and 4.90 for examples.

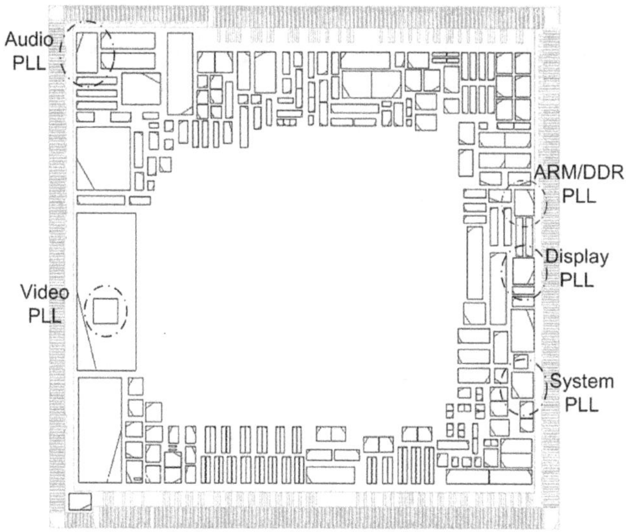

Fig. 6.6. The floor plan that shows the FAPLLs' physical locations in an SoC.

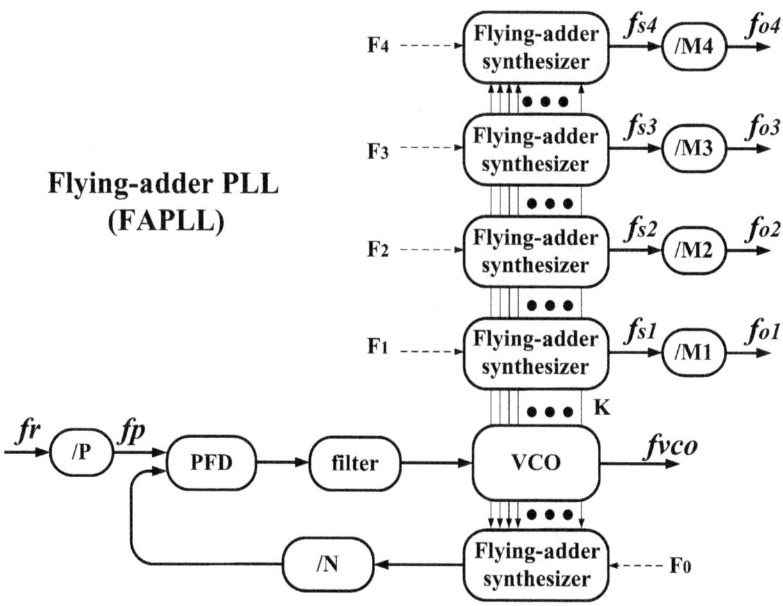

Fig. 6.7. Second-generation FAPLL as clock generator.

For any of the synthesizer, for example, the first one f_{o1}, its output frequency can be calculated by Eq. 6.4. Moreover, PDFR technique can be applied between the pair of (F_0, N) and (F_1, M_1).

$$f_{o1} = ([F_0 \cdot N]/[P \cdot F_1 \cdot M_1]) \cdot f_r \tag{6.4}$$

6.4 FLYING-ADDER AS ADAPTIVE CLOCK GENERATOR

As shown in the left drawing of Fig. 6.8, in digital data processing, there are many cases where data need to be first processed in one time base and then be transferred to another one to be processed again. In this scenario, although the clock rate (frequency) is fixed in circuit-A, there is no guarantee that data are presented in every clock cycle. When this manner of communication is used to pass data between circuit-A and circuit-B, some kind of real-time frequency adjustment is needed in circuit-B if we want a continual data flow in circuit-B (every clock cycle having data). For this application, to ensure both no-information-loss and no-empty-cycle, the clock rate of time-base-B has to follow that of time-base-A in an appropriate way. TAF-based flying-adder technology is suitable for this application since it can generate an arbitrary frequency. Moreover, frequency switching can be accomplished in an instantaneous fashion. The right drawing in Fig. 6.8 characteristically illustrates the operation flow for this application. In here, the flying-adder circuit is used as an adaptive clock generator (FAACG).

Figure 6.9 is a generic example of a time base transfer. In many electronic systems, packet-oriented transmission is used to pass information from one device to another (left drawing). These systems include USB bus, 1394 (FireWire) interface, HDMI interface, MPEG2 transport system, etc. Among these devices, the transmitter sends data packets using its clock and, at the receiving end, the packets are received synchronously on an independent local clock. In some cases, the clock signal is not transmitted along with the data. In others, the clock signal is available but the data might not be presented in every clock cycle. Further, over a long period of time, the transmitter and/or receiver clock sources could have sensible frequency drift. This will cause loss or duplication of data. This is undesirable for real-time audio/video transmission. Therefore, certain synchronization mechanisms have to be established at the receiver side for the smooth operation of downstream processing.

Inside the receiver, FIFO memory is usually employed to accommodate the different data rates at the two ends, as shown in the right drawing of Fig. 6.9. Traditionally, PLL is used to generate the output clock (r_clk). Based on the fullness/emptiness of the FIFO memory, the PLL output clock frequency is adjusted so that there are continuous data flowing out of the receiver at a constant rate. Clearly, the PLL's response speed is critical in this feedback mechanism. For a slow PLL, which requires a long time to respond to the FIFO's status change, the size of the FIFO memory has to be sufficiently large to avoid data loss. Another issue is the frequency resolution. Ideally, the output clock

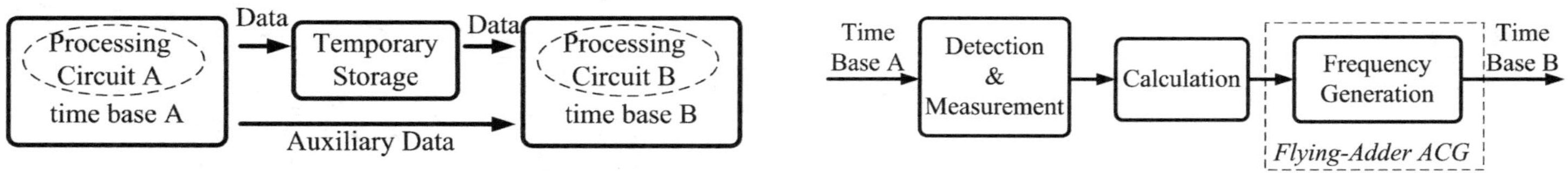

Fig. 6.8. The data flow scenario that requires time-base transfer (left); time-based transfer processing flow (right).

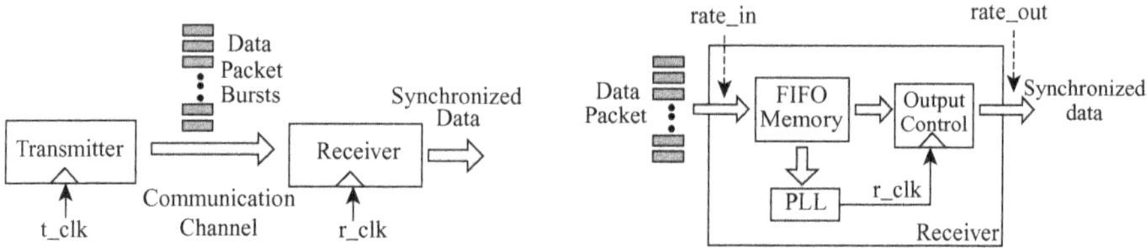

Fig. 6.9. Time-based transfer. A generic example: packet-oriented data transport system (left); synchronization mechanism inside the receiver (right).

frequency should be adjusted only slightly around its center value for small data rate variation. However, conventional PLL has difficulty in achieving fine frequency resolution. Consequently, the frequency change is more or less abrupt. Overall, the slow response and the coarse resolution have prevented system designers from building efficient systems to handle this kind of application.

Figure 6.10 is the simplified block diagram of a USB audio speaker system (left drawing). The USB peripheral interface is used to transfer the digital audio data from a PC to a speaker by isochronous streaming. The CODEC is used to convert the digital data into an analog signal that then drives the speaker. The most commonly used sampling rates are 22.05 KHz for lower-quality PCM and MPEG audio, 44.1 KHz for audio CDs, and 48 KHz for DVDs. However, sending the isochronous USB data directly to the CODEC/speaker could be problematic. In the case of 48 KHz, the isochronous USB mode only guarantees that there is a burst packet consisting of 48 audio samples between two start of frames (SOFs; USB terminology). The placement of this packet could be anywhere within a 1-ms time frame. In other words, The 48 KHz data sent out by the PC are not necessarily aligned with the 48 KHz clock used in the CODEC (plesiochronous system). This could result in some erratic audio with audible clicks and pops. This is very evident to the user if no compensation is made for the alignment of the data and the clock. The problem is even more complex for other sample rates. For example, at 44.1 KHz, the USB host controller sends nine 44-sample bursts followed by one 45-sample burst in a repetitive pattern. It is the responsibility of the USB device to convert this as a continuous stream at a 44.1-KHz rate to the speaker.

The solution to this problem is to insert a stream controller between the PC and the speaker, as shown in Fig. 6.10. The function of the stream controller is to take the isochronous USB data and smooth them or synchronize them before sending them to the CODEC/speaker. The controller stores the data in its memory after receiving them from the USB interface. The center frequency generated by the FAACG is 48 KHz (in the case of a 48-KHz sampling rate), which is used to drive the audio CODEC. As shown, the ACG also receives the USB data. Based on the time stamp (SOF) embedded in the USB packet, the ACG can slightly adjust its output frequency to accommodate the variation in the USB data stream so that the audio noise can be avoided. Because of the finer accuracy and fast response in ACG, the close loop

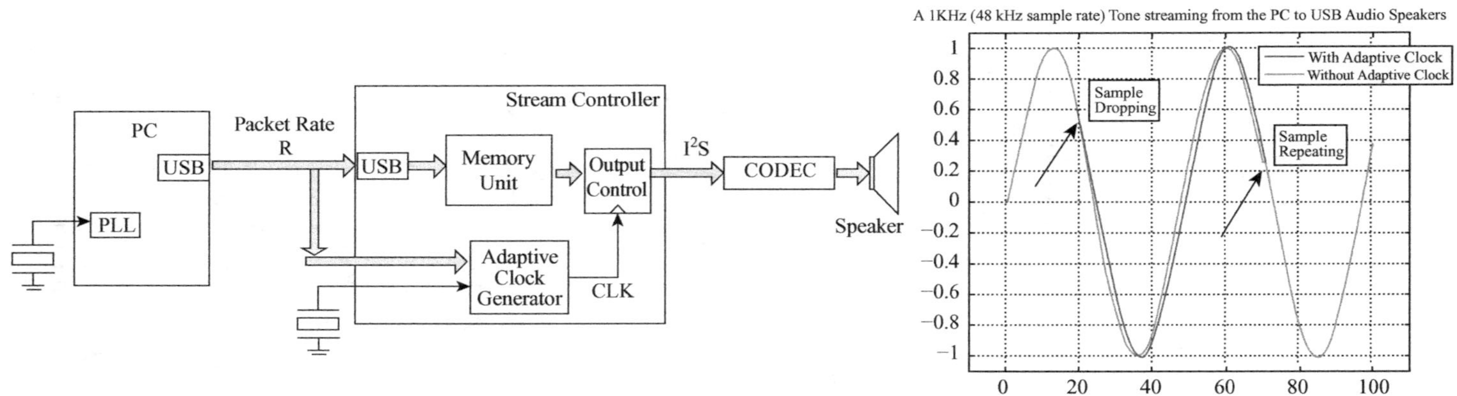

Fig. 6.10. FAACG for USB stream controller: the system (left); the resulting waveforms (right).

parameters (step size, response time) can be adjusted in real time to achieve high fidelity sound. In operation, the USB SOF signal is captured by the clock CLK (or another clock with a known frequency) and stored in a register. This stored value is valid until next SOF occurs (about every 1 ms). The on-chip processor can read this value, do the calculation for generating a new FREQ value, and subsequently update the FREQ control register for applying the new frequency. By locking to the SOF in this fashion, the controller can guarantee the smooth data flow between the PC and the speaker. The right-side drawing in Fig. 6.10 shows the data of 1-KHz tones sent to the speaker by the PC. As shown, when the USB data are directly sent, there are occurrences of sample dropping/repeating, whereas when ACG is used, a clean tone is created (Xiu et al. 2007; http://focus.ti.com/lit/ds/symlink/tusb3200a.pdf).

Figure 6.11 shows another example of using FAACG for MEPG2 data packet synchronization. MPEG2 transport stream (MPEG2-TS) is a data format specified in Part 1, "Systems," of ISO/IEC 13818-1. Its purpose is to allow the multiplexing of digital video and audio data. The basic data unit in MPEG2-TS is the packet, which is usually 188 bytes long but could be 204 bytes as well. The data in the transport system can be transmitted in either packet synchronous mode or packet asynchronous mode. In synchronous mode, the clock rate is the same as the data rate. In asynchronous mode the clock rate is higher than the data rate. At each clock cycle, there may or may not be a valid data. The data are valid only when the DVALID is high as shown in Fig. 6.11a (heterochronous system). The PSYNC signal indicates the start of the packet. Some MPEG2 receivers are only capable of receiving synchronous data. For this reason, the asynchronous MPEG2 data have to be synchronized before being sent to these devices. This process is often referred as packet synchronization or packet smoothing.

Conventionally, the packet synchronizer is realized by the pairing of a PLL and FIFO memory. The clock frequency of the read clock is derived from a PLL that is controlled by the fullness/emptiness of the FIFO and the input data rate. It has to match the average data rate of the input data. Usually, the PLL is designed with low bandwidth to limit the noise from the input side and to maintain high loop stability. As a result, the read clock is unable to follow the input data rate variation quickly. In order to avoid the possibility of data loss, the size of the FIFO memory has to be large enough so that it can hold all the necessary data when the rate shift occurs. If a slow PLL is used, the FIFO size could be as large as several hundreds of packets.

In contrast, if FAACG is used, the memory requirement can be reduced to as small size as two packets. This idea is depicted in Fig. 6.11b. In this scheme, there are two storage units used, each having the size of exactly one packet. They are called ping-pong memory since at any given time there is always one unit being read from and one unit being written to. In the input side that is driven by the higher speed write clock f_w, there is a circuit block of packet counters that constantly counts the number of clock cycles used for transmitting one packet. The structure of this packet counter is depicted in Fig. 6.11c. The *valid data counter* is used to count the number of clock cycles that are

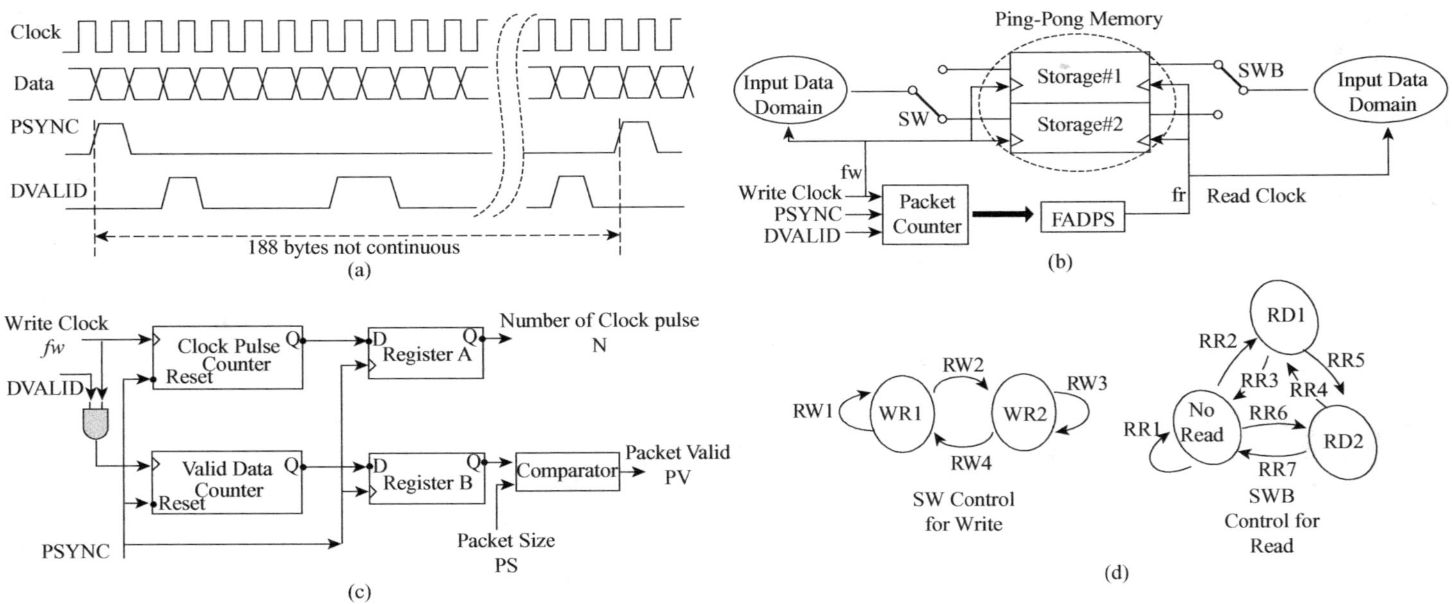

Fig. 6.11. FAACG for MPEG2 packet synchronization: (a) MPEG2 packet asynchronous transmission, (b) packet synchronizer using FAACG, (c) the packet counter, (d) the control states for SW and SWB.

companioned with valid data between two consecutive PSYNC pulses. The *clock pulse counter* is used to count the total number of clock pulses between two consecutive PSYNC pulses. At the beginning of any new packet, the PSYNC becomes active high, which resets the two counters. And at the same time, the contents of the two counters are latched into two registers, *register A* and *register B*, respectively. The content inside *register B* is compared with a preset value *packet size* (e.g., 188 or 204). If equal, the *packet valid* signal will be asserted to indicate that a valid packet has been received. Otherwise, it stays low. The content in *register A* (N) is the number of total clock pulse used for current packet received/stored in memory. Subsequently, the read clock frequency can be calculated as $f_r = (PS/N)*f_w$. Since FAACG can respond instantly, the frequency value will be updated in next read cycle.

Figure 6.11(d) shows the states diagram for the two switches SW and SWB. WR1 and WR2 represent the operations of writing to storage #1 and #2, respectively. The incoming packet size can be 240+ or 400+ (valid plus redundant data) depending on the standards used. The write state machine filters the packet size to 188/204 data per packet. RD1 and RD2 are states for read. The symbols *RW1*, *RW2*, *RR1*, etc. are the rules for state changes indicated by the corresponding arcs. Table 6.1 lists the content of these rules, where *PV1* is the *PV* signal of writing to storage #1. *PV1* = 1 when one packet is successfully written to storage #1. *PV2* is the same signal but for storage #2. Figure 6.12 shows one simulation of this MPEG2 packet synchronizer. In this plot, the signal *wclk* is the write clock. The *dvalid* indicates the validity of the data. The *wclk_count* reports the number of *wclk* cycles used for the packet. When the write operation is finished, an interrupt (*cpu_intrpt*) to the on-chip processor is generated. Based on the *wclk_count*, the CPU will calculate a new frequency control word for the FAACG to adjust the read clock, *rclk*. The resulting *rclk*, along with the data, will be used in the downstream processing. The scheme described in this section is a precise operation. Since only two storage units are needed exactly, the memory requirement has been significantly reduced (Ying and Haider 2006; http://focus.ti.com/lit/ds/slea064a/slea064a.pdf).

The two examples presented in this section are just for illustration purposes. They show that the FAACG can be used wisely for improving system

TABLE 6.1. The Control Rules for Write Switch SW and Read Switch SWB

Rule for SW	Content	Rule for SWB	Content
RW1	NOT *PV1*	RR1	(NOT *PV1*) AND (NOT *PV2*)
RW2	*PV1* AND *PSYNCH*	RR2	*PV1*
RW3	NOT *PV2*	RR3	(RD1 Done) and NOT *PV2*
RW4	*PV2* AND *PSYNCH*	RR4	(RD2 Done) and *PV1*
		RR5	(RD1 Done) and *PV2*
		RR6	*PV2*
		RR7	(RD2 Done) and NOT *PV1*

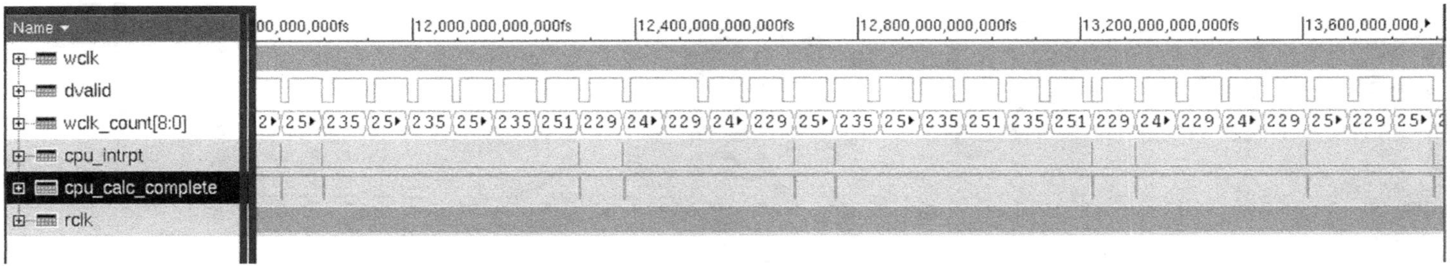

Fig. 6.12. One simulation snapshot of an MPEG2 packet synchronizer.

efficiency. There are many other applications that can utilize this FAACG for optimization at higher levels. All those are due to the two important features that are not available in conventional PLL: the fine frequency resolution and the fast response speed.

6.5 FLYING-ADDER AS ON-CHIP VCXO

One of the major advantages of digital transmission is the reliable delivery of information from one location to another with high fidelity. In this transmitting/receiving process, the timing information can be embedded in the data stream. This technique of embedding timing information within data has been used in many modern applications, such as telecommunications, digital TV broadcasting, digital audio, ADSL, and set-top boxes. The extraction of this embedded timing information from the data is often referred as timing recovery. The key issue in timing recovery is the clock synchronization between the transmitter and the receiver. It requires the constant adjustment of the local clock (receiving side) based on the timing information extracted. One of the methods is to use a VCXO (voltage controlled crystal oscillator), which can be an external VCXO chip (Lee et al. 1996; Logan et al. 1998; Qiuting et al. 1988; Watanabe et al. 2006) or an on-chip VCXO/DCXO module (Balan and Pan 2002; Huang and Basedau 1996; Lee and Bulzacchelli 1992; Lin 2005; Mujica et al. 2003).

The top left drawing in Fig. 6.13 shows the block diagram of an MPEG2 data stream transmission and reception used in a digital TV satellite broadcast. On the transmitter (TX) side, pictures are converted into a digital data stream by a video encoder. Then the data stream is transmitted through a communication channel to the receiving (RX) side, such as a step-top box or a TV set. The data stream on the encoder side is generated based on its 27-MHz reference clock (with some multiplication). This data stream is then latched by the decoder with its own local clock. Without frequency matching between the two sides, the pictures generated from the video decoder will not be synchronized with that of the source. In the MPEG2 specification, the maximum allowable variation is 27 MHz ± 810 Hz. Typically, the local clock frequency adjustment is achieved by an external VCXO chip. It is separated from the main processing chip as shown in the bottom left drawing in Fig. 6.13.

In the MPEG2 transport mechanism, a time stamp called a program clock reference (PCR) is inserted into the data stream. This PCR time stamp can be used by the receiver to retrieve the source clock information. The feedback from this time-stamp processing is used to drive the VCXO, to slightly adjust the local 27-MHz reference. More details are shown in the right drawing of Fig. 6.13. The cornerstones in this timing recovery method are the system time clock (STC) and the PCR. When a video program is encoded, a 27-MHz encoder clock drives the encoding process. When the encoded data packets are packed into the transport system, a time stamp (PCR) is inserted into the data stream at the rate of about once in every 100 ms (>10 Hz). On the

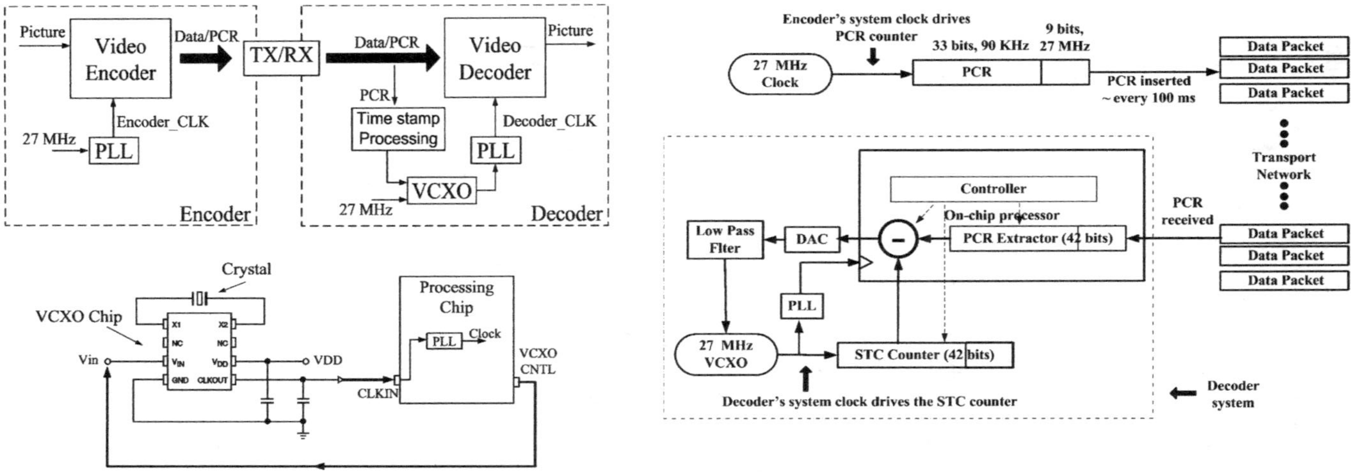

Fig. 6.13. The MPEG2 data transport system: the system (top left); using an external VCXO chip (bottom left); the working principle (right).

decoder side, when the PCR is received, it is compared with a local PCR value that is stored in an STC counter that is driven by a VCXO. The difference is fed into a DAC, which then drives the off-chip VCXO. The goal of this feedback mechanism is to adjust the VCXO output so that the difference in PCRs gradually becomes zero. In other words, the remote PCR is used to direct the local PCR stored in the STC, which is the time reference of the MPEG2 decoder, into a value intended by the encoder.

Another approach to achieving local clock synchronization is to build a VCXO block on-chip. An on-chip integrated DCXO module has been reported in Lin (2005) that uses an array of on-chip capacitors to tune the Colpitts crystal oscillator. The accumulation of the on–off state of each capacitor in the array will determine the output frequency. It is controlled digitally by the information extracted from the communication channel. This approach can eliminate the off-chip VCXO chip and reduce system costs. A flying-adder synthesizer, owing to its fine frequency resolution and fast response, can be used to serve this purpose even more efficiently (Xiu 2008; Xiu et al. 2012). The digital control information extracted from the time stamp can be directly applied at the synthesizer's control word FREQ to generate the VCXO capable clock. This idea is illustrated in the left drawing of Fig. 6.14.

From the discussion of Section 4.19 and Fig. 4.58, we know that the flying-adder synthesizer's (FADPS) frequency transfer function is *1/x* as illustrated in the middle plot of Fig. 6.14. It is a single value function. For any given control word FREQ, there is one and only one corresponding frequency. This is essential for VCXO function. Moreover, this *1/x* characteristic is valid for the synthesizer's entire operation range ($2 \le FREQ \le 2*K$, where K is the number of VCO outputs). This feature guarantees the wide VCXO pulling range. Furthermore, *1/x* function is monotonic, which is another crucial requirement for VCXO operation. The above three characteristics make FADPS ideal for functioning as a VCXO module. Since the control variable is a digital word, a flying-adder VCXO is better termed FADCXO.

In VCXO/DCXO-related applications, the required frequency pulling range is usually small. Typically, the range is about ±300 ppm, or less, of the center frequency. As depicted in the right drawing of Fig. 6.14, in small area,

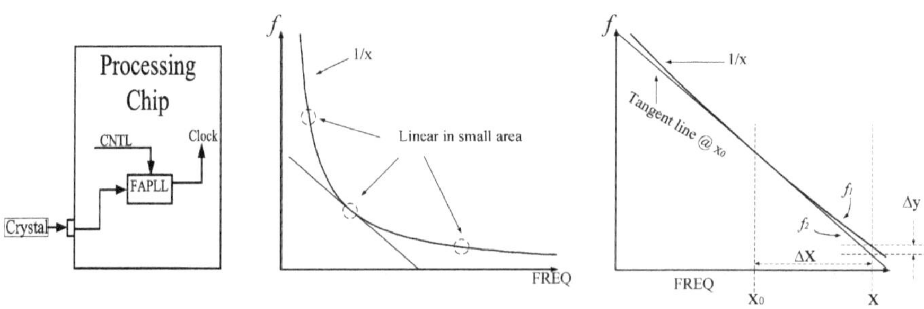

Fig. 6.14. FAPLL as an on-chip VCXO. The principle idea (left), FADPS transfer function (middle), FADPS transfer function in a small range (right).

the *1/x* curve can be approximated by a straight line. Therefore, in real applications, the FADCXO can further be improved to a linear DCXO. If we define a variable z as $z = (FREQ - FREQ_0)/FREQ_0$, where $FREQ_0$ is a fixed value, then FREQ can be expressed as $FREQ = FREQ_0 * (1 + z)$, and we have Eq. 6.5. When operating in a small area around $FREQ_0$, we have $|z| \ll 1$. In such cases, a Taylor series can be used, and Eq. 6.5 can be approximated by Eq. 6.6. This equation clearly shows the linear characteristic of the FADCXO.

$$f = \frac{1}{FREQ * \Delta} = \frac{1}{\Delta * FREQ_0 * (1 + z)} \tag{6.5}$$

$$\begin{aligned} f &= \frac{1}{\Delta * FREQ_0 * (1+z)} = \frac{1}{\Delta * FREQ_0}(1 - z + z^2 - z^3 + z^4 - z^5 + \dots) \\ &\approx \frac{1}{\Delta * FREQ_0}(1 - z) \end{aligned} \tag{6.6}$$

The performance of the FADCXO can be investigated from these perspectives: pulling range, linearity, frequency resolution, modulation rate, slope polarity, slope sensitivity, and stability.

- **Pulling Range:** The FADCXO frequency transfer function is $1/(\Delta * FREQ)$. This is valid for the entire operation range: $2 \le FREQ < 2*K$. Usually, this is a very wide range. In a typical FADPS implementation, the frequency span is on the order of hundreds of MHz. Therefore, the pulling range of the FADCXO is very large. In most timing recovery applications, we only use it in a small range (several hundreds Hz).
- **Linearity:** As shown in the right drawing of Fig. 6.14, at any given point x_0, the *1/x* curve can be approximated by the tangent line associated with this point. This tangent line can be expressed in Eq. 6.7, where c is a constant. If Δy is defined as the difference between the two functions at any given point x, Eq. 6.8 is the error between the two functions (see Appendix 6.A for details) and can be used as the measurement for linearity in VCXO specification.

$$f_2(x) = -\frac{c}{x_0^2}x + \frac{2c}{x_0} \tag{6.7}$$

$$\frac{\Delta y}{y} = \frac{(f_1(x) - f_2(x))}{f_1(x)} \approx \left(\frac{\Delta x}{x_0}\right)^2 \tag{6.8}$$

- **Frequency Resolution:** The frequency resolution of the FADCXO is defined as the frequency step achieved when FREQ is advanced or retreated by one LSB. From Eq. 4.7, the frequency resolution δf can be derived as in Eq. 6.9, where x is the number of fractional bits in FREQ, and f is the frequency at synthesizer's output.

$$\delta f = -2^{-x} * \Delta * f^2 \tag{6.9}$$

- **Modulation Rate:** It is defined as the rate at which the control voltage (or other variable) changes that can result in corresponding frequency changes. Currently, commercial VCXO chips and on-chip DCXO blocks [such as the one reported in Lin (2005)] all require a certain amount of time to respond to a control change, but the FADCXO can response instantly to the FREQ change. This fast response can help achieve fast lock in the timing recovery process.
- **Slope Polarity:** The slope polarity denotes the direction of frequency change versus control voltage (or other control variable). The FADCXO has negative slope polarity since frequency decreases when FREQ increases. It has mathematically guaranteed monotonicity, which is essential for a clock recovery system.
- **Slope Sensitivity or Slope Linearity:** This measures the smoothness of the VCXO operation. For the FADCXO, the slope at any given point x can be mathematically derived as in Eq. 6.10. This is a continuous function. It guarantees that the DCXO frequency will not change abruptly at any point.

$$f_1'(x) = -\frac{c}{x^2} \tag{6.10}$$

- **Stability:** This concerns the VCXO frequency dependence on temperature variation, aging, etc. Since the FADCXO is an oscillator made of CMOS transistors, it does not impose any additional stability issues. In this regard, it is just a regular circuit component like a frequency divider. In addition, the temperature variation and aging affect can be easily compensated by periodical reprogramming (calibration) the FADCXO.

One real FADCXO example is implemented using the following configuration (in a 90-nm process): $f_r = 27$ MHz, $N = 32$ ($f_{VCO} = 864$ MHz, $K = 8$, $\Delta = 144.68$ ps), FREQ = 8 ($f_s = 864$ MHz), and $M = 32$. Under these conditions, from Eq. 4.8, the output is 27 MHz. For the VCXO function, FREQ is allowed to vary in the range of 7.997038134 to 8.002874021. This results in FADCXO frequencies of 27 MHz ± 10 KHz, or about ±370 ppm. There are 21 bits in the fraction. Using Eq. 6.9, the frequency resolution of f_s around the central 864 MHz is ~51 Hz, and the resolution at the final output is 1.6 Hz.

$$\delta f_s = -2^{-k} * \Delta * f_s^2 = -2^{-21} * (144.68e-12) * (864e6)^2 = 51.50Hz$$

$$\delta f_o = \frac{\delta f_s}{M} = 51.50Hz / 32 = 1.61Hz$$

Figure 6.15 shows the lab measurement result. The one on the top left is the pulling range. Within the FREQ range of 7.997038134 to 8.002874021, the calculated and measured frequencies are displayed together. The small frequency offset is due to the crystal error (f_r is not precisely 27 MHz). This is not a problem for the FADCXO since an offset can be programmed into the FREQ to compensate this. From these data, the linearity defined in Eq. 6.8

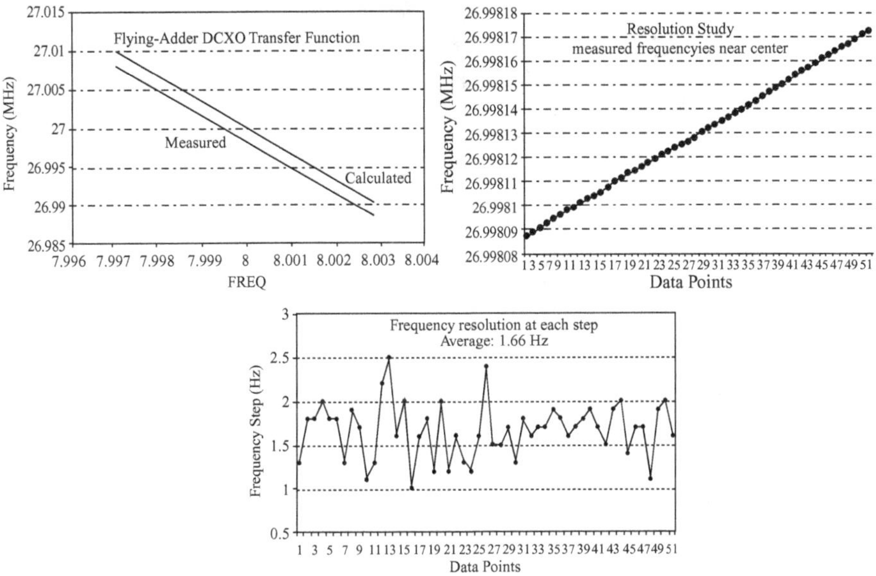

Fig. 6.15. The FADCXO measurement results: pulling range (left); resolution (center); frequency difference between adjacent points (right).

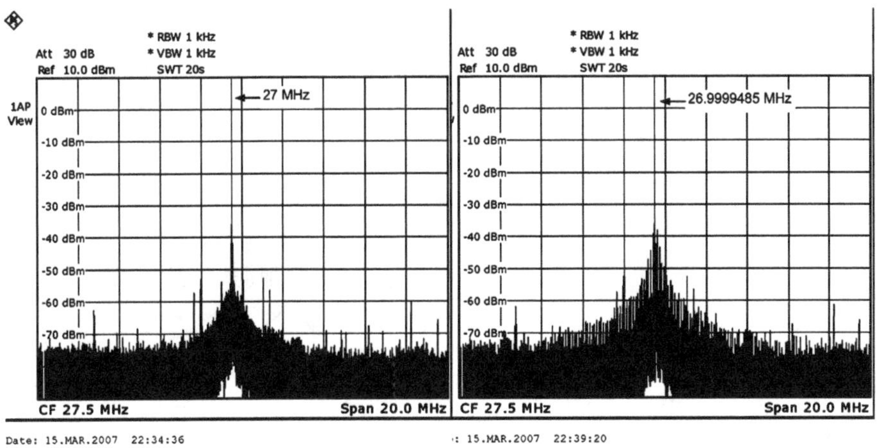

Fig. 6.16. The frequency spectrum of 27 MHz (left) and 26.9999485 MHz (right).

can be calculated as 0.001%. The top right plot shows the frequency resolution. In this measurement, one LSB is advanced each time. The resulting frequency change is plotted. The plot at bottom shows the same information in another way: the frequency difference between any two adjacent points is calculated and displayed. The average step is about 1.66 Hz, which agrees with the calculation favorably. Figure 6.16 shows the frequency spectrum of 27 MHz and 26.9999485 MHz. In the 26.9999485-MHz case, the dense spurs are caused by the fraction 0.000015259. Figure 6.17 presents a control flow diagram that

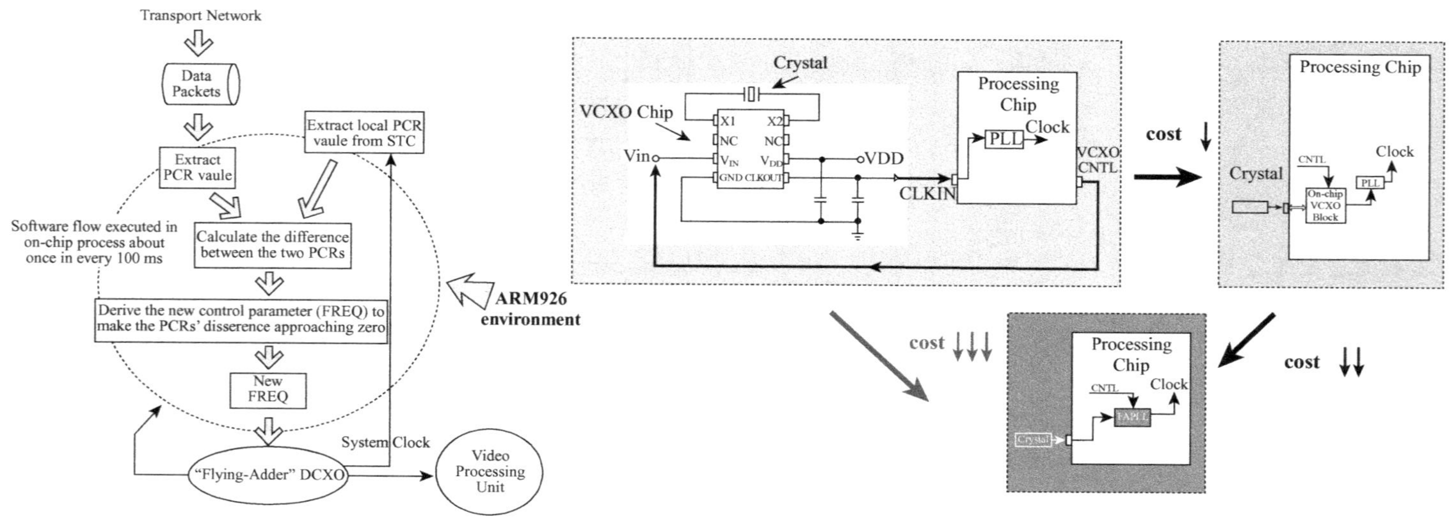

Fig. 6.17. The control flow chart of an FADCXO in a real environment (left) and the benefit of cost savings (right).

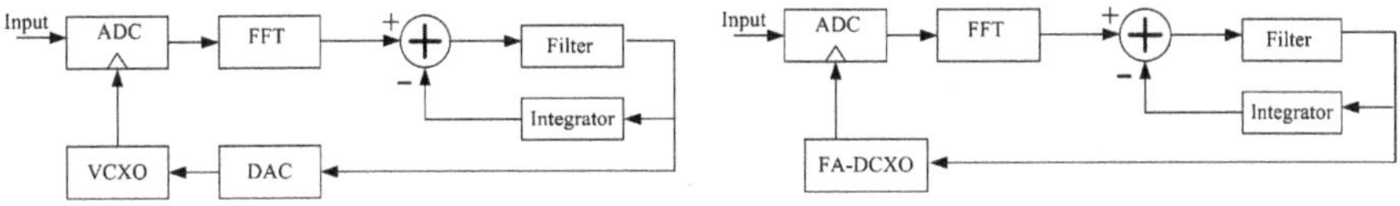

Fig. 6.18. ADSL timing recovery (left) and FADCXO-based ADSL timing recovery (right).

shows how this FADCXO is used in the real environment (left). On the right, the benefit of cost saving is graphically demonstrated.

The VCXO-based timing recovery scheme has been used in many communication applications. Figure 6.18 (left) shows another example of its usage in ADSL applications. Information about the remote clock (frequency and phase) is embedded in the input signal. The loop is constructed to extract this information and direct the VCXO to adjust its frequency so that the ADC can be synchronized with the remote clock. A method of using a numerically controlled oscillator and tapped delay line to jitter the ADC clock is proposed in Mujica et al. (2003) for serving this purpose. The goal is cost reduction. The same goal can be achieved more efficiently with the FADCXO as illustrated in the drawing on the right of Fig. 6.18.

6.6 FLYING-ADDER FOR FRAME RATE SYNCHRONIZATION AND DISPLAY MONITOR ACCOMMODATION

The left-side drawing in Fig. 6.19 is a simplified HDTV video display system. Starting from a video source, the video frames sequentially pass through several processing units inside the decoder chip before being displayed on a display device. The video content is processed and displayed frame by frame. Within the decoder, usually there is a frame buffer between the video processor and the display unit to accommodate their different processing speeds. The display unit and display device are all driven by the pixel clock generated from the on-chip PLL. In a standard TV system, the digital video content is displayed pixel by pixel on the display device as shown in the right drawing of Fig. 6.19. The rate at which the pixels are displayed is controlled by a pixel clock. Its frequency is determined by Eq. 6.11, where *F_rate* is the frame rate, or number of video frames per second. *F_size* is the frame size, or number of video lines per frame (scan size), and *L_size* is line size, or the number of pixels per line (scan size).

$$f_{pixel\_clock} = F\_rate * F\_size * L\_size \tag{6.11}$$

In some cases, the frame rate of the video signal is not constant but varies depending on how the video source is generated. Different video contents are derived from different video sources that bear their own frame rates. For

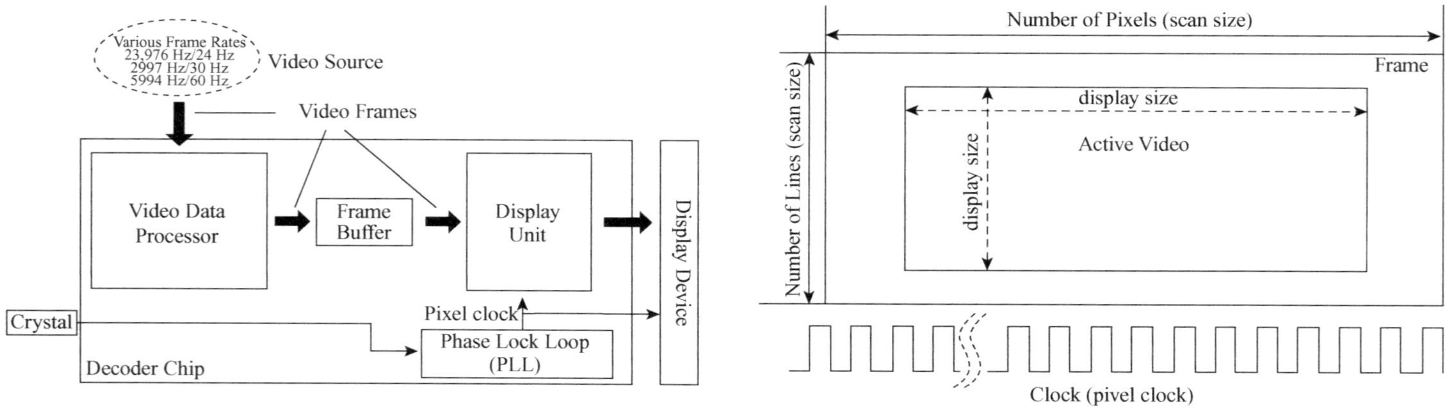

Fig. 6.19. HDTV video display system (left), and pixel clock and display device (right).

example, when a movie is broadcasted in HDTV 720p mode, it could use a 59.94-Hz frame rate. However, during the commercial breaks, the frame rate used for advertisement program might be in 60 Hz. In the receiving system, the hardware must consequently be designed to handle this frame rate variation. Traditionally, a PLL is designed only for several commonly used frequencies, and these frequencies often cannot satisfy the need to display the video content seamlessly under all different frame rates. In the case of the 720p mode, the display frame size is 1650 (pixels per line) by 750 (lines per frame). The pixel clock frequency required for the 60-Hz frame rate is then calculated as 74.25 MHz by using Eq. 6.11. In most decoders, the PLL is designed for this frequency. However, when the broadcasting source changes to a frame rate of 59.94 Hz, the corresponding pixel clock frequency is 74.17575 MHz. If the same hardware (74.25 MHz) is used to display the video, there will be a frame buffer underflow problem with unpleasant visual results.

To compensate for this problem, a technique has been used that repeats or drops a video frame in the video stream from time to time. This can make the incoming video rate pseudo match that of the displaying hardware. In the above example, roughly for every 1000 frames of video, there will be a shortage of one frame of data for display. In operation, for every 16 seconds, one extra video frame can be added to compensate this problem. However, viewer will experience the "video jitter" of repeated images at such a high rate of occurrence. Another approach is to adjust the length of each video line (but keep the active video length the same as defined in the display standards) to compensate for the frame rate difference. In the previous 720p example, if the line's length is changed to 1651 pixels (from the original 1650) and 74.25 MHz is still used as the pixel clock frequency, the new frame rate can be calculated as $F\_rate = 74.25\ \text{MHz}/(1651 * 750) = 59.96366$. Now, the error between 59.96366 and 59.94 is 0.0394%. In other words, for every ~2500 frames, the system will require a repeated picture frame. The video jitter is improved from the previous 16 seconds to 42 seconds. The problem can also be solved by using multiple crystals in the system, each for its dedicated frame rate. The drawback of course is the higher cost.

A FAPLL can be used to solve this problem elegantly. The display PLL in Fig. 6.5 can be configured as such to generate the 74.25 MHz: $P = 1$, $N = 33$, $FREQ = 12$, $M = 8$ ($f_r = 27$ MHz and $K = 8$). If $FREQ$ is changed to 12.012012012 (from the original 12), the resulting frequency becomes 74.17575 MHz, which meets the requirement of the 59.94 Hz frame rate. In this implementation, since the $FREQ$ uses a finite size of 21 bits for the fraction, the 12.012012012 cannot be exactly represented due to the truncation error. The actual value achieved is 12.0120120049, which will be translated into a frame rate of 59.9400000355 Hz. Now the error between 59.94 Hz and 59.9400000355 Hz is 0.000592 ppm. In this approach, for about every 1.7 billions frames, the frame buffer underflow problem happens once. In other words, the event of frame repeat/drop only needs to occur once when the decoder chip has continuously operated for 326 days. By using the FAPLL, not only is the hardware cost significantly reduced but the visual artifact is virtually eliminated. The FAPLL is flexible enough to accommodate all the HDTV display modes with all different frame rates.

Just as with frame rate synchronization, the FAPLL's superior frequency generation capability can also be used to accommodate the various monitors and digital TV displays that the decoder chip has to support. A typical example is the WXGA display mode. Wide XGA (WXGA) is a set of nonstandard resolutions derived from the XGA display standard by widening it to a wide screen aspect ratio. WXGA is generally understood as referring to the resolution of 1366 × 768, with an aspect ratio of 16:9. This is the most popular resolution for liquid crystal display (LCD) televisions and HD plasma flat panel displays. However, other resolutions have also been labeled as WXGA, ranging from 1920 × 1080 to 720 × 800. The following list is the other resolutions bearing the WXGA label.

- 1280 × 720, monitor, 16:9
- 1280 × 768, monitor, 5:3 (16:9.6)
- 1280 × 800, monitor, 8:5 (16:10)
- 1360 × 768, LCD TV, 16:9
- 1366 × 768, LCD TV, 16:9
- 1920 × 1080, projector, 16:9

From Eq. 6.11, it is clear that the pixel clock frequency required for these different resolutions varies greatly. Typically, a monitor or flat panel display vendor provides three key parameters for its device: pixel clock frequency, vertical scan size, and horizontal scan size (refer to the right-side drawing of Fig. 6.19). These parameters are all provided in the style of suggested range and typical value. It is up to the HDTV chip to configure the display device for the intended resolution. Due to the limited number of frequencies that are available from a conventional PLL, the appropriate *F_size* and *L_size* have to be selected to accommodate the pixel clock frequency that is synthesizable from the PLL. In many cases, this is simply unachievable. An FAPLL can be used to solve this problem easily. From the typical values of *F_size* and *L_size* provided by the display device vendor, the pixel clock frequency can be readily derived from Eq. 6.11 and consequently be generated from the FAPLL. For other graphic modes, such as WSXGA (1600 × 900 or 1680 × 1050) and WUXGA (1920 × 1200), the FAPLL can easily be reused as well without any hardware modification.

6.7 FLYING-ADDER FOR FREQUENCY SYNCHRONIZATION IN DIGITAL COMMUNICATION: A PREVIEW

The features of instantaneous frequency switching and fine frequency resolution make the flying-adder an ideal technology for the digital communications discussed in Section 6.1.3. Specifically, the flying-adder DLL (FADLL)

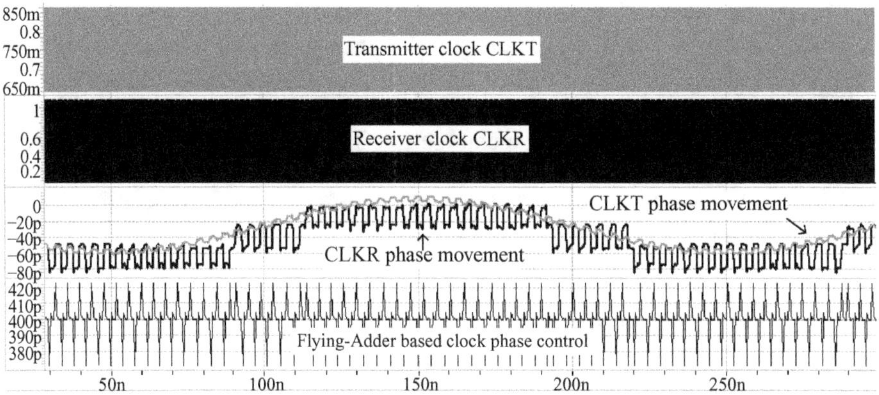

Fig. 6.20. Time-average-frequency-based synchronization for a mesochronous system.

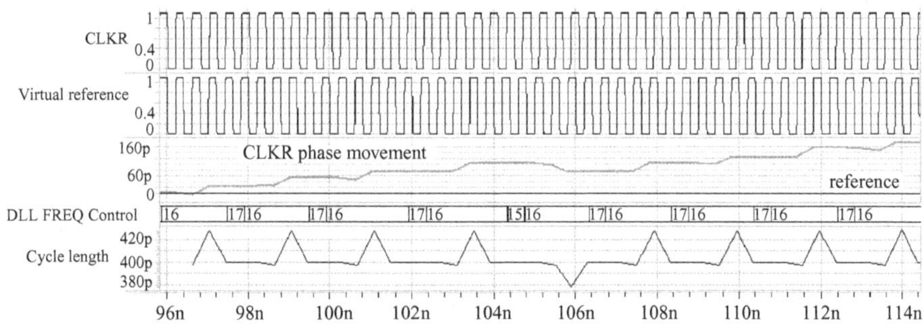

Fig. 6.21 Time-average-frequency-based synchronization for plesiochronous and heterochronous systems.

introduced in Section 4.25 is the enabler for a new approach: *TAF-based frequency synchronization for digital communications*. The FADLL capability demonstrated in Fig. 4.95 can be used for phase compensation in a mesochronous system and for frequency compensation in plesiochronous/heterochronous system. It is a DLL with the capability of infinite-long-delay generation.

Figure 6.20 is a demonstration of this approach in a mesochronous system. The transmitter clock CLKT (refer to Fig. 6.2) has nonzero phase movement as shown in the third plot from the top (the red trace). In the fourth plot, the FADLL is shown to adjust its cycle length to effectively move the clock phase. The nominal frequency (period) is 2.5 GHz (400 ps). Two additional periods available from the FADLL are 375 ps and 425 ps. They are used to track the target. As a result, the receiver clock CLKR is able to track the CLKT's phase movement (the black trace in the third plot).

Figure 6.21 demonstrates the case of using a TAF-based synchronization method for plesiochronous and heterochronous systems where a sensible frequency offset existed between the transmitter and the receiver. In this case, the FADLL can move its cycle length in one particular direction consecutively to track the frequency offset. There are two crucial design parameters in this TAF-based frequency synchronization approach: step size and loop latency. Loop latency defines the time required from the receiver between when it senses the incoming data's phase change to a FADLL sending out the CLKR phase correction. Step size is the time resolution of the method. Section 6.8 will present an application of TAF-based clock data recovery as an example of this frequency synchronization approach.

6.8 FLYING-ADDER FOR CLOCK DATA RECOVERY

Serialization and deserialization (SeDes) is a technology commonly used in chip-to-chip communication for its low cost and high reliability. Its principle idea is illustrated in the left-hand drawing of Fig. 6.22. Clock data recovery (CDR) is a key technique in this application (Horowitz et al. 1998; Razavi 2002). It plays critical role in multi-Gbps datalink standards, such as SATA, PCI, SONET, XAUI, USB, DisplayPort, etc. The task of CDR is brought out because the clock signal is not transmitted during the information exchange between the transmitter (TX) and the receiver (RX). Instead, the clock signal is embedded in the electrical waveform transmitted. It has to be extracted from the transmitted data and subsequently used in the receiver in order for the transmitted electrical waveform to be correctly interpreted. In most modern datalink standards, the TX and RX usually do not share a common frequency reference. Consequently, there is strong possibility (almost 100%) that some amount of offset exists between the TX and RX clocks' frequencies as shown in right-side drawing of Fig. 6.22. Further, the magnitude of this offset most likely varies from time to time. If care is not taken, it could result in errors when interpreting the transmitted data. Therefore, some kind of mechanism must be employed to adjust the RX clock's frequency to match that of TX's clock (in the long-term sense).

This kind of system is the plesiochronous system discussed in Section 6.1.3. In a plesiochronous system, the signals' significant instants occur at nominally the same rate. Any variation in rate is constrained within a specified limit. In other words, a TX and RX operate plesiosynchronously if they operate at the same nominal frequency but may have slight short-term frequency mismatches from time to time (which leads to phase drifting). USB 3.0 is a typical example of a plesiochronous system. In general, a plesiochronous system behaves similarly to a synchronous system except that it must be equipped with certain ways to cope with the synchronization slip that happens at intervals because of the plesiochronous nature of the system. Time-average-frequency can be used effectively to handle this synchronization slip since it can use different clock pulse lengths to generate the desired patterns. Moreover, the underline circuit (flying-adder) can carry out this pulse-length adjustment very quickly.

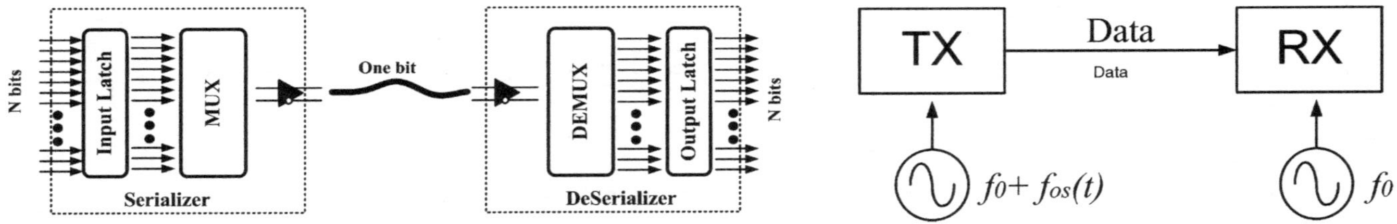

Fig. 6.22. SeDes in wired communication (left); frequency offset between transmitter and receiver (right).

Referring to Fig. 6.22, assume that the TX clock frequency is f_t (=1/P_t). Also assume that, on the RX side, there is a group of discrete frequencies* that can be generated from a TAF-based DCO (FADCO): f_{r1}(=1/P_{r1}), f_{r2}, f_{r3}, . . . , f_{rn}. Then, it is theoretically possible to use this group of RX frequencies to track the TX frequency as expressed in Eq. 6.12.

$$P_t = a_1 * P_{r1} + a_2 * P_{r2} + a_3 * P_{r3} + \cdots + a_n * P_{rn} \tag{6.12}$$

This frequency matching is achieved through many cycles, where a_1, a_2, a_3, . . . , a_n are the probabilities of respective frequencies' occurrence, and $a_1 + a_2 + a_3 + \cdots + a_n = 1$. It is interesting to point out that the conventional frequency-matching approach in CDR is, in principle, using the same mechanism as stated in Eq. 6.12. The only difference is that the n is infinite, or a very large number.

The motivation of using time-average-frequency for CDR is the potential cost saving. In theory, we can use only two or three types of cycles on the RX side to match the frequency of the TX. The real-time dynamic characteristic of the transmitted data determines the particular type of cycle needed in the receiver's clock. This mechanism is feasible since most modern datalink standards have very stringent constraints on the transmitter's frequency accuracy. It makes the TAF-based receiver clock circuitry easy to be constructed. The spirit behind the TAF-based CDR is "dynamic balancing," or achieving the balance by constantly moving/dancing around a central target using two or three discrete steps. The SKP code used in USB 3.0 is for adjusting the rate mismatch between the TX and RX. Its addition and removal are performed on the data side. The operation of adding or removing one extra Δ (switching from one frequency to another) in TAF-based frequency matching functions similarly, but it is done in the clock side.

The benefits can also be investigated from the circuit perspective. Modern datalink CDR designs all use a binary PD due to its speed advantage. But it requires a digital-to-analog process to convert the PD's digital output to the oscillator's analog input control. During this process, (1) it uses a lot of resources, (2) it degrades the RX clock's performance since the frequency generator is based on a noisy reference, and (3) the slow digital-to-analog process (which takes many RX clock cycles) makes it difficult to track any rapid TX phase trajectory variation. This motivates the use of TAF to build the CDR circuit. Since the DCO's (digital controlled oscillator) input control is digital, it is naturally suitable to work with the binary PD. A high-speed DCO, such as the FADCO (refer to Section 4.19), can generate high-quality discrete frequencies. To be precise, it produces clock cycles of different lengths at various times based on the current control status. The combination of these clock cycles can achieve any average frequency desired, to match the

* Here, frequency means the instantaneous frequency (the inverse of instantaneous period; refer to Section 3.2.3).

TX frequency as shown in Eq. 6.12. In this TAF-based CDR implementation, fewer resources are required. The RX clock can be designed in a high-quality manner. Furthermore, fast TX phase trajectory variation can be followed by the RX clock since the switching between the discrete frequencies can be done quickly.

Figure 6.23 shows an example of the TAF-based CDR (TAF–CDR). The left drawing is the loop structure. The loop only has three components: the binary PD, the loop control, and the FADCO. The PD is the conventional Alexander detector, which only reports three states: *clock-late*, *clock-early*, or *data-no-transition*. Based on the PD's output, the control block creates the instruction to instruct the FADCO to act accordingly. This FADCO only produces three possible frequencies f_1, f_2, and f_3. This loop will use these three frequencies to track any data rate variation (within a certain range) in the TX side. The table in the middle lists the PD outputs and their corresponding desirable actions from the DCO. The control block's state flow is depicted in the drawing on the right. At the system level, all variables in the whole loop are digital. No analog voltage is involved. To apply this TAF–CDR for a USB 3.0 application that has a data rate of 5 Gbps, an FADCO can use 16 inputs of 2.5 GHz ($f_r = 125$ MHz, $N = 20$, $f_{vco} = 2.5$ GHz, $K = 16$, $\Delta = 25$ ps [refer to Fig. 4.27]). The valid values for the control word of this DCO are FREQ = 15, 16, and 17, which results in three discrete frequencies: 2.35 GHz (425 ps), 2.5 GHz (400 ps), and 2.67 GHz (375 ps). This structure is a half-rate CDR where both the rising and falling clock edges are used to sample the data.

A TAF–CDR circuit is designed in a 55-nm process. Figure 6.24 shows the SPICE simulation result using data downloaded from a real environment as its input. In Figure 6.24a, the upper plots contain both the clock signal and the input data (the differential pair of 200-mv amplitude in the middle). As shown, the clock rate is half the data rate. The sampling clock is this clock's 90° and 270° phase delayed versions (not shown). The bottom plot is the period (frequency) trend plot. It is clear that, based on the incoming data, the clock rate is constantly jumping around its central value of 400 ps (2.5 GHz). The number below the period plot is the current FREQ settings (produced by the control block). Between the numbers and the actual periods, there are several cycles of delay. It is fixed and is caused by the latency of the loop. The plot in Figure 6.24b is the eye diagram (width of two-bits time: 400 ps). Both the data (200 ps, −0.2 v to 0.2 v) and clock (~400 ps, 0 v to 1.1 v) are displayed.

$$x = A * \sin(2\pi f_s t + J_m * \sin[2\pi f_j t]) \tag{6.13}$$

To test this circuit further, known jitter and frequency offset can be added into the data. In standard practice, the jitter effect is studied by injecting a sinusoidal wave into the phase part of a signal as shown in Eq. 6.13, where f_s is the signal rate, f_j is the jitter frequency, and J_m is the jitter amplitude.

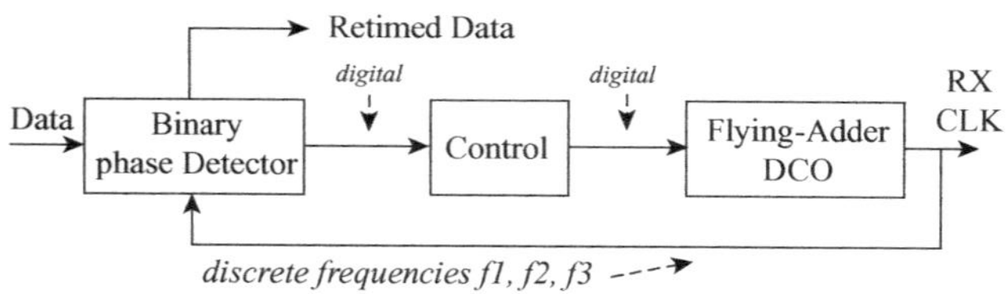

X	Y	Relationship	Action
0	0	Data no transition	Hold the Rx clk
0	1	CLK is late	Speed up Rx clk
1	0	CLK is early	Slow down Rx clk
1	1	CLK is late	Speed up Rx clk

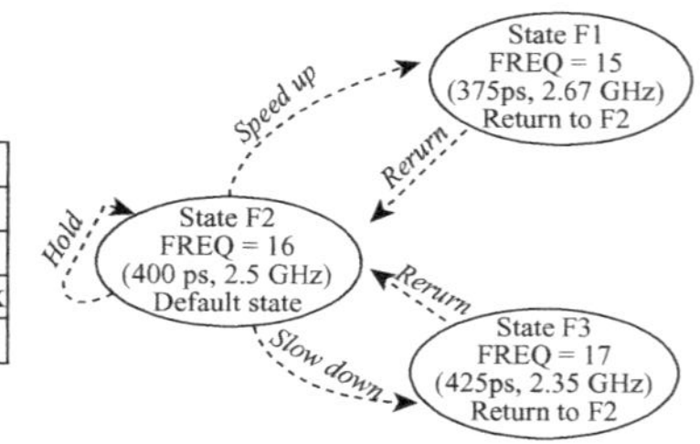

Fig. 6.23. The TAF–CDR structure (left); binary PD output (middle); control state machine (right).

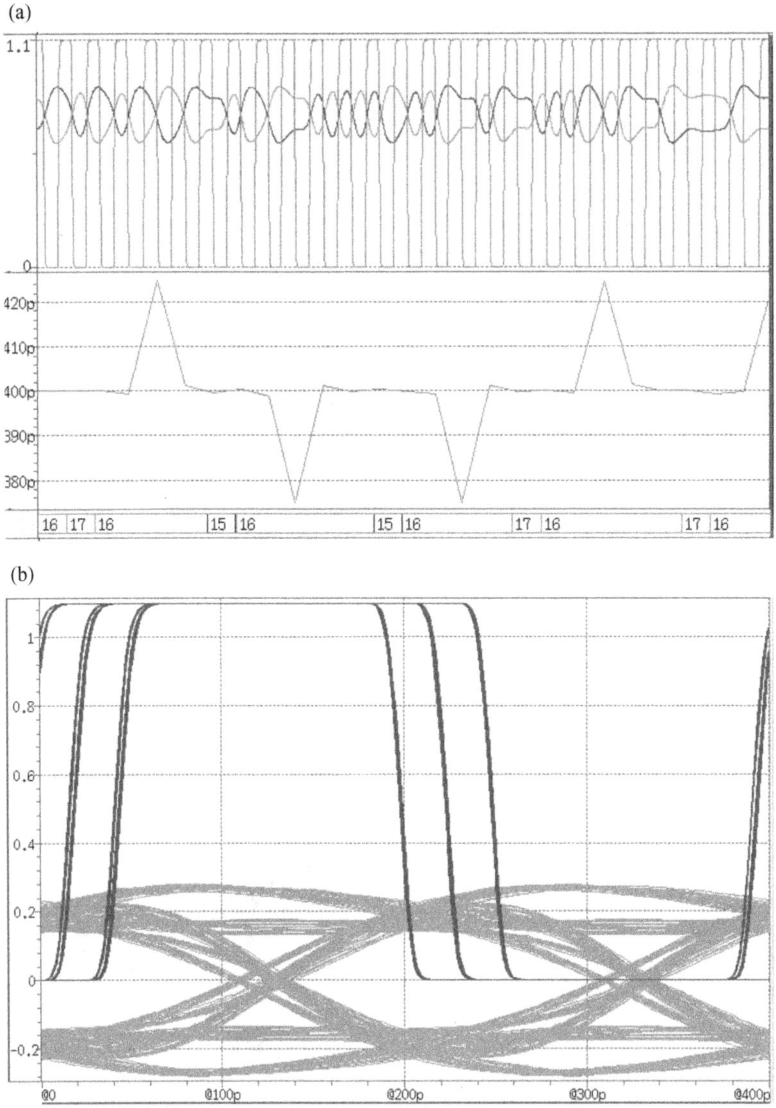

Fig. 6.24. SPICE simulation result of TAF–CDR using real transmitted data.

Figures 6.25 and 6.26 show some interesting simulation results. In these simulations, the input data is a pattern of repetitive "10." Four scenarios are simulated: case #1 of clean data; case #2 of 5 MHz (f_j) jitter with 67 ps amplitude (J_m); case #3 of 50 MHz jitter with 67 ps amplitude. and case #4 of 5 MHz jitter with 34 ps amplitude plus ~10,000 ppm frequency offset (f_s is 2.475 GHz, instead of 2.5 GHz). Figure 6.25 is the eye diagram. Figure 6.26 is the clock period time trend plot. Figure 6.27 is the period histogram. In all four cases,

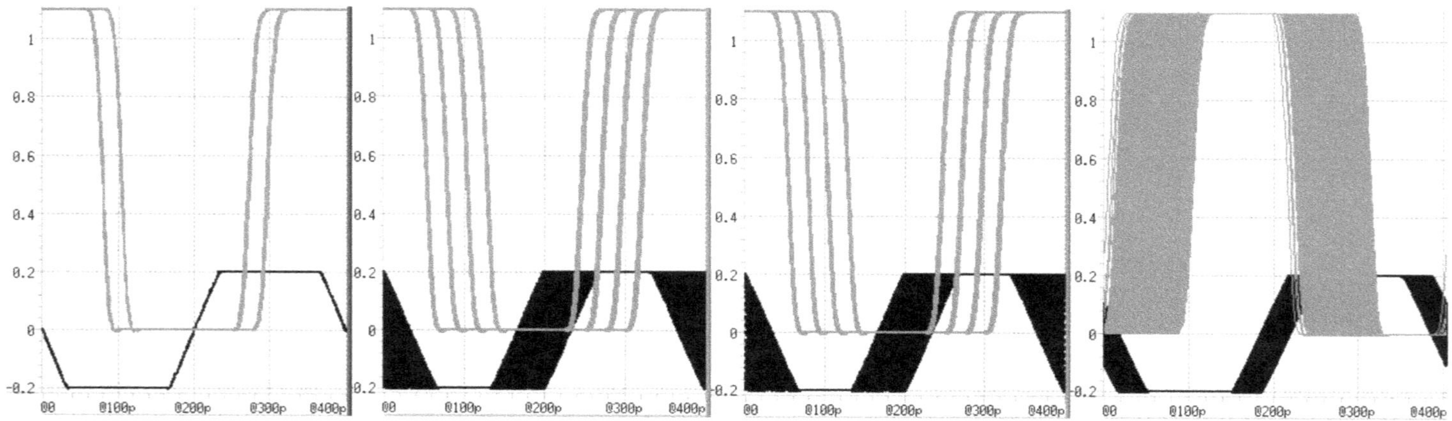

Fig. 6.25. Eye diagrams: case #1, case #2, case #3, and case #4 (from left to right).

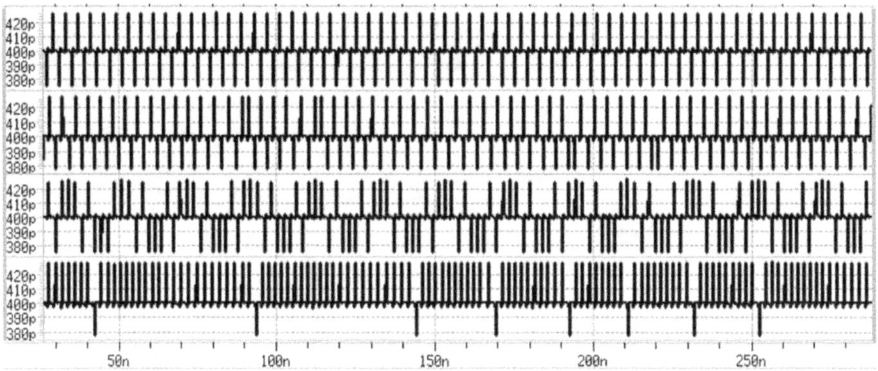

Fig. 6.26. Clock period time trend plot: case #1, case #2, case #3, and case#4 (from top to bottom).

the data have been correctly interpreted by the TAF–CDR. From these simulations, several observations on TAF–CDR characteristics are made:

(1) The TAF–CDR constantly moves its clock edge to follow the data movement (jitter). This is done in cycle-to-cycle base (Fig. 6.26).
(2) The tracking is done with only three frequencies (Fig. 6.27).
(3) The dynamic of the tracking depends on jitter frequency (the second and third plots in Fig. 6.26).
(4) The frequency offset is compensated by appropriately assigning weights to the three frequencies (the fourth plot in both Figs. 6.26 and 6.27).

The TAF–CDR's characteristics can be explored in more detail using Fig. 6.28. In the top plot, the data are clearly seen with large jitter (67 ps). If a fixed-rate clock of 2.5 GHz is used to sample these data as shown in the third plot of this figure, there is chance that data could be interpreted incorrectly if the clock edge is not in the center. This can happen when the CDR circuit's bandwidth is low compared to the jitter frequency. In other words, when jitter frequency is high enough, it is very difficult for a conventional CDR circuit to always place the clock edge in the data center. On the other hand, this is not a problem for TAF–CDR since it can move its edge very quickly. In the middle plot of Fig. 6.28, it can be seen that the clock edge moves around several positions (they are separated by one $\Delta = 25$ ps). From a circuit perspective, the TAF–CDR can be viewed as an FADLL (the left drawing in Fig. 4.94, Section 4.25). This DLL dynamically moves its phase forward or backward from looking at the data (through the PD). Most of the time, the TAF–CDR stays at its central frequency 2.5 GHz (see the plots in Figs. 6.26 and 6.27). When needed, it changes to other frequencies (2.35 or 2.67 GHz) once and immediately returns back to 2.5 GHz. In case #1, where the data are clean with

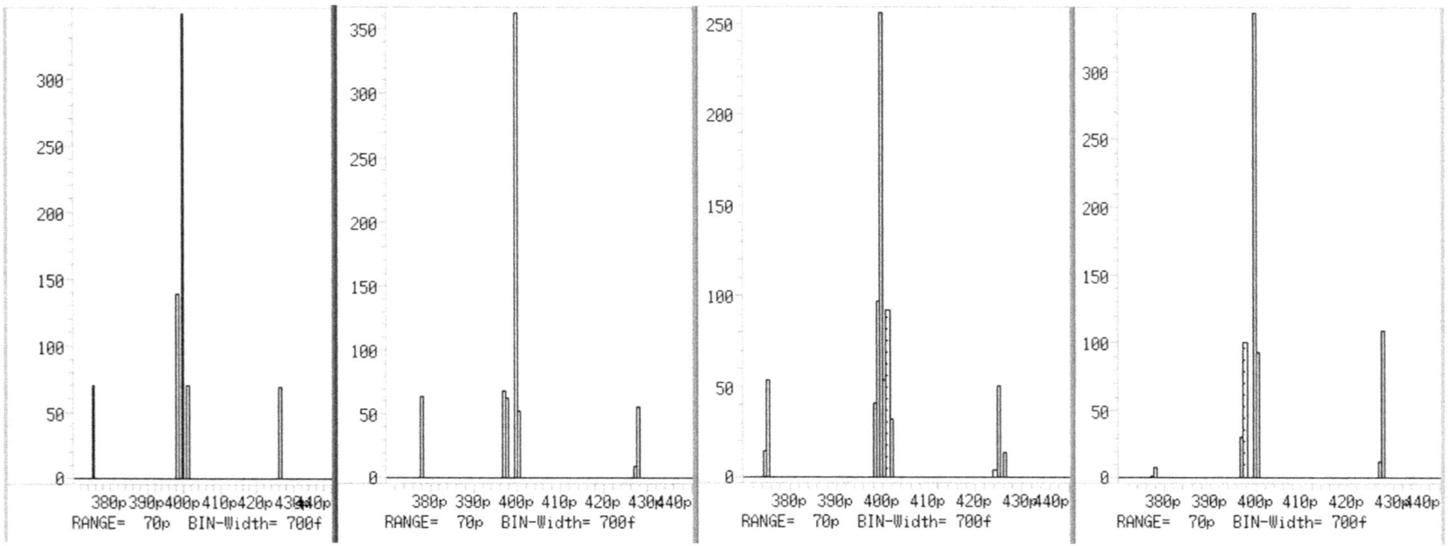

Fig. 6.27. Clock period histogram: case #1, case #2, case #3 and case #4 (from left to right).

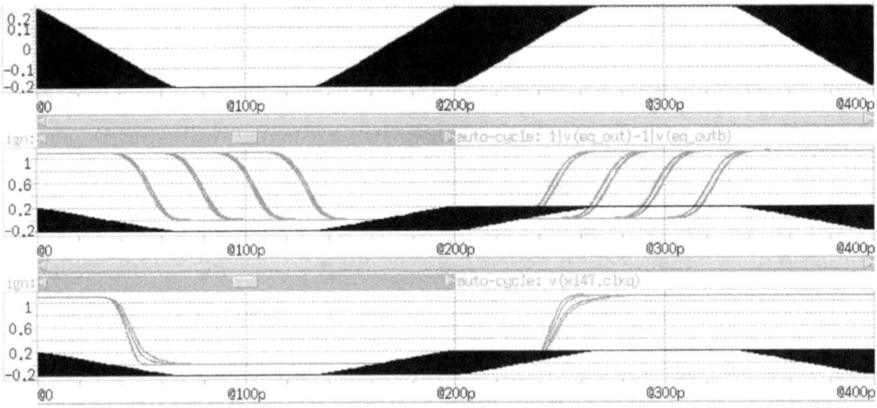

Fig. 6.28. Eye diagram of case #3 (f_j = 5 MHz, J_m = 67 ps): data (top), data and TAF–CDR clock (middle), data and a fixed 2.5-GHz clock (bottom).

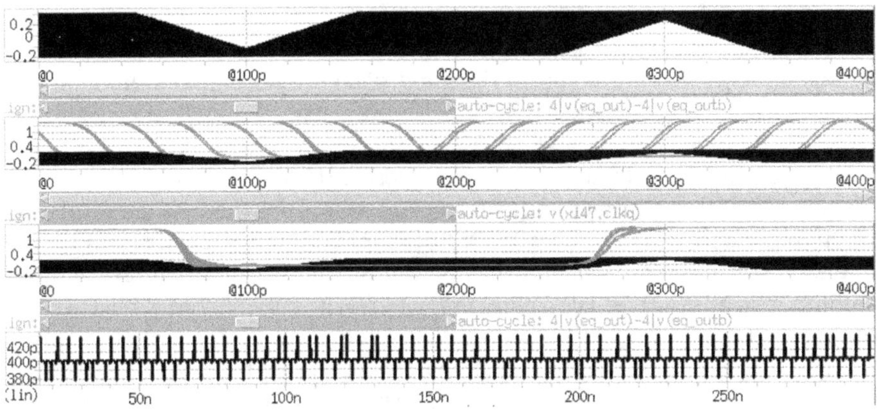

Fig. 6.29. Eye diagram of f_j = 5 MHz, J_m = 160 ps: data (first from the top), data and TAF–CDR clock (second), data and a fixed 2.5-GHz clock (third), FADLL output's period trend (fourth).

zero jitter, the movement is balanced. In other words, the number of forward and backward movements are equal (dynamic balancing even when there is no noise). As a result, there are only two possible positions in the eye diagram (Fig. 6.25, leftmost plot). In the jittery cases (cases #2–4), the FADLL can move in the same direction for more than once before it changes direction. As a result, we see more clock edge positions in the eye diagram. In general, the more steps are accumulated in one direction, the more edge positions we will see in the eye diagram. This is supported by Fig. 6.29 where the jitter amplitude is increased to 160 ps (80% of bit time). In this figure, the period time trend plot is added at the bottom to show that more consecutive movements in the same direction are needed to track this larger jitter (compare it to the second plot in Fig. 6.26).

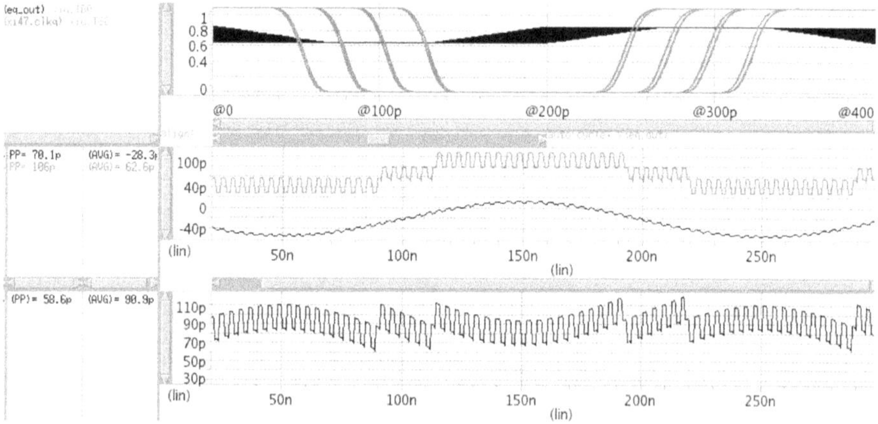

Fig. 6.30. TIE measurements on data and TAF–CDR clock of case #2.

For TAF–CDR, an eye diagram is not the right tool to measure performance since TAF–CDR has very fast dynamic characteristics that cannot be revealed by this static tool. A more appropriate approach is by using the time-interval-error concept (TIE; see Section 1.2). If we use a fixed 2.5 GHz clock as reference, the effectiveness of TAR–CDR TIE following the data TIE is a much better measurement. Figure 6.30 shows these TIE measurements of case #2. The top plot is the eye diagram (instead of a differential, the data are displayed as the single end of 0.65 v to 0.85 v). In the middle plot, the curve in black is the data TIE. As can be seen, it has period of 200 ns (5-MHz jitter) with an amplitude of about 70 ps (PP value on the left). The curve in red is the FADLL output clock's TIE. Its trend follows that of the data. It is interesting to see that this curve is made of four discrete values. This agrees with the number of edges in the eye diagram. The bottom plot is more interesting. It is also a TIE measurement of the TAF–CDR clock, but using data as the reference. This is equivalent to doing a subtraction operation between the two curves in middle plot. It measures how far the clock edge is away from the data edge. As can be seen, the distance varies with time. Its average is 91 ps, the maximum is 119 ps, and the minimum is 61 ps. Since the USB 3.0's bit-time is 200 ps, the ideal sampling point is 100 ps so that clock edge can be placed at the middle position of the data. This TIE shows that, although the data edge is moving constantly, the TAF–CDR clock is able to follow it with a reasonable sampling margin.

Figure 6.31 shows a similar measurement with a larger jitter of 160 ps (still at 5 MHz). The top plot is the eye diagram. The second from the top is the same eye diagram but displayed with a 200-ps time window. It is used to show that the jitter is so large that the eye is now very small. Both these eye diagrams show that there are eight edge positions. The third plot from the top is the TIE measurements of the data and the clock. As can be seen, the clock follows the data. There are eight discrete values in the clock TIE plot (the red curve), which corresponds to the eight edges in the eye diagram. The fourth plot is the distance between the data and the clock. As can be seen, its average is

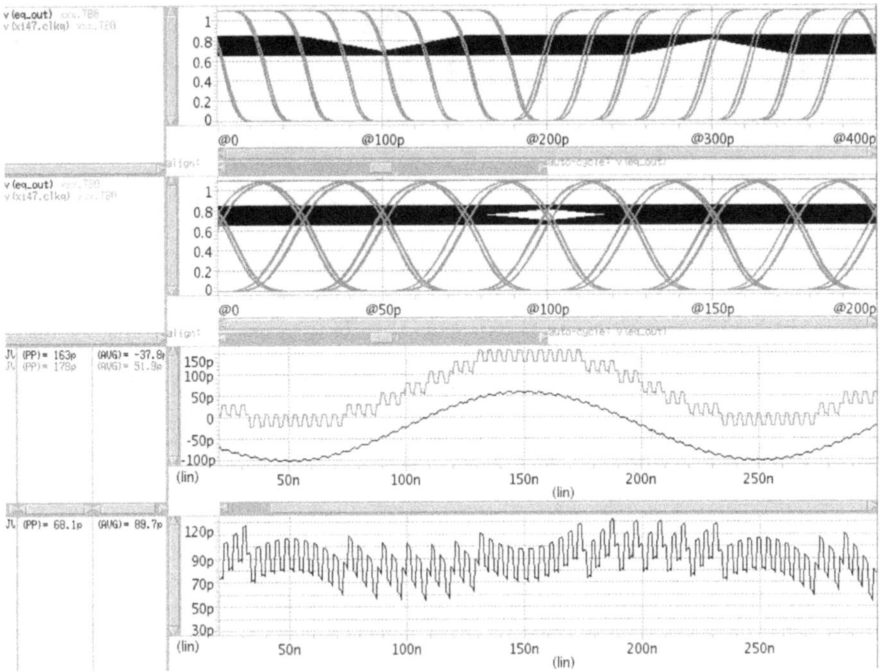

Fig. 6.31. TIE measurements on data and TAF–CDR clock of jitter 160 ps, 5 MHz.

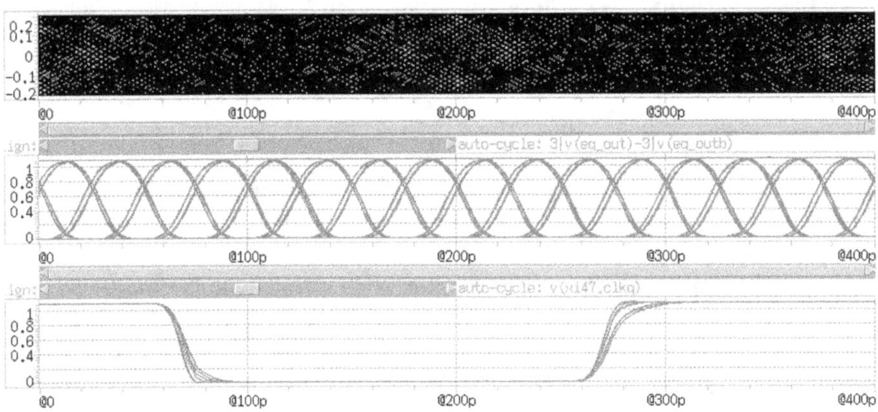

Fig. 6.32. Eye diagram of case #4: frequency offset (2.475 GHz) and jitter (5 MHz, 34 ps) in data; displayed in 400-ps time window.

90 ps, with a minimum of 60 ps and a maximum of 110 ps. This sampling margin is decent enough. Figures 6.30 and 6.31 reveal the fact that eye diagram, due to its static nature, is not appropriate for TAF–CDR, whose loop dynamic is very fast. This fast dynamic is certainly favorable for tracking jitter.

Figures 6.32 and 6.33 are simulation results from case #4 of frequency offset ~10000 ppm (2.475 GHz) and jitter 34 ps at 5 MHz. Figure 6.32 displays the

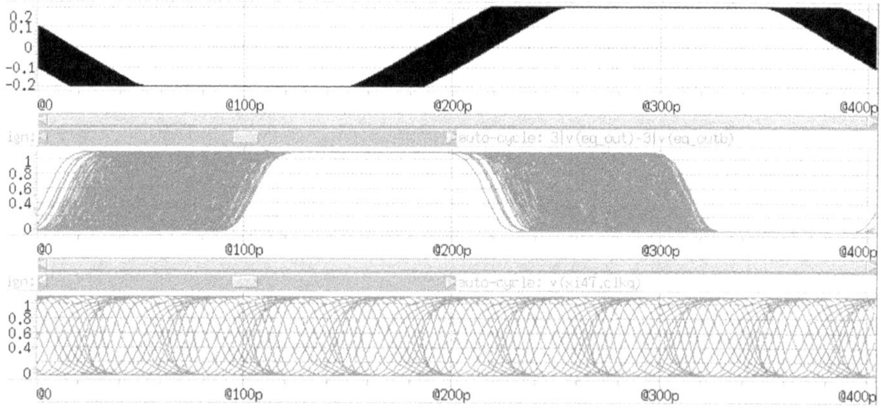

Fig. 6.33. Eye diagram of case #4: frequency offset (2.475 GHz) and jitter (5 MHz, 34 ps) in data; displayed in 404-ps time window.

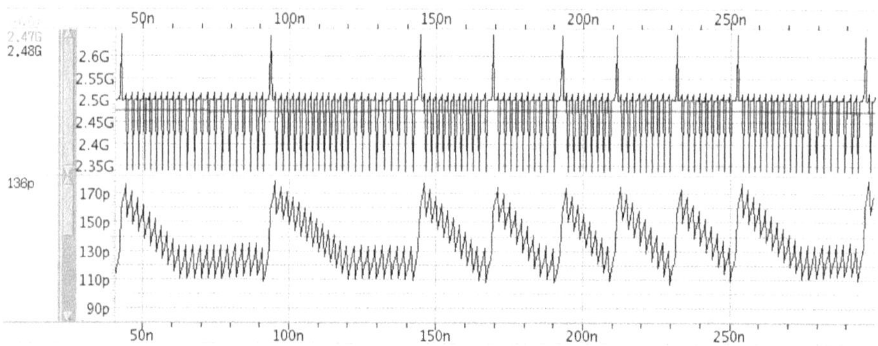

Fig. 6.34. Frequency measurement (top) and TIE measurement (bottom) for case #4.

eye using a 400-ps time window. The top plot is the data, the middle plot is the TAF–CDR clock, and the bottom one is a fixed 2.5 GHz clock. As seen, due to the 10000 ppm frequency offset, the data eye is not recognizable at all. Figure 6.33 displays the same information using a time window of 404 ps. The data eye is visible now (two-bits time displaying a repetitive "10"). It can be seen that the TAF–CDR clock edges lean heavily toward one side due to the frequency offset. Figure 6.34 reveals this fact from another angle. The top plot is the frequency trend plot. The red curve belongs to the data whose average is 2.475 GHz with a slight variation due to the 5-MHz jitter. The black curve belongs to the TAF–CDR clock. Clearly, it uses more of the lower frequency to compensate the –10000 ppm offset in data. The bottom is the TAF–CDR clock TIE with respect to the data. The average distance is 136 ps, with a maximum of 180 ps and a minimum of 110 ps. The simulation confirms that data can still be interpreted correctly at this small margin.

Figure 6.35 is the measurement data from FAPLL hardware. The configuration is K = 8 and f_{vco} = 1 GHz (Δ = 125 ps). The FREQ is configured to

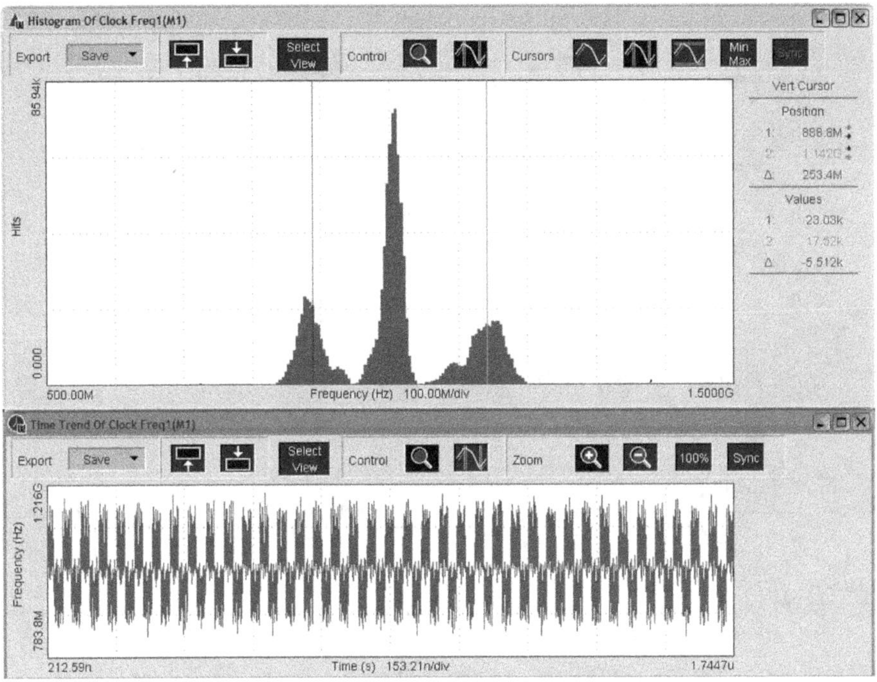

Fig. 6.35. FAPLL output of three discrete frequencies.

alternate among three values: 7, 8, and 9. It results in three discrete frequencies: 889 MHz, 1 GHz, and 1.14 GHz (refer to Eq. 4.9). The top plot is the measured histogram for frequency. The bottom plot is the frequency time trend. This real measurement data corresponds to the simulation plots in Figs. 6.26 and 6.27.

6.9 FLYING-ADDER DLL FOR DESKEW

Clock skew is an important phenomenon in clock distribution. It describes the fact that a clock signal can arrive at different cells/devices/components at different times. This term has been discussed in Section 1.2.3 within the scope of on-chip clock tree synthesis. Clock skew can also be an issue at the module, chip, and even board level. At these higher levels, skew can be caused by many different things, such as interconnect wire length difference, temperature variations, variation in intermediate devices, capacitive coupling, material imperfections, and differences in input capacitance, etc. Deskew is a technique that uses delay compensation to alleviate or eliminate the skew. In Fig. 6.36, DLL and PLL are used for this purpose. In the left-hand drawing, a delay line is inserted between the input CLKI and the distribution network. Its purpose is to add an appropriate amount of delay so that the edges of CLKO and CLKI align.

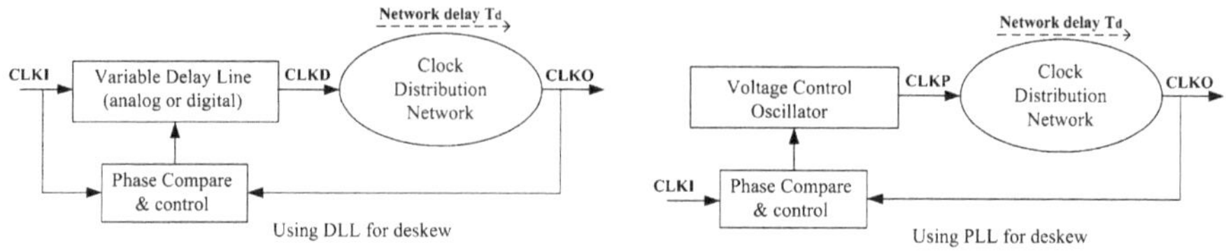

Fig. 6.36. Using a DLL and PLL for deskew.

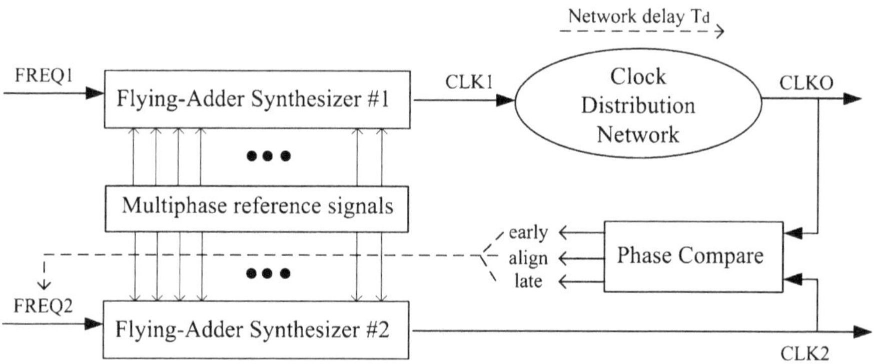

Fig. 6.37. Using a flying-adder DLL for deskew.

This effectively removes the network delay T_d associated with the distribution network. In the right-hand drawing, a PLL is used to achieve the same goal.

Flying-adder DLL can be used for deskew as shown in Fig. 6.37. In this application, two synthesizers are needed. Normally, FREQ1 = FREQ2 so that CLK1 and CLK2 have the same frequency. At the output of the clock distribution network, CLKO is compared with CLK2 by using a binary phase detector. The result will be fed to FREQ2 and direct the synthesizer #2 to move its output phase forward or backward (refer to Fig. 4.95). When the edges are aligned, FREQ2 will stay at its nominal value (=FREQ1). The advantages of FADLL-based deskew are: (1) all the frequency generation flexibility associated with a flying-adder synthesizer are available; (2) there is no risk of harmonic locking; (3) no duty cycle correction circuit is needed (as in analog DLL); (4) the delay compensation can be achieved in arbitrary amount (more than one cycle), and multiple-cycles compensation is possible; and (5) circuit overhead is small compared to a PLL or conventional DLL.

6.10 FLYING-ADDER FOR DIGITAL FREQUENCY-LOCKED LOOP (FLYING-ADDER DFLL)

A frequency-locked loop (FLL) is a circuit that can generate a clock signal using frequency conversion from a reference source of a difference frequency, as shown in the left drawing of Fig. 6.38. It is similar to PLL except that there

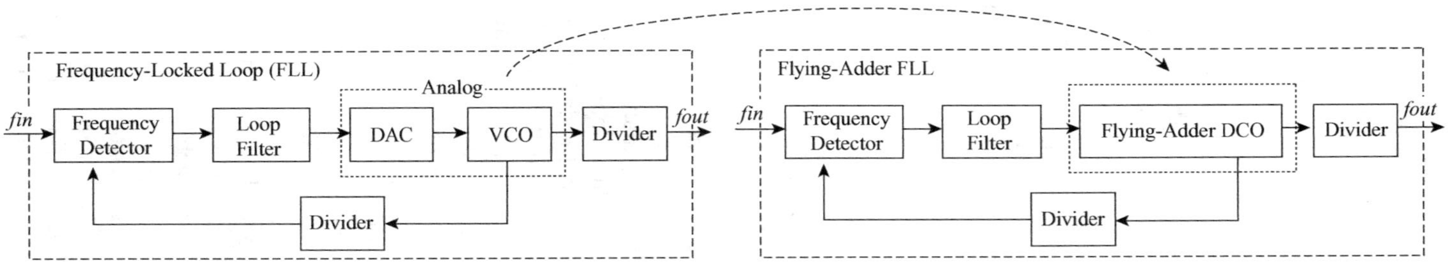

Fig. 6.38. Frequency-locked loop (left) and flying-adder digital frequency-locked loop (right).

is no known phase relationship between the input and output. When lock is reached, the output frequency is in certain multiples of the input frequency. The frequency detector is often implemented in a digital manner (e.g., using counters). Thus, its output is a digital value. Consequently, the loop filter can be digital as well. Since the oscillator is an analog voltage controlled oscillator, a DAC is needed between the filter and the oscillator. One of the FLL's advantages is programmability. Since the loop filter is in the digital domain, it can be constructed in a highly adaptive fashion: low gain for narrow bandwidth and high gain for wide bandwidth (depending on the input's quality). Further, the loop's update/sample rate can be user adjustable to fit the needs of different applications. A commercial example of FLL is available at (http://www.wolfsonmicro.com).

Since a flying-adder circuit can function as a digital oscillator (FADCO; see Section 4.19), the DAC–VCO pair in the conventional FLL can be replaced by the FADCO as illustrated in the right-hand drawing of Fig. 6.38. Unlike the case in a conventional FLL, where the signal between the DAC and the VCO is analog, all the loop variables in an FADCO-based FLL are digital (FADFLL). Further, the filter control can be implemented in software to make it a software FLL (Xiu et al. 2004). A real example of an FADFLL used in video decoders is presented below.

Video decoders are one of the critical components in the TV-related consumer electronics industry. Color video information is often transmitted and stored as a composite signal that includes a luminance component (Y), a chrominance component (C), a blanking signal, and horizontal and vertical synchronization signals. The luminance component expresses the intensity (i.e., black to white) whereas the chrominance component represents the color and its intensity. Furthermore, the chrominance component is created by modulating a subcarrier signal with two color difference components (U and V, or I and Q). The phase and amplitude of the modulated signal determines the color and its intensity. The chrominance component is then superimposed over the luminance component to generate the active portion of the composite signal. The left drawing in Fig. 6.39 shows one composite video line waveform of a color bar test signal. The composite video signal facilitates the transmission and storage of video information. However, it has to be decomposed, or decoded, into (Y, U, V) or (R, G, B) format before it can be displayed on a television set or a monitor. The right drawing in Fig. 6.39 is the simplified functional block diagram of a generic video decoder. The major tasks include sync separation, Y/C separation, and color demodulation.

For a composite signal that conforms to a standard (NTSC, PAL, or SECAM), digital decoding is usually achieved by first sampling the composite video signal with a clock that is either locked to the line frequency f_H (horizontal sync, or HS) or the subcarrier frequency f_{sc} by a PLL. The sampling frequency is selected to be $k*f_H$ ($k = 910$ for an NTSC-compliant signal) for line-locked architecture, and $m*f_{sc}$ for burst-locked architecture. The line-locked architecture can generate a fixed number of samples for each video

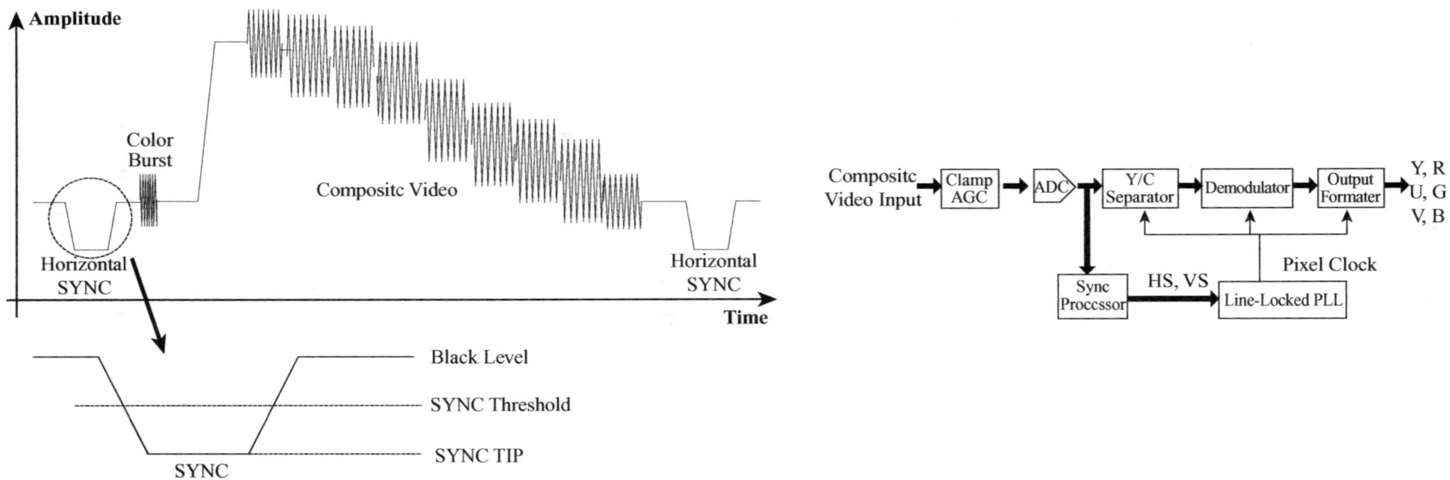

Fig. 6.39. The waveform of one composite video line (left) and the simplified block diagram of a video decoder (right).

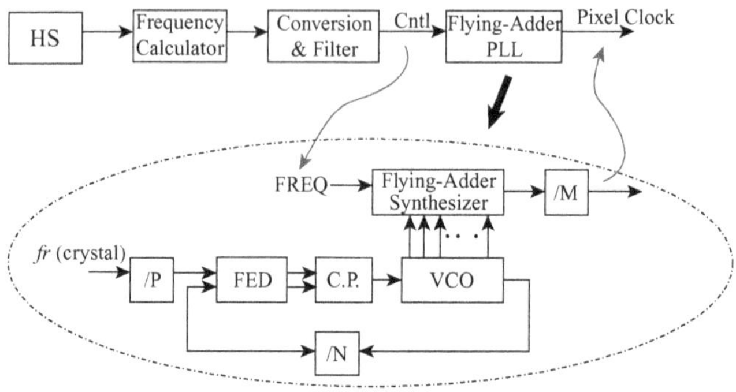

Fig. 6.40. The flying-adder-based line-locked PLL: an example of an FADFLL.

line, and it is often chosen over the burst-locked architecture. The key component of the video decoder, the line-locked PLL, can be constructed by FADFLL with some advantages. Figure 6.40 is the block diagram of flying-adder-based line-locked PLL. The digitized HS pulse (by the ADC) is first processed to find the time stamp that crosses the sync threshold (Fig. 6.39). Then, based on the ADC sampling frequency, the HS's absolute frequency f_H can be calculated. After the multiplication of $k*f_H$ and some filter functions, this information can be passed to the frequency control word of the flying-adder synthesizer, which subsequently generates the pixel clock. In this architecture, the pixel clock has a fixed multiple ratio to input HS, but no known phase relationship. Thus, this is a frequency-locked loop.

There are several advantages of using this FADFLL as a video line-locked PLL. First of all, as shown in Fig. 6.40, the pixel clock is generated from a FAPLL, which references a crystal of tens of MHz (e.g., 14.31828 MHz). Compared to the conventional analog PLL approach of referencing directly to the HS, this input is much cleaner and several orders of magnitude higher in frequency. The flying-adder synthesizer isolates itself from this noisy HS signal electrically. As a result, better jitter performance can be expected on the resulting pixel clock. Secondly, the feature of instantaneous response of FAPLL is especially helpful in tracking the line length variation. In a real video environment, the HS frequency is not a constant but varies continuously from line to line. The FADFLL-based line-locked PLL can track this variation quickly owing to its fast response. Based on the HS input, the operation associated with frequency calculation and control conversion can be carried out within one video line. The new frequency can be applied immediately to the pixel clock in the next video line. In comparison, the response time for a conventional PLL is more than several video lines, which could result in significant video quality degradation. Third, for nonstandard video signals, especially those from a VCR or TV game, the video line length could have very large deviations from the standard. This is problematical for the conventional PLL since its construction depends heavily on the divider ratio (equal to the number of pixels per line) inside the loop. This is not a problem for FAPLL since it

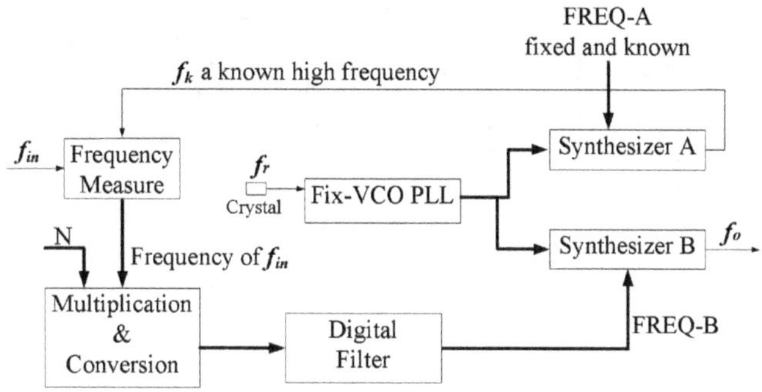

Fig. 6.41. A generic FADFLL architecture.

can synthesize frequency in very wide range. This FADFLL-based line-locked PLL architecture has been used in many commercial video decoders with great success (Xiu and Meiners 2008; http://focus.ti.com/docs/prod/folders/print/tvp5160.html).

A more general FADFLL architecture is illustrated in Fig. 6.41. As shown, a simple fixed frequency PLL is included in this system. Based on this PLL/VCO, two flying-adder frequency synthesizers can be constructed. Synthesizer A is used to produce a predefined high frequency f_k by using a fixed and known frequency control word *FREQ-A*. This known high frequency signal (e.g., in the GHz range) can be used to measure the absolute frequency of the input f_{in} (in KHz or low MHz range). This operation can be readily carried out by a counter. The result, after being multiplied by the multiplication ratio N, can be converted into the frequency control word for synthesizer B so that the output f_o and input f_{in} have the desired frequency relationship $f_o/f_{in} = N$. This scheme has the following advantages:

(1) The f_o and f_{in} are electrically isolated. The potential large noise associated with f_{in} will not be passed to f_o since it references to a clean crystal.
(2) The fixed frequency multiplier does not have to be low bandwidth since the crystal could be in high tens of MHz. And this simple PLL is a fixed frequency design with much reduced analog complexity.
(3) N could be arbitrarily large since it is not bandwidth related. It is only limited by the highest synthesizable frequency from the flying-adder synthesizer.
(4) No large R and C are required in the system as in the conventional solution for this problem.
(5) All the processing tasks (multiplication, conversion, and filtering) use digital values, and they can be easily implemented in the digital domain, or even in software.
(6) Synthesizer A can be a simplified version since it only needs to generate a known high frequency. Furthermore, it can even be completely

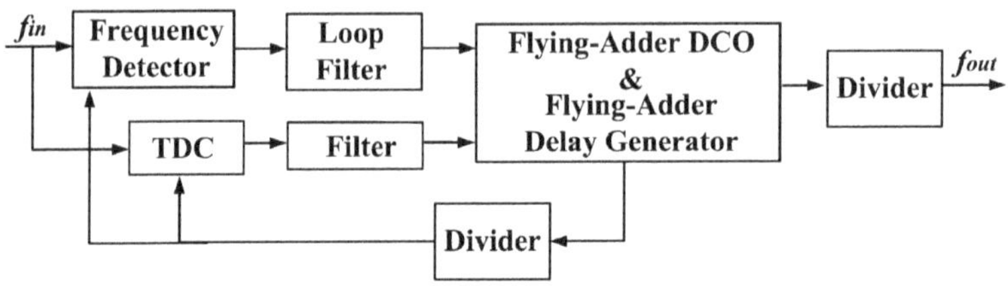

Fig. 6.42. Flying-adder digital phase-locked loop (FADPLL).

eliminated and be replaced by an output from the VCO since its sole purpose is to measure the f_{in} (through a known high frequency).

6.11 FLYING-ADDER FOR DIGITAL PHASE-LOCKED LOOP (FLYING-ADDER DPLL)

Using the flying-adder phase synthesis feature discussed in Section 4.25, the FADFLL structure in Fig. 6.38 can be further improved to the flying-adder digital phase-locked loop (FADPLL). This idea is depicted in Fig. 6.42. After the desired frequency is obtained through the FLL loop, a time-to-digital converter (TDC) is used to detect the phase difference between the input and output. Its output is used to direct the flying-adder delay generator to adjust its output phase so that the phase alignment between the input and output can be achieved. In Fig. 6.42, the FADCO and the flying-adder delay generator are the same circuit. The frequency generation and the delay generation are activated through the different usage of the frequency control word F. To make the structure even simpler for cost reduction, instead of a TDC, a binary phase detector can be used to bring the FADPLL output phase aligned with the input f_{in}'s phase. Overall, the strong digital nature of FADFLL and FADPLL make them attractive for many potential applications.

6.12 FLYING-ADDER TECHNOLOGY FOR DYNAMIC FREQUENCY SCALING

As discussed in Section 4.23, in a typical flying-adder clock waveform, there are two types of cycles used: $T_{TAF} = (1 - r)*T_A + r*T_B$. By adjusting r or T_A and T_B, we can change the T_{TAF} (frequency). This could be applied to the technique of dynamic frequency scaling that is useful in many applications (such as low power operation). Figure 4.92 illustrates the FAPLL dynamic frequency scaling characteristic. The plot at left shows the fine-tuning mode. The plot at right shows the large step change. In both cases, the frequency is instantly changed after the command is received (two cycles later). In the fine-tuned case, it takes a longer time before the new frequency can be sensed (a small frequency

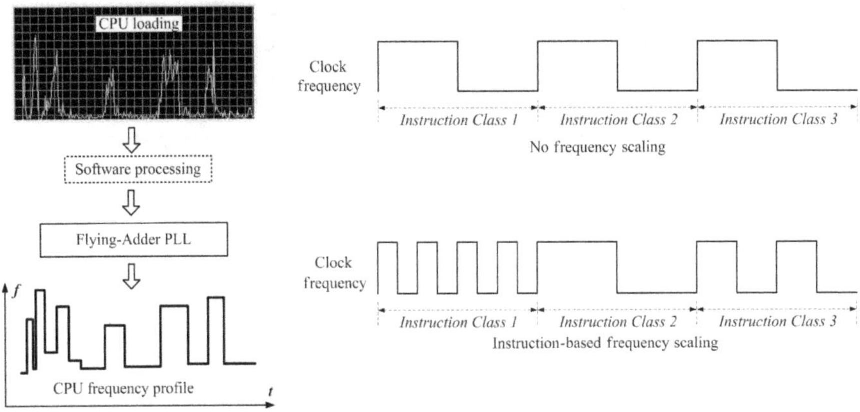

Fig. 6.43. Activities-based dynamic frequency scaling (left) and instruction-based dynamic frequency scaling (right).

difference requires a long observation time). The fine-tune mode is especially applicable to spread spectrum clock generation (SSCG). For processor chips (CPU, DSP, microcontroller), the spread spectrum function can even be achieved purely using software by programming the on-chip FAPLL in real time. For example, we can write a simple program to let the FAPLL's control word F alternate between I and $I+1$ at certain rates. As a result, its clock energy will be spread out. This is called software-controlled dynamic frequency scaling.

The large step mode could be useful for CPU dynamic frequency scaling to achieve low power operation. In principle, based on system activities, the system clock's frequency profile could be instructed to exactly follow the loading profile as illustrated in the right drawing of Fig. 6.43. Masakatsu et al. (2005) is an example of applying this principle where the activities from the CPU, the internal data bus, and embedded memory are continually monitored. The resulting information is used to control a "clock thinning circuit" to adjust the clock frequency in real time. FAPLL can be the more powerful and efficient version of this "clock thinning circuit." Furthermore, the control mechanism can potentially be realized in software.

More than a decade ago, the concept of instruction-based dynamic frequency clocking was proposed (Ranganathan et al. 1998), which allows for "on-the-fly" adjustment of processor clock speed based on current CPU instruction. This approach starts from the observation that different functional modules (adder, multiplier, etc.) have different clock requirements. Hence, they can be driven by different frequencies based on their critical path delays. For example, in low-level image processing, the "filtering" operation involves mostly *add* and *multiply*; "connected component labeling" often requires *comparison*, while "thinning" only needs a simple logic function. Therefore, the same processor can operate at different speeds while performing all these algorithms in the same application environment. This clocking scheme is illustrated in the right drawing of Fig. 6.43, where the clock control unit is responsible for varying the clock frequency based on the current instructions. In the past, the only available circuit component that could be used for performing this task was the frequency divider. As a result, the frequency step is coarse (a divisor of 2 is mostly used). Due to its flexible frequency generation capability, a flying-adder synthesizer can be an enabler for fully exploring this low power clocking scheme.

6.13 FLYING-ADDER AS 1-BIT DDFS

An NCO- (numerical controlled oscillator) based DDFS provides a flexible architecture that enables easy programmability. It has been used in many applications for on-the-fly frequency and phase synthesis, for example, as digital up-/ down-converters used in 3G wireless and software radio systems, as a digital PLL used in radar systems and in drivers for optical or acoustic transmissions, and as multilevel FSK/PSK modulators/demodulators. A standard cell based flying-adder implementation is actually a 1-bit DDFS. Its implementation cost is very low. It can potentially replace the NCO + DDFS hardware in some of the digital-based timing-recovery schemes, such as the carrier frequency recovery for a MIMO–CDMA system, or Doppler-effect compensation in a GPS receiver.

Figure 6.44 illustrates the circuit block diagram of a standard cell-based 1-bit DDFS system. The multiphases reference can be generated from a chain of k flip-flops, which are driven by a clock of frequency f_r. The flip-flops can be initiated with certain values (such 11,110,000 for $k = 8$) to form the

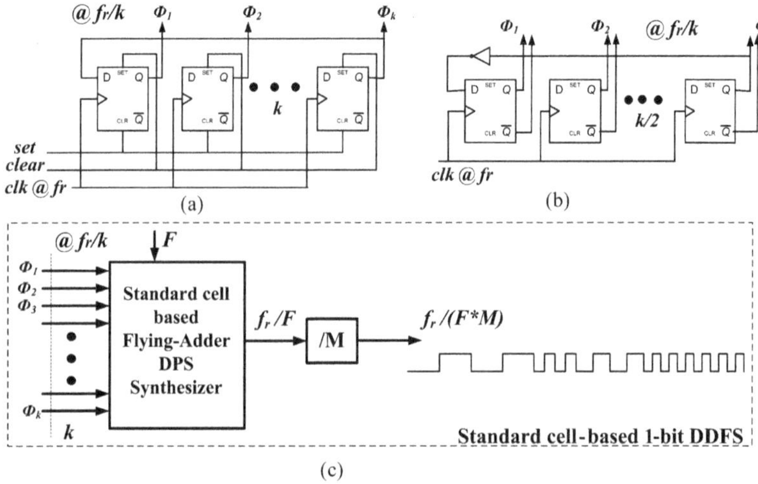

Fig. 6.44. Flying-adder DPS as standard cell-based 1-bit DDFS.

oscillation of frequency f_r/k (Fig. 6.44a). The same oscillation can also be achieved by configuration of Fig. 6.44b with half the number of flip-flops (refer to Section 4.23, Fig. 4.10). The resulting k outputs are used in the following standard cell-based flying-adder DPS synthesizer as shown in Fig. 6.44c. In this approach, the flying-adder synthesizer can also be viewed as a programmable divider since its output is f_r/F, where F can take any integer in [2, 2K]. Furthermore, it is more than just a programmable divider since fractions can be used in the control word F. It is the fraction divider described in Section 4.8. Overall, Fig. 6.44 is one example of the low-cost implementation discussed in Section 4.16. It could be a powerful yet very low cost solution to many applications. It can even be implemented entirely in FPGA.

6.14 FLYING-ADDER FOR SPREAD SPECTRUM CLOCKING

The technique of spread spectrum clock generation (SSCG) is used in electronic systems to reduce the effect of electromagnetic interference (EMI). Clock signal, because of its periodic nature, has sharply focused frequency tones in its spectrum. A perfect clock signal would have all its energy concentrated at the desired frequency and its odd harmonics, and would therefore radiate energy with very high efficiency that can exceed the regulatory limit for EMI. Spread spectrum clocking is a technique that can be used to alleviate this problem. Instead of one frequency, this technique generates a group of frequencies around a center value and therefore reshapes the system's electromagnetic emission profile. It can help make the system comply with EMI regulation as illustrated in Fig. 6.45. As demonstrated in Section 5.7, time-average-frequency naturally spreads the clock energy since it uses two frequencies (periods) to mimic one virtual average frequency (period) by assigning appropriate weights to these two frequencies. The open-loop style ensures the operation precision. One benefit of this approach is that the clock spectrum can be manipulated easily by digitally adjusting the weight and the occurrence pattern of the two frequencies in a dynamic fashion. No sophisticated VCO-related analog operation is required.

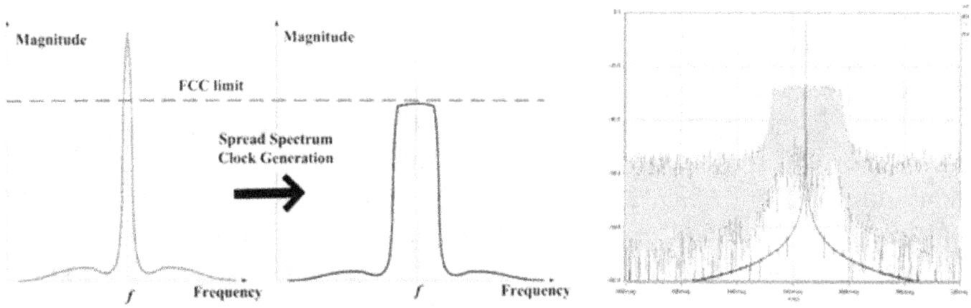

Fig. 6.45. Spread spectrum clocking for reducing EMI.

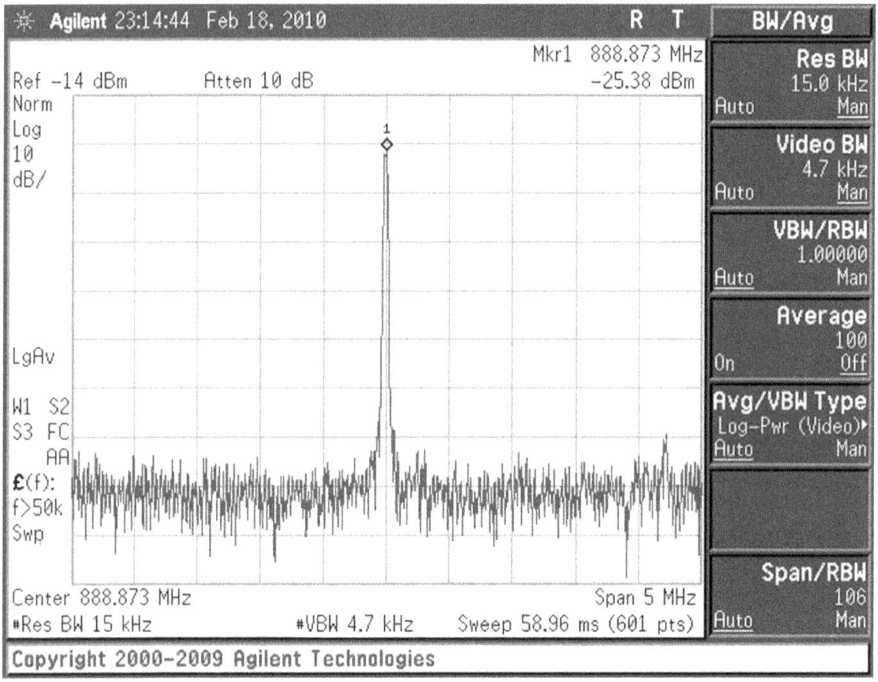

Fig. 6.46. The measured spectrum of 889 MHz from a FAPLL.

One of the principal concerns among the many issues associated with SSCG is the risk that modifying the system clock runs the danger of clock and logic circuit misalignment. In other words, with the conventional VCO-adjustment-based SSCG, the short-term jitter (period jitter) could be out of control since the VCO is a complex nonlinear component. With flying-adder technology, only two types of periods are used whether the spread spectrum feature is on or off. The PLL/VCO is fixed. When spread spectrum is needed, only the weight and the occurrence pattern of the periods need to be adjusted (You and Xiu 2007). Therefore, no additional timing risk is added.

The principle of time-average-frequency-based spread spectrum has been explained in Section 5.7 in detail. Presented below are a few cases obtained from real lab data. Accompanying them are the simulation results obtained from using the program in Appendix 5.C. Figure 6.46 is the measured spectrum of 889 MHz from a FAPLL (f_{vco} = 1 GHz, K = 8, F = 9, f_s = 888.89 MHz). Figure 6.47 is the spectrum obtained using a triangular modulation (refer to Fig. 5.25) on this frequency of 889 MHz. In this measurement, the modulation magnitude is 2^{-8} = 0.0039. The step is 2^{-18} = 0.000003815. The clock used for controlling the modulation block is 20 MHz. From this setup, the calculated frequency spread thus is (0.0039/9)*888.89*2 = 0.77 MHz (refer to Section 5.7). The plot at the left of Fig. 6.47 is the measured result, which shows about a 0.8-MHz spread. The plot in the middle is the simulation result. The two plots

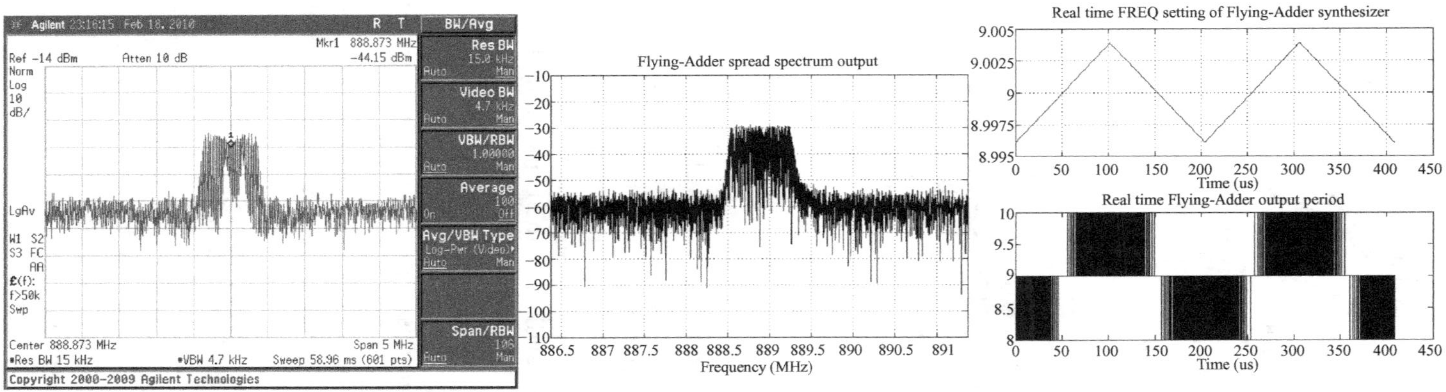

Fig. 6.47. The spectrum obtained using a triangular modulation of modulation magnitude $2^{-8} = 0.0039$ and step $2^{-18} = 0.000003815$, measured (left), simulated (center), FREQ and period trends (right).

are displayed using exact the same scales both in X and Y axes for ease of comparison. Clearly, the measured and the simulated results agree favorably with each other. The plot at right is the time trends of the FREQ and the instantaneous period (from simulation). We can see that it takes about 200 us for the modulation to finish one full cycle. FREQ varies between 8.9961 and 9.0039 linearly with time. As a result, three discrete periods are used: 8Δ, 9Δ, and 10Δ.

Figure 6.48 is the case with magnitude $2^{-10} + 2^{-12} = 0.00122$ and step $2^{-20} = 9.54\text{e-}7$. The calculated spread is 0.241 MHz. The measured result shows about the same amount of spread as expected. Figure 6.49 is the case of magnitude $2^{-12} = 0.000244$ and step $2^{-20} = 9.54\text{e-}7$. The calculated spread is about 48 KHz. The measured one agrees with it.

Figures 6.47–6.49 are three cases with different frequency spreads. The modulation strength varies greatly among the three cases. In all of them, the calculated and the simulated results are in good agreement with the measured ones. The clock energy has been effectively spread to a broader range based on the modulation strength. They clearly illustrate the fact that flying-adder-based spread spectrum can be precisely controlled; time-average-frequency is an ideal vehicle for spread spectrum.

6.15 FLYING-ADDER FOR DRIVING SAMPLING SYSTEM

As discussed in Chapter 5, a TAF-based clock concentrates most of its energy on its average frequency (period). At the same time, there is some portion of energy leaks to its fundamental frequency (period). This portion of energy shows its present as spurious tones that could be harmful to the signal sampling system (e.g., ADC) or to the signal reconstruction system (e.g., DAC) (refer to Section 1.4). Depending on the bandwidth of the signal of interest, these spurious tones can be no harm to system operation (Gui et al. 2010). When they do, the profile of these spurious tones could be reshaped by one of these two methods: (1) Convert the spurious tones to noise (refer to Section 5.5), or (2) move the spurious tones to other frequency locations (refer to Section 5.6).

Theoretically, the operation of converting the spurious tones to noise is equivalent to destroying the fundamental period (actually prolonging the fundamental period by a large extension) but keeping the average period (the wanted frequency) unaltered. This can be understood by examining the equation of $T_{TAF} = (1 - r)*T_A + r*T_B$, where T_{TAF} is the desired frequency (period). T_A and T_B are the component frequencies (periods). The weight factor is r, and it can be expressed as $r = p/q$, where the greatest common divisor (GCD) between p and q is 1. If T_{FD} is used to represent the fundamental period, it is proven that $T_{FD} = q*T_{TAF}$ (refer to Section 5.4). Therefore, the spacing between the spurs of the TAF clock is $f_{FD} = 1/T_{FD} = 1/(q*T_{TAF}) = f_{TAF}/q$.

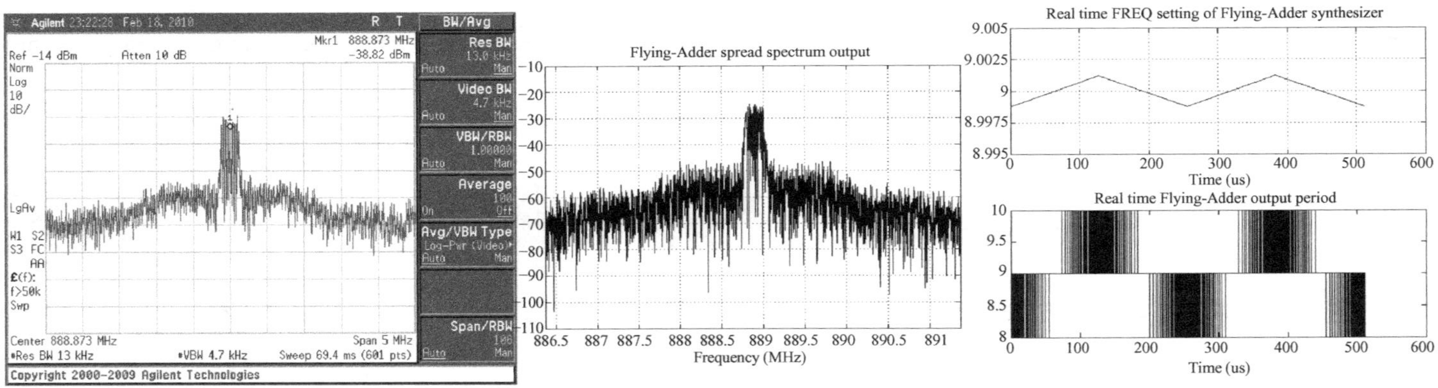

Fig. 6.48. The spectrum obtained using a triangular modulation of modulation magnitude 0.00122 and step $2^{-20} = 9.54e\text{-}7$, measured (left), simulated (center), FREQ and period trends (right).

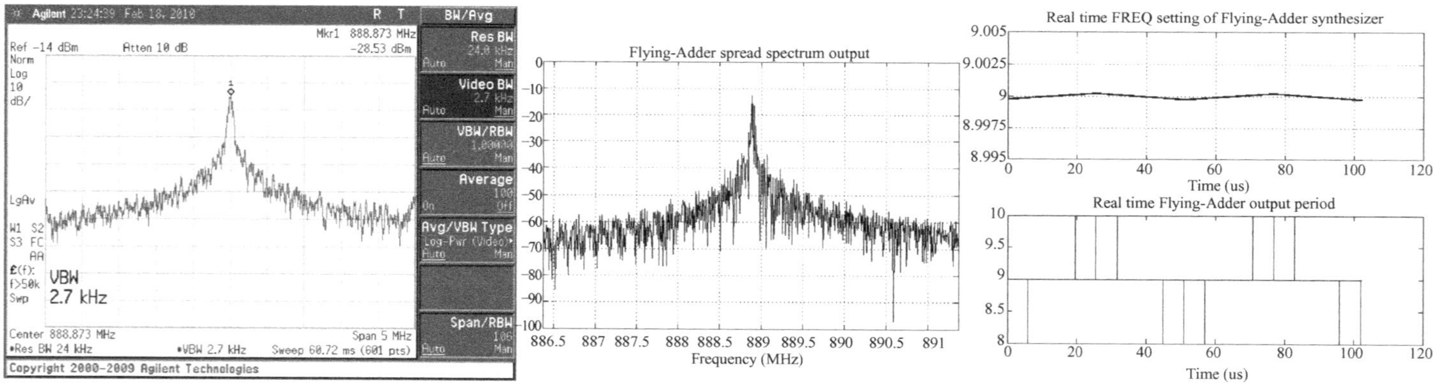

Fig. 6.49. The spectrum obtained using a triangular modulation of modulation magnitude 0.000244 and step $2^{-20} = 9.54\text{e-}7$, measured (left), simulated (center), FREQ and period trends (right).

The three parameters T_A, T_B, and r determine both the average frequency (period) and the fundamental frequency (period). Under the constraint of keeping the average frequency fixed, we can select T_A, T_B, and r to ensure that the waveform pattern (T_A and T_B occurrence pattern) does not repeat itself for long time. Furthermore, we can dynamically change the r from time to time to make the waveform take even longer time to repeat. Both of these methods enlarge the q (or the *effective* q, in the case of dynamically adjusting the r). Consequently, the T_{FD} is prolonged and f_{FD} becomes smaller. In the extreme case when q is very large, f_{FD} is so small that the spurs tones appear like noise (refer to Section 5.5 and Xiu et al. [2011]). Similarly, it is also possible to chose T_A, T_B, and r so that the fundamental period is changed to another value and, at the same time, the average period is untouched (refer to Section 5.6). This will result in the spurious tones being moved to other locations where it might have no impact on system operation or can be filtered by a filter of some sort.

Besides converting the spurs to noise and moving spurs around, there is another approach of compensating the TAF-induced timing irregularity beforehand (Runner et al. 2009). As illustrated in Fig. 6.50, a time-average-frequency precorrection filter can be inserted before the DAC to "precorrect" the input data. The filter is constructed in the form of variable delay that compensates the expected variations in the period of the time-average-frequency clock. As a result, the filtered data reflect the values that the sampled data would have if all the clock cycles had an identical duration. This is possible since the time-average-frequency clock's period variation is deterministic. It is known beforehand.

6.16 FLYING-ADDER FOR NON-UNIFORM SAMPLING

Non-uniform sampling is natural in many real-world applications. This is due to, for example, event-triggered phenomena, mismatched clocks, or imperfect sensors. In non-uniform sampling, unlike the case of the conventional sampling system, where only signal amplitude is outputted, both the signal amplitude and time stamp are delivered. Figure 6.51 shows a continuous-time signal that

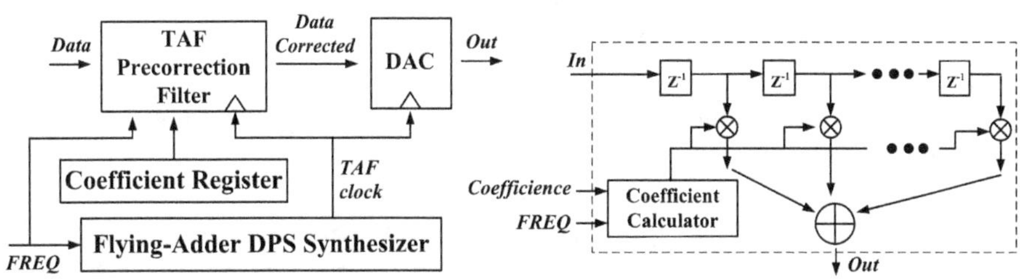

Fig. 6.50. TAF precorrection filter for compensating the delay: the system (left) and the TAF precorrection filter (right).

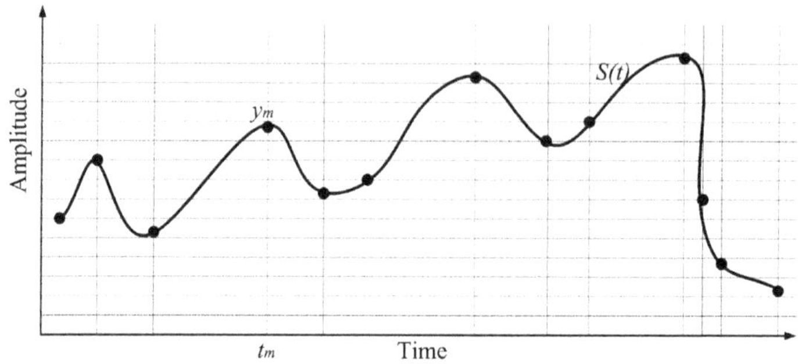

Fig. 6.51. A continuous-time signal $s(t)$ is non-uniformly sampled.

is non-uniformly sampled. The sampling is done at time t_m to get the sample value y_m from the continuous-time signal $s(t)$. In the field of non-uniform sampling, there are three major issues (Eng 2007):

- Given measurements (y_m, t_m), how do we best characterize the frequency content in the original signal $s(t)$?
- Given measurements (y_m, t_m), how do we find the best approximation of the original signal $s(t)$?
- Given a signal $s(t)$, how do we optimally place the sampling instant t_m and what is the optimality?

Among the three, the last one of optimal sampling is of importance to many applications, such as for anti-jamming in radar, for suppression of alias frequencies in frequency transform, and for placement of sensors in spatial sampling. In these cases, optimal sampling is hopefully achieved by the predeterminately chosen sampling points. Optimal sampling is useful for digital alias-free signal processing and ADC alias-free sampling. The idea here is to choose the placement of sampling points for reducing or removing aliasing, which allows a high-frequency analog signal to be sampled at a much lower sample rate and yet avoid the addition of any aliases in its digital spectra. Bland and Tarczynski (1997),; Papenfuss et al. (2003),, and Shapiro and Silverman (1960) provide some motivations for non-uniform sampling and some user guidelines for sampling point placement. An example of ADC alias-free sampling is given in Liu (2003).

TAF-based flying-adder DPS is suitable for the application of optimal sampling since the sampling points can be easily placed by adjusting the flying-adder clock period.

1. The non-uniform sampling points can be placed in an integer multiple of a base unit.
2. The non-uniform sampling points can be placed in a fraction multiple of a base unit.

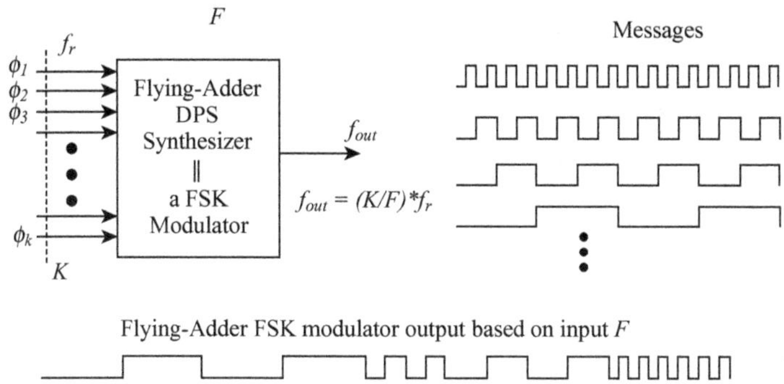

Fig. 6.52. A flying-adder DPS synthesizer as a digital FSK modulator.

3. The non-uniform sampling points can be placed with any desirable pattern (such as random) since the control of flying-adder DPS is in digital domain.
4. The non-uniform sampling points can be predetermined. They can also be resolved dynamically in real time based on the sampled data. This is possible since the flying-adder DPS can adjust its output within two cycles.

6.17 FLYING-ADDER AS DIGITAL FSK MODULATOR

Frequency-shift keying (FSK) is a frequency modulation scheme in which digital information is transmitted through the discrete frequency changes of a carrier. Since flying-adder architecture is an open-loop structure, it can change the pulse length (instantaneous period) instantly at any desirable moment. This technology is a natural fit for FSK applications. Figure 6.52 depicts the idea of using a flying-adder DPS as s digital FSK modulator. There are K inputs (Φ_1, Φ_2, . . . , Φ_K) of frequency f_r fed to the flying-adder DPS synthesizer. As discussed in Section 4.8, F is the frequency (period) control word. The valid range for F is [2, 2K]. For any given F, the output frequency is given by $f_{out} = (K/F)*f_r$.

From a communications point of view, F can be viewed as a message since there is a one-to-one relationship between the F and the output waveform. In this application, the flying-adder output is the FSK modulated signal. As shown in Fig. 6.52, there are multiple distinguished messages that can be coded into this modulator. This modulator can be implemented either in pure digital fashion by using standard cells for low cost, or in mixed-signal fashion for higher performance. Compared to the technique presented in Nemer (1998), this flying-adder FSK modulator is much more powerful and efficient in terms of throughout and resources requested.

We can go one step further to use time-average-frequency for constructing the FSK modulator as shown in Fig. 6.53. In this scenario, a fraction a is used

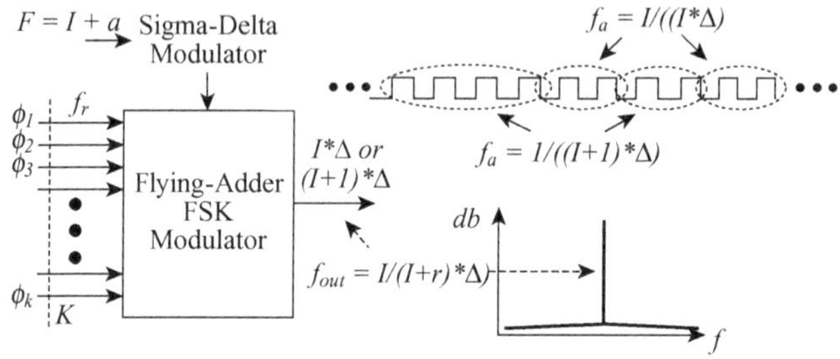

Fig. 6.53. A time-average-frequency-based flying-adder FSK modulator.

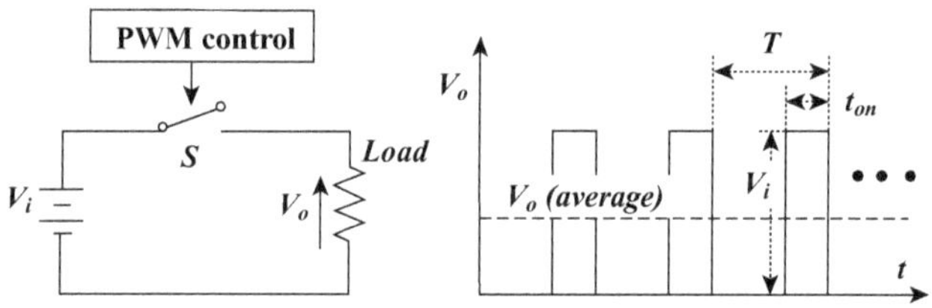

Fig. 6.54. The principle idea of a switching DC-DC converter.

in the control word F. Now, the output is composed by either the cycle of $I*\Delta$ or the cycle of $(I + 1)*\Delta$. The average output frequency is $f_{out} = 1/([I + a]*\Delta)$. A message can be coded into a. The sigma-delta modulator is used to randomize the input for reducing or eliminating the spurs. A similar idea has been explored in Liu and Lin (2009).

6.18 FLYING-ADDER FOR PWM/PFW DC-DC POWER CONVERSION

A DC-DC converter is an electronic device that is used for transforming DC electrical power from one voltage level to another. One important type of DC-DC converter is switching power supply. Its principle idea is illustrated in Fig. 6.54, where the power transfer from source to load is achieved by controlling the average voltage applied to the load. This can be done by opening and closing a switch in a rapid fashion. Pulse width modulation (PWM) is commonly used to control the switch. The average voltage seen by the load is $V_o(avg) = (t_{on}/T)*V_i$. Compared to linear power supply, this kind of switching supply has much higher power efficiency.

In the switching DC-DC converter circuit, one of the key components is the PWM modulator. Thanks to its capability of generating arbitrary period and adjustable duty cycle, the flying-adder DPS synthesizer is a natural fit for this application. This idea is shown in the top drawing of Fig. 6.55. For most of the DC-DC converter applications, the flying-adder PWM/PFM (pulse frequency modulation) modulator can be implemented in an all-digital fashion (100% standard cell based) at a very low cost. Furthermore, for modern large

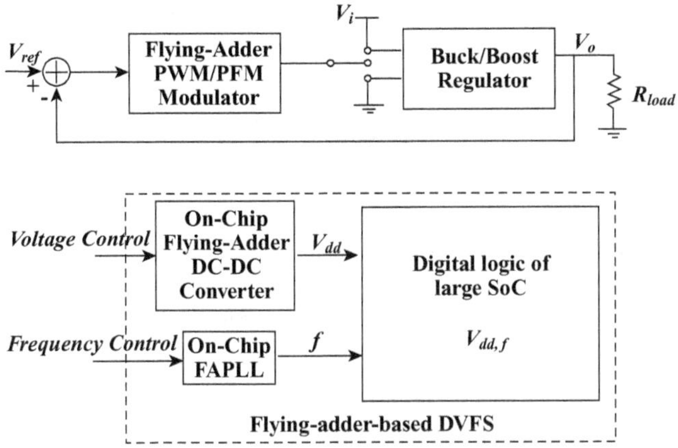

Fig. 6.55. The flying-adder-based DC-DC switching converter (top) and the flying-adder-based DVFS for low power operation (bottom).

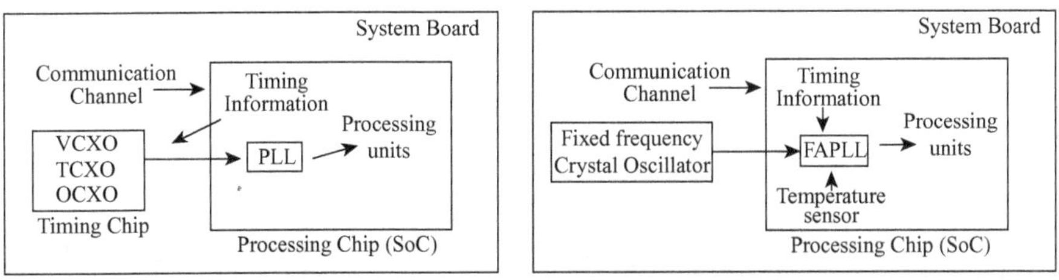

Fig. 6.56. Using an FAPLL to integrate timing chips into a processing chip: VCXO/TXCO/OCXO chip on board (left), FAPLL on chip (right).

SoC designs, FAPLL and flying-adder-based DC-DC converters can be used together to create a powerful system for implementing the dynamic voltage frequency scaling (DVFS) methodology for low power operation as shown in the bottom drawing of Fig. 6.55.

6.19 INTEGRATE CLOCKING CHIPS INTO PROCESSING CHIPS

In modern electronic system design, information (data) is often moved between chips through various communication channels. Inside the chips, data are processed under the control of local clocks. As discussed in Section 6.1.3, there are two methods to coordinate the clocks in involved parties: clock data recovery and timing extraction. In timing extraction, a VCXO is often used, as shown in the example of Section 6.5. In certain applications, TCXO (temperature-controlled VCXO) or even OCXO (oven-controlled VCXO) chips are required to maintain high temperature stability for frequency accuracy. This scenario is depicted in the left drawing of Fig. 6.56. Those TCXO/OCXO chips are significant costs in the BOM (bill of material) of the final system. Sometimes, their costs can be comparable to the main processing chip. One of the potential cost-saving solutions is to integrate those timing chips into the main processing chip.

As illustrated in the right drawing of Fig. 6.56, instead of TCXO/OCXO, a much cheaper fixed frequency crystal oscillator is used to provide the reference to the on-chip clock circuitry. Any communication-related frequency deviation or temperature-induced frequency instability can be compensated by the on-chip FAPLL. This solution is practical since the FAPLL can produce a very fine frequency step and can do it rapidly. Moreover, the processing units are very likely made of digital circuits that can accept time-average-frequency readily.

BIBLIOGRAPHY

"AM389x Sitara ARM Microprocessors Technical Reference Manual," http://focus.ti.com/lit/ug/sprugx7/sprugx7.pdf, Texas Instruments Inc., 2011.

Balan, V. and T. Pan. 2002. "A Crystal Oscillator with Automatic Amplitude Control and digitally Controlled Pulling Range of ±100 ppm." *Proc. IEEE Int. Symp. Circuit and System*, vol. 5, Scottsdale, AZ, May, pp. 461–464. A 200 uA, 78 MHz CMOS Crystal-Oscillator Digitally Trimmable to 0.3 ppm.

Bland, D. M. and A. Tarczynski. 1997. "Optimum nonuniform sampling sequence for alias frequency suppression," ISCAS'97, pp. 2693–2697.

Eng, F. 2007 "Non-Uniform Sampling in Statistical Signal Processing," Ph.D dissertation, Linkopings University.

Gui, P., C. W. Huang, and L. Xiu. 2010. "The Effect of Flying-Adder Clock on Digital-to-Analog Converter," *IEEE Trans. Circuit Syst. II*, vol. 57, no. 1, pp. 1–5.

Horowitz, M., C. K. K. Yang, and S. Sidiropoulous. 1998. "High-Speed Electrical Signaling: Overview and Limitation," *IEEE Micro.*, vol. 18, no. 1, pp. 12–14.

Huang, Q. and P. Basedau. 1996. "A 200 μA, 78 MHz CMOS crystal-oscillator digitally trimmable to 0.3 ppm," *Proc. Int. Symp. Low Power Electronic and Design*, Monterey, CA, Aug, pp. 305–308.

Lee, T. H. and J. F. Bulzacchelli. 1992. "A 155 MHz Clock Recovery Delay and Phase-Locked Loop," ISSCC92, pp. 160–161.

Lee, Y., Y. Sasaki, H. Otsuka, and Y. Sekine. 1996. "A Wide Variable Range VCXO for IC," *Proc. 50th IEEE International Frequency Control Symposium*, pp. 722–727.

Lin, J. 2005. "A Low-Phase-Noise 0.004 ppm/Step DCXO With Guaranteed Monotonicity in the 90 nm CMOS process," *JSSC*, vol. 40, no. 12, pp. 2726–2734.

Liu, H. Q. 2003. "ADS82x ADC with non-uniform sampling clock," Texas Instrument Analog Applications Journal, 4Q, http://www.ti.com/sc/analogapps.

Liu, Y. H. and T. H. Lin. 2009. "A Wideband PLL-based G/FSK Transmitter in 0.18 um CMOS," *IEEE J. Solid-State Circuits*, vol. 44, pp. 2452–2462.

Logan, S. M., D. M. Embree, R. E. Sheehey, and D. S. Stevens. 1998. "A Wide Frequency Range Surface Mountable Voltage Controllable Crystal Oscillator Family," *Proc. 42nd Annual Frequency Control Symposium*, pp. 276–278.

Nakai, M., S. Akui, K. Seno, T. Meguro, T. Seki, T. Kondo, A. Hashiguchi, H. Kawahara, K. Kumano, and M. Shimura. 2005. "Dynamic Voltage and Frequency Management for a Low-Power Embedded Microprocessor," *Solid-State Circuits IEEE J.*, vol. 40, pp. 28–35.

Messerschmitt, D. G. 1990. "Synchronization in Digital System Design," *IEEE J. Sel. Areas Commun.*, vol. 8, no. 8, pp. 1404–1409.

Mujica, F. A., U. Dasgupta, and M. Ali. 2003. "Digital Timing Recovery for Communication Systems," *GLOBECOM*, vol. 4, pp. 2130–2135.

Nemer, J. C. 1998. "Digital FSK *Modulator*," US patent US5712878, Jan..

Papenfuss, F., Y. Artyukh, E. Boole, and D. Timmermann. 2003. "Optimal sampling functions in nonuniform sampling driver designs to overcome the Nyquist limit," ICASSP'03, vol. 6, pp. 257–260.

Qiuting, H., W. C. Sansen, M. J. Steyaert, and P. M. Peteghem. 1988. "Design and Implementation of a CMOS VCXO for FM Stereo Decoder," *JSSC*, vol. 23, no. 3, pp. 784–793.

Ranganathan, N., N. Vijaykrishnan, and N. Bhavanishankar. 1998. "A Linear Array Processor with Dynamic Frequency Clocking for Image Processing Applications," *IEEE Trans. Circuits Syst. Video Tech.*, vol. 8, pp. 435–445.

Razavi, B. 2002. "Challenges in the Design of High-Speed Clock and Data Clock Recovery Circuits," IEEE Communication Magazine, Aug..

Runner, K., W. Walter, and L. Xiu. 2009. "Jitter Precorrection Filter in Time-Average-Frequency Clocked Systems," US patent pending, 12/628339.

Shapiro, H. and R. A. Silverman. 1960. "Alias-free sampling of random noise," *J. Soc. Ind. Appl. Math.*, vol. 8, no. 2, pp. 225–248.

"TMS320C6A8x Integra DSP+ARM Processors Technical Reference Manual," http://focus.ti.com/lit/ug/sprugx9/sprugx9.pdf, Texas Instruments Inc., 2011.

"TMS320DM816x DaVinci Digital Media Processors Technical Reference Manual," http://focus.ti.com/lit/ug/sprugx8/sprugx8.pdf, Texas Instruments Inc., 2011.

"TUSB3200A USB Streaming Controller," Data Manual, Texas Instruments inc., http://focus.ti.com/lit/ds/symlink/tusb3200a.pdf, 2010.

"TVP5160 NTSC/PAL/SECAM/Component 2x10-bit Digital Video Decoder (Rev. E)," Texas Instrument Inc, 2011, http://focus.ti.com/docs/prod/folders/print/tvp5160.html.

"TVP9900 VSB/QAM Receiver Data Manual," Texas Instruments Inc., http://focus.ti.com/lit/ds/slea064a/slea064a.pdf, 2007.

Watanabe, M., M. Umeki, and M. Okazaki. 2006. "High Performance VCXO with 622.08 MHz Fundamental Quartz Crystal Resonator," International Frequency Control Symposium and Exposition, IEEE, pp. 54–57.

"Wolfson Frequency Looked Loop (FLL)," Wolfson microelectronics, WAN_0209, http://www.wolfsonmicro.com.

Xiu, L. 2007. "A 'Flying-Adder' On-chip Frequency Generator for Complex SoC," *IEEE Trans. Circuit Syst. II*, vol. 54, pp. 1067–1071.

Xiu, L. 2008. "A Novel DCXO Module for Clock Synchronization in MPEG2 Transport System," *IEEE Trans. Circuit Syst. I*, vol. 55, pp. 2226–2237.

Xiu, L. and J. Meiners. 2008. "Flying-Adder Frequency Synthesizer-Based Digital-Controlled Oscillator and Video Decoder Including the Same," Patent US7356107, April.

Xiu, L., W. Li, J. Meiners, and R. Padakanti. 2004. "A Novel All Digital Phase Lock Loop with Software Adaptive Filter," *IEEE J. Solid-State Circuit*, vol. 39, no. 3, pp. 476–483.

Xiu, L., S. Clynes, S. Gurrapu, T. Haider, F. Ying, and W. Mohammed. 2007. "Flying-Adder PLL Based Synchronization Mechanism for Data Packet Transport," DCAS 2007, Dallas, TX, USA, Nov..

Xiu, L., M. Ling, and H. Jiang. 2011. "A Storage Based Carry Randomization Techniques for Spurs Reduction in Flying-Adder Frequency Synthesizer," *IEEE Trans. Circuit Syst. II*, vol. 58, no. 6, pp. 326–330.

Xiu, L., et al. 2012. "Method and Apparatus for On-Chip Voltage Controlled Oscillator Function," US patent no. 8165199.

Ying, F. and T. Haider. 2006. "MPEG 2 Transport Stream Packet Synchronizer," US patent pending, serial no. 11/558519.

You, Z. and L. Xiu. 2007. "System and Method for Generating a Spread-Spectrum Clock Signal," US patent pending, serial no. 11/960840.

CHAPTER 7

LOOKING INTO FUTURE: THE ERA OF "TIME"

7.1 THE FOUR FUNDAMENTAL TECHNOLOGIES IN MODERN CHIP DESIGN

In modern VLSI circuit design, there are four fundamental technologies that support the whole chip design superstructure: (1) processor technology, (2) memory technology, (3) analog technology, and (4) clock technology, as illustrated in the left drawing of Fig. 7.1. Processor technology focuses on the task of computation. Within a system, the processor performs the basic arithmetical and logical operations. From an implementation perspective, processor technology is a pure digital technology. Information is represented through two levels of 1 and 0. Computation is carried out in binary fashion. Analog technology deals with systems of time-continuous multilevel signals, in contrast to the time-discrete two levels of digital signals. The term *analog* describes the proportional relationship between a signal and a voltage (or current) that represents the signal. An analog circuit is needed in VLSI chips because the information processed by the chip has to eventually interface with humans. Our five human senses—sight, hearing, taste, smell, and touch—are based on proportional relationships. Memory technology specializes in storing and retrieving information at a fast pace and in large amounts. In memory systems, the target (the information) is in binary style, but the circuit implementation requires some degree of analog processing.

Nanometer Frequency Synthesis Beyond the Phase-Locked Loop, First Edition. Liming Xiu.

Published 2012 by John Wiley & Sons, Inc.

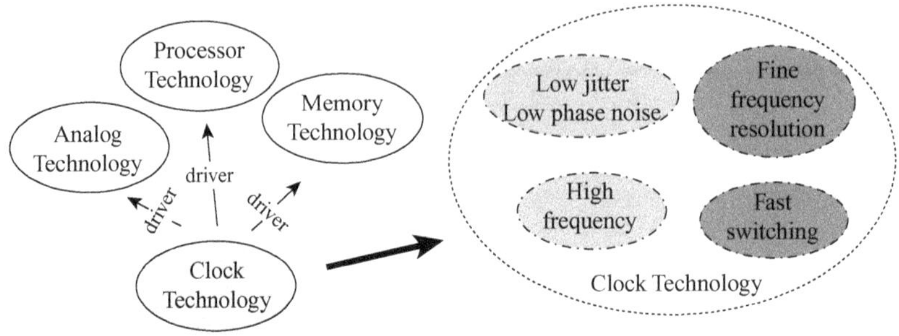

Fig. 7.1. The four fundamental technologies in modern chip design (left) and the four important issues in clock technology (right).

These three technologies directly deal with information. To be functional, however, they need a crucial signal called "clock." Clock circuit is unique because it does not handle information directly; its role is to support the others. Clock technology focuses on how to create good and reliable clock signal. Using a human body as an analogy, the processor is the brain—it makes decisions. Memory is the memory. The five senses are the sensors, and the limbs are the actuators. The clock generator is the heart, and the clock pulse is the heart beat. The clock signal is the blood. The clock distribution network is the vessel. From outside the body (chip), you cannot see the blood (clock signal), but it is there. Without it, there will be no life (the chip will not move).

As depicted in the drawing on the right of Fig. 7.1, there are four key issues in clock technology: low noise, high frequency, fine frequency resolution, and fast switching. Among them, the first two (low noise and high frequency) have been studied thoroughly. Decent solutions exist. However, the last two problems (fine resolution and fast switching) have not been solved to our complete satisfaction, regardless of all the great techniques developed so far. When all applications are investigated, it is found that only in very high end applications (a small portion of all applications), that extreme low noise and extreme high frequency are needed. In contrast, fine frequency resolution and fast switching are always preferred in almost all applications. Therefore, these two remaining problems present a great opportunity for us to improve things, to invent better things.

Looking at the development history of the processor, memory, and analog technologies, and comparing this history to clock technology history, it is believed that the lack of a theoretical breakthrough is probably the very reason of that no good solutions have developed for these two problems. In the clock field, most of the theoretical work lies within the scope of (1) fine-tuning the feedback theory (for PLL), (2) the spectrum analysis of direct digital synthesis, (3) and the understanding of the noise mechanism in various devices. The characteristics of the clock signal itself are hardly investigated. The fundamental question of "what is frequency?" is never questioned. The time-average-frequency concept, in the author's belief, is the first attempt in this direction. It could provide the foundation for us to solve these two decades-long problems.

Processor technology, memory technology, and analog technology have all enjoyed tremendous advances in the past several decades. They have made today's chip sophisticated enough to compete with human brain. To leapfrog further, the next push should come from the clock side.

7.2 "TIME"-BASED ANALOG PROCESSING

Unlike the other three technologies, the clock circuit is traditionally not used for processing information since it deals with "time." Up until now, time has not been used as information in VLSI signal processing. In the interaction between human and chip, information is represented through level. Levels are natural for our body's processing systems since our five senses are built on proportional relationships. Level is also inherently compatible in electronic systems because in a circuit, the level (voltage or current) is proportional to the number of electrons flowing inside the devices. However, with the trend of reducing power consumption, the supply voltage has been forced to decrease in every process advance. As a result, the useful room left for information processing keeps shrinking. Another fact in modern CMOS digital processes is that most advanced processes are digitally oriented for the benefit of a high degree of integration. These processes mainly target three metrics for tradeoffs: area, speed, and power dissipation. Many analog-related metrics, such as gain, linearity, input/output impedance, noise, and voltage swings are not seriously addressed. This makes the voltage-level-based analog processing very challenging in this environment. Therefore, to continue to improve processing efficiency (in term of power per bit), "time"-based processing should be considered as another option.

The word *analog* (or *analogue*) has its roots from a Greek word that means "proportional." In circuit design, analog signal processing is built on proportional relationships. When "time" is used as a medium, similar to voltage level, the proportional relationship can be built on it as well. In this approach, the time is not treated as a *moment* but as a *window*. Given a window of known size, information can be created based on the number of events that occur within this window. Mathematically, a proportional relationship can be established between the information and the number of events. This idea is illustrated in Fig. 7.2. In the left drawing, traditional voltage-level-based analog processing is demonstrated. The window is the available voltage range to be used for processing signal. A particular voltage level is used to represent the actual information. In this method, two proportional relationships exist. The first one is the relative position of this voltage level within this window (the information). The second one is the one-to-one relation between this voltage level and the number of flowing electrons (a scaling factor). In the right-hand drawing, the time-based approach is depicted. A window is created with the

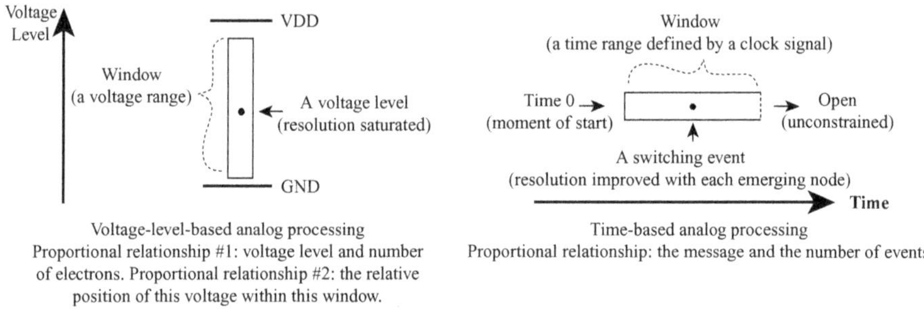

Fig. 7.2. Analog signal processing: voltage level based (left) and time based (right).

assistance of a clock signal. The number of events that occur in this window is the information. There is only one proportional relationship here: the information and the number of events. The role of the clock signal is different in these two approaches. In the voltage-level-based case, the clock signal is only used for marking the information (identifying event sequence). In the time-based approach, a clock is used both for marking the information (the clock cycle index, a number, for marking when events happen) and defining the window (based on the clock cycle's size).

When semiconductor technology moves forward, the pros and cons of these two different analog processing approaches can be analyzed from the following perspectives: the available range for processing, the processing resolution, and the processing efficiency.

- **Range.** Voltage-(level)-based approach is constrained by VDD (supply voltage), which decreases with each emerging node (for the reason of lowering power consumption). The time-based approach has virtually no up-limit.
- **Resolution.** The resolution in the voltage-based approach is saturated (a few uV) since the noise level does not go down as the geometry shrinks. The resolution in the time-based method is compatible with process advances. For each emerging node, the transistor is switching faster (nowadays, a few ps for an inverter). The resolution is continuously enhanced.
- **Efficiency.** In the voltage-based method, the proportional relationship needs to be revealed through an analog-to-digital device so that the result can be used in the following processing unit (which is in binary format). In the time-based method, the information is already represented in binary format. This is because in the voltage-based system, the clock signal functions only as an index, while in the time-based system, the clock is used both as an index and as a measuring tool (the window). Thus, efficiency is improved.

This time-based analog processing method is the "rate-of-switching" approach described in Section 5.1. In our five senses of sight, hearing, taste, smell, and touch, all of them are level (or strength, amplitude) based. Among them, at least three (hearing, touch, and sight) are capable of sensing the rate-of-switching. Therefore, time-based analog processing has ample justification to be useful.

7.3 "TIME" AND FREQUENCY: ENCODING MESSAGES THROUGH MODULATION

In communications, modulation is often used to convey a message signal (e.g., a digital bit stream or an analog audio signal) inside another signal called a carrier that is more suitable to be physically transmitted. In practice, a sine waveform of a certain frequency is used as a carrier. A baseband message signal can be transformed into a passband signal through modulating this sinusoidal wave. For example, an audio signal can be attached to a radio frequency (RF) signal for long-distance transmission. Three modulation techniques are commonly used: amplitude modulation, phase modulation, and frequency modulation. These three modulation methods can be carried out in either an analog or digital domain. In digital modulation, the aim is to transfer a digital bit stream over an analog bandpass channel. Instead of modulation, these three approaches are referred as *keying* in digital modulation: amplitude shift keying, phase shift keying, and frequency shift keying.

As the process advances, the "time" can be more readily manipulated. Since time is closely related to frequency, frequency and phase shift keying is expected to be facilitated by this trend. As a pulse waveform synthesis technology, direct period synthesis can synthesis almost arbitrary lengths and variable duty-cycle pulse waveforms in high frequency (GHz range). The resulting time-average-frequency-based waveform is very rich in frequency content. This provides us with the possibility to directly encode the message signal into the carrier signal. In this approach, the frequency control word of the direct period synthesizer can function as the message. In other words, the tasks of the "frequency shift keying" and the carrier generation are accomplished in one single step. It is expected that this approach can reduce system cost (both the transmitter and the receiver), and, at the same time, make the system more resilient to noise.

7.4 MANIPULATE "TIME": THE TOOLS

In level-based processing, the tools needed are quantization and sampling. They are used to transform the proportional-based continuous analog signal into a discrete binary signal for processing. The representative hardware is the analog-to-digital converter (ADC). In time-based processing, the principle of quantization and sampling still holds. But the implementation is different. The proportional-to-binary conversion is accomplished by a time-to-digital converter (TDC) or voltage-to-digital converter (VDC). The TDC, whose role is

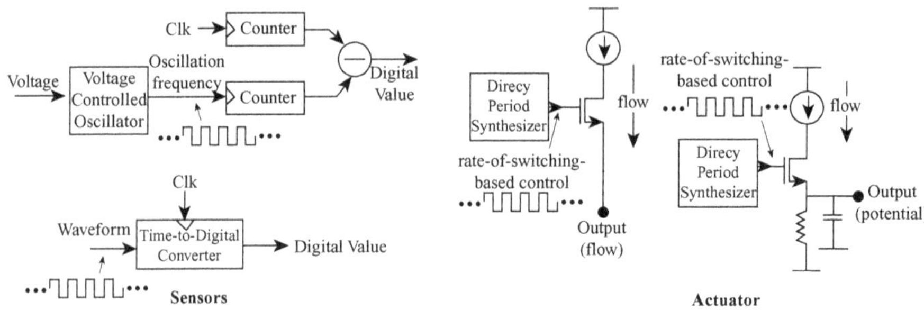

Fig. 7.3. In time-based processing, the tools for analog-to-digital conversion (left) and digital-to-analog conversion (right).

depicted in the lower left-hand drawing of Fig. 7.3, is a mature technique. The VDC can be built around a voltage controlled oscillator (VCO), which converts a voltage to an oscillation frequency (the upper left drawing of Fig. 7.3).

On the other side of the spectrum (refer to Fig. 5.3), the digital-to-analog converter (DAC) is used in level-based processing as a bridge between the binary domain and the analog world. In time-based processing, a rate-of-switching-based actuator is the corresponding counterpart. The output of this actuator can either be a *flow* or a *potential* to interact with the outside world, as depicted in the drawings on the right side of Fig. 7.3. The enabler for the rate-of-switching-based actuator is the direct period synthesizer.

7.5 IT IS TIME TO USE "TIME"*

In the field of semiconductor circuit design, level has played the dominant role since information is carried by this media. "Time" only plays supporting role. Its purpose is mainly for bookkeeping: recording what happened when and scheduling what needs to happen in the future. Today, we have reached the crucial point where "time" needs to be seriously considered as another information carrier. In this new approach, level will play the supporting role. It will only be used to differentiate the two states of "on" and "off": Is something happening? Is something changing? "Time," or rate-of-switching, answers the more important question: What happened? What shall happen in future?

7.5.1 But, Does This Make Sense?

This world is virtually made of numbers. All human activities are based on numbers. Numbers do not have a physical presence. They only exist in our spiritual world. Numbers are created to help us to reason. Humans sense the

* To author's best knowledge, this phrase is first used by R.B. Staszewski in 2010 IEEE CAS summer school.

physical world through the proportional relationships that are naturally built within our body. However, the physical world is more manageable only after we organize it through our mental product: the number. In the past, level has being effectively used to materialize the number. Just like the five basic senses, the "sense of time" is also inherently built within our body. Thus, it is reasonable for us to quantify this "sense of time" through numbers and utilize the result to convey information, beyond just for bookkeeping (sequencing).

7.5.2 And, Is It Worth It?

Time-based information processing is increasingly feasible since the transistor is switching faster. At the same time, level-based processing is continuously losing playable room. This trend will continue.

Time-based information processing is more efficient. In circuit practice, "time" is realized through a special signal called a clock. Unlike level, which has only one useful parameter, a clock cycle (time) bears two useful parameters: the *size of the cycle* and the *index of the cycle*. The size of the cycle is used to carry messages; the index can be used for sequencing. In the process of creating this special signal of clock, resources are already spent. Making additional use of this signal improves efficiency.

7.5.3 Will It Replace Level?

No. In certain applications, the time-based approach is efficient. In other applications, the level-based approach is better.

The selection between time-based or level-based processing only happens in the boundary between the human world and the electronic world (see Fig. 5.3). Once the information is inside the chip, where it is already in binary-level format, and where computation and logic operations will happen, using the binary system to process it is the most efficient way.

7.5.4 Finally, Is It Ready?

Yes.

- **The Concepts.** Time-average-frequency, rate-of-switching.
- **The Theory.** Time-average-frequency theory.
- **The Tool.** Direct period synthesis architecture, digital-to-frequency converter.

APPENDICES

APPENDIX 4.A: THE VHDL CODE FOR FLYING-ADDER SYNTHESIZER (PLEASE REFER TO FIG. 4.17)

```
======================================================================
----- This block is used to generate the inputs for
flying-adder, 8 inputs, K = 8 -----
LIBRARY ieee;
USE ieee.std_logic_1164.ALL;

ENTITY ticks IS
  PORT (TICK_OUT: OUT std_logic_vector(7 DOWNTO 0) );
END ticks;

ARCHITECTURE behavior OF ticks IS

signal tick_int: std_logic_vector(7 DOWNTO 0) :=
"00000000";
signal CLK: std_logic := '0';
```

Nanometer Frequency Synthesis Beyond the Phase-Locked Loop, First Edition. Liming Xiu.

Published 2012 by John Wiley & Sons, Inc.

```
BEGIN
    CLK <= NOT CLK after 0.3125 ns;    -- frequency =
    1.6 GHz, 0.625 ns

    TICK_GEN : FOR i IN 0 TO 7 GENERATE
        tick_int(i) <= TRANSPORT CLK after i*0.078125
        ns;
    END GENERATE TICK_GEN;

    TICK_OUT <= tick_int;
END behavior;



================================================================
----- This block is used to model the 8 → 1 MUX, MUX8 -----
LIBRARY ieee;
USE ieee.std_logic_1164.ALL;

ENTITY mux8 IS
 PORT (A:     IN std_logic_vector(7 DOWNTO 0);
       sel:   IN std_logic_vector(2 DOWNTO 0);
       Y:     OUT std_logic);
END mux8;

ARCHITECTURE behavior OF mux8 IS

SIGNAL output_int: std_logic := '0';

BEGIN
    MUX_UPDATE : PROCESS(A, sel)
    BEGIN
        IF  ( sel = "000") THEN output_int <= A(0);
        ELSIF ( sel = "001") THEN output_int <= A(1);
        ELSIF ( sel = "010") THEN output_int <= A(2);
        ELSIF ( sel = "011") THEN output_int <= A(3);
        ELSIF ( sel = "100") THEN output_int <= A(4);
        ELSIF ( sel = "101") THEN output_int <= A(5);
        ELSIF ( sel = "110") THEN output_int <= A(6);
        ELSIF ( sel = "111") THEN output_int <= A(7);
        ELSE          output_int <= 'X' ;
        END IF;
    END PROCESS;
```

```
    Y <= output_int after 0.05 ns when sel'event else
output_int after 0.02 ns;  -- the values can be
changed, based on your design.
END behavior;



==============================================================
----- This block is used to model the 2 → 1 MUX, MUX2 -----
LIBRARY ieee;
USE ieee.std_logic_1164.ALL;

ENTITY mux2 IS
 PORT (A:     IN std_logic;
       B:     IN std_logic;
       SEL:   IN std_logic;
       Y:     OUT std_logic);
END mux2;

ARCHITECTURE behavior OF mux2 IS

SIGNAL output_int: std_logic := '0';

BEGIN
    MUX_UPDATE : PROCESS(A, B, SEL)
    BEGIN
        IF ( SEL = '0') THEN output_int <= A;
        ELSIF ( SEL = '1') THEN output_int <= B;
        ELSE          output_int <= 'X' ;
        END IF;
    END PROCESS;

    Y <= output_int after 0.015 ns when sel'event else
output_int after 0.02 ns; -- the values can be
changed, based on your design.
END behavior;



==============================================================
----- This block is used to model the toggle flip-flop,
TFF -----
LIBRARY ieee;
USE ieee.std_logic_1164.ALL;
```

```
ENTITY tff IS
  PORT (CLK:      IN std_logic;
        Q:      OUT std_logic);
END tff;

ARCHITECTURE behavior OF tff IS
    SIGNAL Q_int: std_logic := '0';
BEGIN
    PROCESS (CLK)
    BEGIN
        IF ( CLK'event and CLK='1' and CLK'last_value
        = '0' ) THEN
            Q_int <= NOT Q_int after 0.03 ns; -- the
value can be changed, based on your design.
        END IF;
    END PROCESS;
    Q <= Q_int ;
END behavior;
```

```
=====================================================================
----- This block is used to model the adder in PATH_B,
3-bits -----
LIBRARY ieee;
USE ieee.std_logic_1164.ALL;
USE ieee.std_logic_arith.all;
USE IEEE.std_logic_misc.all;
USE IEEE.std_logic_unsigned.all;
```

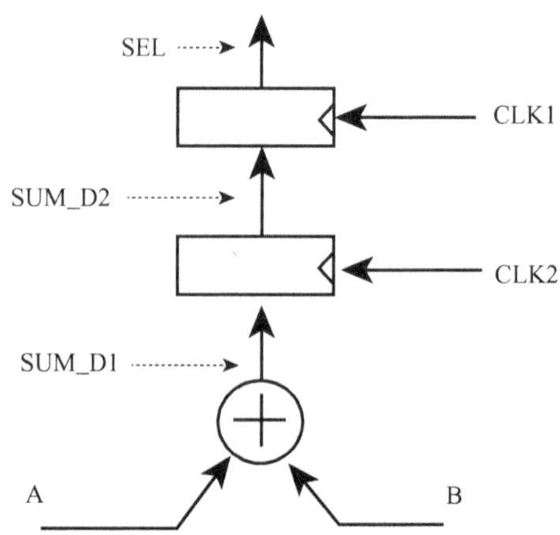


```
ENTITY adder IS
 PORT (     CLK2:      IN std_logic;
```

```
  CLK1:      IN std_logic;
  A:      IN std_logic_vector(2 DOWNTO 0);
  B:      IN std_logic_vector(2 DOWNTO 0);
  SEL:   OUT std_logic_vector(2 DOWNTO 0));
END adder;

ARCHITECTURE behavior OF adder IS

  SIGNAL SUM_D1, SUM_D2 : std_logic_vector(2 DOWNTO 0)
  := "000";

BEGIN
        SUM_D1 <= A + B ;

        process
        begin
            wait until CLK2'event and CLK2 = '1';
                SUM_D2 <= SUM_D1 ;
        end process;

        process
        begin
            wait until CLK1'event and CLK1 = '1';
                SEL <= SUM_D2 ;
        end process;

END behavior;




=====================================================================
----- This block is used to model the accumulator in
PATH_A, 31-bits -----
LIBRARY ieee;
USE ieee.std_logic_1164.ALL;
USE ieee.std_logic_arith.all;
USE IEEE.std_logic_misc.all;
USE IEEE.std_logic_unsigned.all;

ENTITY accumulator IS
  PORT ( A:          IN std_logic_vector(30 DOWNTO
0);    -- 3 bits for integer, 28 bit for fractional
part, total 31 bits
       CLK2:         IN std_logic;
```

```
        SEL:          OUT std_logic_vector(2 DOWNTO 0);
        TO_PATH_B:    OUT std_logic_vector(2 DOWNTO 0));
END accumulator;

ARCHITECTURE behavior OF accumulator IS
```

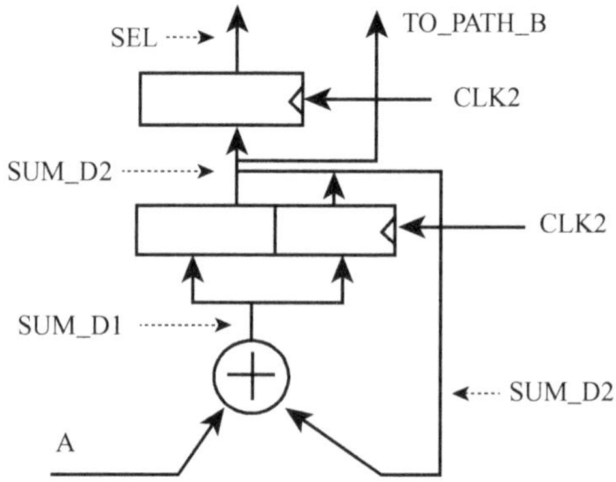


```
  SIGNAL SUM_D1, SUM_D2 : std_logic_vector(30 DOWNTO 0)
  := ( others => '0' );

BEGIN
        SUM_D1 <= A + SUM_D2 ;
        TO_PATH_B <= SUM_D2(30 DOWNTO 28) ;    -- top
3 bits

        process
        begin
            wait until CLK2'event and CLK2 = '1';
            SUM_D2 <= SUM_D1 ;
            SEL <= SUM_D2(30 DOWNTO 28) ;     -- top
            3 bits
        end process;
END behavior;



==============================================================
--- Flying-adder synthesizer, please refer to Fig.
4.17 -----

LIBRARY ieee;
USE ieee.std_logic_1164.ALL;
USE IEEE.std_logic_misc.all;
USE IEEE.std_logic_unsigned.all;
```

```
ENTITY Flying-adder IS
 -- 32 bits cntl word, 4 bits for integer since there
 are 8 inputs, but 2 paths
 PORT( FREQ:          IN std_logic_vector(31 DOWNTO 0);
       CLK_OUT:       OUT std_logic);
END Flying-Adder;

ARCHITECTURE structure OF flying-adder IS

    COMPONENT ticks          -- This is the block for
    generating the 8 inputs
        PORT( TICK_OUT:      OUT std_logic_vector(7
        DOWNTO 0) );
    END COMPONENT;

    COMPONENT accumulator     -- This is the
    accumulator in PATH_A
        PORT(A:              IN std_logic_vector(30 DOWNTO
                             0);
             CLK2:              IN std_logic;
             SEL:             OUT std_logic_vector(2
                              DOWNTO 0);
             TO_PATH_B:        OUT std_logic_vector(2
                               DOWNTO 0));
    END COMPONENT;

    COMPONENT adder     -- This is the adder in PATH_B
        PORT(CLK2:             IN std_logic;
             CLK1:             IN std_logic;
             A:           IN std_logic_vector(2 DOWNTO
                          0);
             B:           IN std_logic_vector(2 DOWNTO
                          0);
             SEL:            OUT std_logic_vector(2 DOWNTO
                             0) );
    END COMPONENT;

    COMPONENT mux8
        PORT(A:      IN std_logic_vector(7 DOWNTO 0);
             SEL:    IN std_logic_vector(2 DOWNTO 0);
             Y:      OUT std_logic);
    END COMPONENT;

    COMPONENT mux2
        PORT(A:      IN std_logic;
```

```
                B:      IN std_logic;
                SEL:    IN std_logic;
                Y:      OUT std_logic);
    END COMPONENT;

    COMPONENT tff
        PORT(CLK:      IN std_logic;
                Q:     OUT std_logic);
    END COMPONENT;

    SIGNAL FREQ_up, low_to_up: std_logic_vector(2
    DOWNTO 0) := (others => '0');
    SIGNAL FREQ_low: std_logic_vector(30 DOWNTO 0) :=
    (others => '0');
 -- Initial values have to be set on "sel_up" and
"sel_low" to avoid "X"s. This is important. Any values
are fine.
    SIGNAL sel_up, sel_low: std_logic_vector(2 DOWNTO
    0) := (others => '0');
    SIGNAL ticksout: std_logic_vector(7 DOWNTO 0);
    SIGNAL muxout_low, muxout_up, muxout: std_logic;
    SIGNAL clk1, clk2: std_logic;

BEGIN
    FREQ_low <= FREQ(30 DOWNTO 0);              -- The
lower 31 bits is sent to PATH_A
FREQ_up <= FREQ(31 DOWNTO 29) + FREQ(28) ;    -- The
                                upper 3 bits is
                                shifted right (/2) and
                                sent to PATH_B.
                                Additionally, FREQ(28)
                                is used for balancing
                                the duty cycle.

    CLK_OUT <= clk1 ;
    clk2 <= not clk1 after 0.01 ns;          -- The
value could be changed, based on your design.

    tickgen : ticks    -- block for generating the
    flying-adder inputs
        PORT MAP( TICK_OUT => ticksout );

    low_adder : accumulator          -- The accumulator
    in PATH_A
        PORT MAP(
```

```
                A            => FREQ_low,
                CLK2           => clk2,
                TO_PATH_B      => low_to_up,     -- See
                Fig. 4.16
                SEL            => sel_low );

    low_mux : mux8            -- The MUX8 in PATH_A
        PORT MAP(
                A            => ticksout,
                SEL           => sel_low,
                Y            => muxout_low );

    up_adder : adder              -- The adder in PATH_B
        PORT MAP(
                A            => FREQ_up,
                B            => low_to_up,
                CLK2          => clk2,
                CLK1          => clk1,
                SEL           => sel_up );

    up_mux : mux8             -- The MUX8 in PATH_B
        PORT MAP(
                A            => ticksout,
                SEL           => sel_up,
                Y            => muxout_up );

    interlock_mux : mux2      -- The MUX2 for
                              interlocking
        PORT MAP(
                A            => muxout_up,
                B            => muxout_low,
                SEL           => clk1,
                Y            => muxout );

    toggle_dff : tff          -- The final toggle
                              flip-flop
        PORT MAP(
                CLK            => muxout,
                Q            => clk1 );
END structure;
```

APPENDIX 4.B: HOW CLOSE CAN IT REACH AN INTEGER?

$$F = N/(\gamma \cdot f_o) = \eta \cdot N \qquad (4.B.0, 4.11)$$

Section 4.9.1, Eq. 4.11 (which is listed here as Eq. 4.B.0) presents a mathematic problem: given a real number η and an integer N from the range of $[N_L, N_H]$, by doing the operation of $\eta \cdot N$, how close can we reach another integer F? N has to be chosen from $[N_L, N_H]$. This problem can be graphically illustrated with an analogue presented in Fig. 4.B.1. There is a rack with evenly spaced steps inside. Given a stick of certain length, can the stick's end reach one of the steps if we keep stacking the sticks one after another? Assume that the rack is infinitely long and also assume that we can stack the stick however many times we like; then it is understandable that the stick's end will reach one of the steps sooner or later regardless the length of the stick and no matter where the stick first starts from. In this analogue, each step in the rack corresponds to an integer F. The integer N is used for indexing the stacking operations. The length of the stick represents the real number η. Now the flying-adder synthesizer related question becomes:

> Given a stick of certain length (a desired frequency $\eta = K \cdot f_r / f_o$), starting from the first step in the rack region of (F_L, F_H), you are allowed to do maximum $N_H - N_L$ times of stacking, can the stick's end reach one of the steps between (F_L, F_H)? If not, what is the guaranteed closest distance to one of the steps?

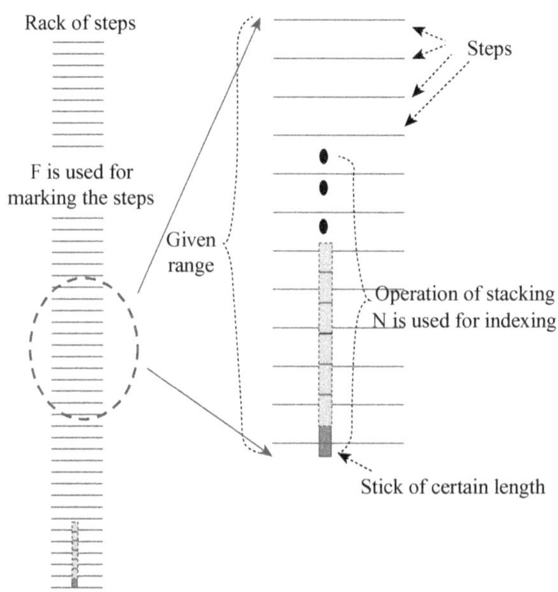

Fig. 4.B.1. How close can the stick's end reach a step?

In an FAPLL circuit, it has been arranged in such way that the $N_H - N_L$ times of stacking operation will cover the rack region of (F_L, F_H). Depending on the length of the stick (the asked frequency), it is possible that the stick's end could hit one of the steps [successfully locate one of the integers in (F_L, F_H)]. When that happens, this frequency can be directly generated. Otherwise, we have to choose the closest step (closest integer). The difficulty of this problem is due to the fact that we need to find the guaranteed smallest frequency error when the asked frequency is unknown (the stick's length is a variable). One solution is presented below.

There are $(N_H - N_L)\cdot(F_H - F_L)$ possible combinations if one N and one F make a pair of (F, N). A sequence can be built based on the value of F/N. This sequence can be sorted in ascending order as shown in Eq. 4.B.1 below where integer p is used for indexing.

$$FN\_seq\_sorted = \left\{ \left(\frac{F}{N}\right)_1, \left(\frac{F}{N}\right)_2, \ldots, \left(\frac{F}{N}\right)_p, \left(\frac{F}{N}\right)_{p+1}, \ldots \right\} \quad where \qquad (4.B.1)$$
$$\left(\frac{F}{N}\right)_p < \left(\frac{F}{N}\right)_{p+1}$$

This sequence of F/N is used to mark the axis in which the real number η moves, as illustrated in the left drawing of Fig. 4.B.2. Any η can be identified by its two boundary values $(F/N)_p$ and $(F/N)_{p+1}$ as shown in Eq. 4.B.2.

$$\left(\frac{F}{N}\right)_p \leq \eta \leq \left(\frac{F}{N}\right)_{p+1} \qquad (4.B.2)$$

Define that $F' = \eta\cdot N$ (F' is not guaranteed to be an integer) and F is the closest integer that we are searching for. Then the error $e_{p,p+1}$ can be defined by Eq. 4.B.3 below.

$$e_{p,p+1} = \frac{F - F'}{F} = 1 - \eta\left(\frac{N}{F}\right) \qquad (4.B.3)$$

It can be easily proven that Eq. 4.B.2 is equivalent to Eq. 4.B.4. Therefore, when η is in the range defined by Eq. 4.B.2, the trend of $e_{p,p+1}$ can be illustrated by the drawing on the right in Fig. 4.B.2.

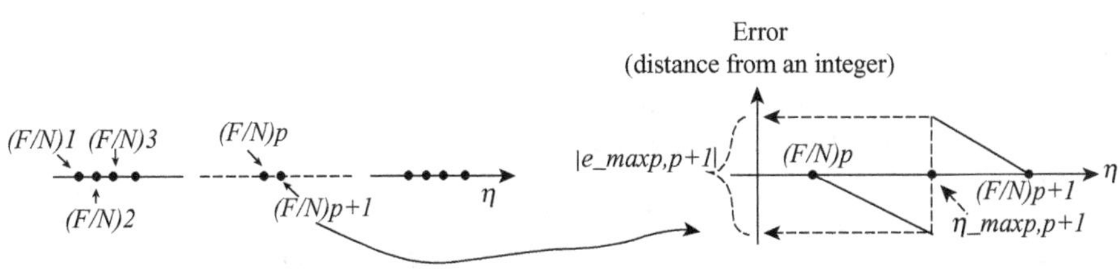

Fig. 4.B.2. The local maximum distance (local maximum approximation error).

$$1-\eta\left(\frac{N}{F}\right)_p < 0 < 1-\eta\left(\frac{N}{F}\right)_{p+1} \tag{4.B.4}$$

The size of $e_{p,p+1}$ is proportional to the distance of F' to an integer. This distance changes piecewise linearly when η moves. For any η in this region defined by Eq. 4.B.2, we can use either $(F/N)_p$ or $(F/N)_{p+1}$ to approximate it. However, the maximum distance (the maximum approximation error) is obtained when Eq. 4.B.5 is satisfied. This is true since we can always move η to increase the error on one side and decrease the error on the other side if η does not make the two equal.

$$1-\eta\left(\frac{N}{F}\right)_p = -\left\{1-\eta\left(\frac{N}{F}\right)_{p+1}\right\} \tag{4.B.5}$$

From Eq. 4.B.5, it is concluded that the maximum error can be found by Eq. 4.B.7 and it occurs in the location of Eq. 4.B.6. Equations 4.B.6 and 4.B.7 are what are needed for estimating the frequency error. A simple algorithm can be created to first build the sorted sequence of Eq. 4.B.1. Then $\eta\_max_{p,p+1}$ and $e\_max_{p,p+1}$ can be calculated using Eqs. 4.B.6 and 4.B.7.

$$\eta\_\max{}_{p,p+1} = \frac{2}{(N/F)_{p+1} + (N/F)_p} \tag{4.B.6}$$

$$e\_\max{}_{p,p+1} = \frac{(N/F)_{p+1} - (N/F)_p}{(N/F)_{p+1} + (N/F)_p} \tag{4.B.7}$$

For an integer-only FAPLL example of $f_r = 26$ MHz and $K = 8$, we assume that a VCO circuit is designed for a range of 1.3 GHz to 2.6 GHz. As a result, the valid range for N is (50, 100). Using those parameters, Fig. 4.B.3 (left) shows the plot of $e\_max_{p,p+1}$ envelop versus η (the X axis has been transformed into a real frequency by $\eta = K{\cdot}f_r/f_o$). Figure 4.B.4 is its magnified plot around the area of 1492 MHz. As can be seen, the frequencies of 1490.667 MHz and 1493.818 MHz can be directly generated with ($F = 12$, $N = 86$) and ($F = 11$, $N = 79$), respectively. At these two frequencies, the approximation error is 0. Any frequencies in between have to be approximated by either one of them. The further the asked frequency is away from these two frequencies, the larger the error is. The error increases linearly verses the distance. The maximum error occurs at frequency of 1492.242 MHz. To improve the performance (reduce error magnitude), we can enlarge the N's range. This is equivalent to doing more steps of stacking with a smaller stick. While doing so, we assume that the VCO circuit is not modified; the previous range of 1.3 GHz to 2.6 GHz is still valid. In a PLL design, this can be achieved by reducing the reference frequency. The middle plot in Fig. 4.B.3 is the case of $f_r = 13$ MHz (N:[100, 200]). The plot at the right is the case of $f_r = 6.5$ MHz (N:[200, 400]). Clearly, the approximation error is significantly reduced with a large N range (the size of the step in the rack becomes smaller). It is also interesting to see that in all

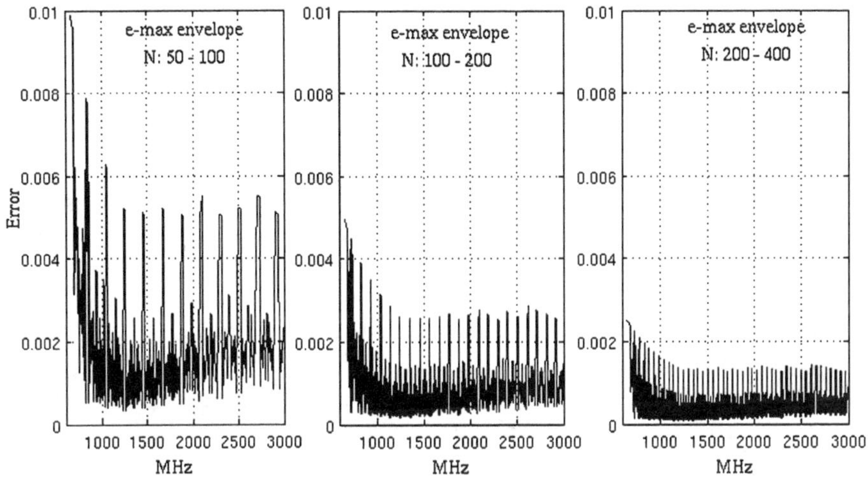

Fig. 4.B.3. The $e\_max_{p,p+1}$ envelop for three different N ranges.

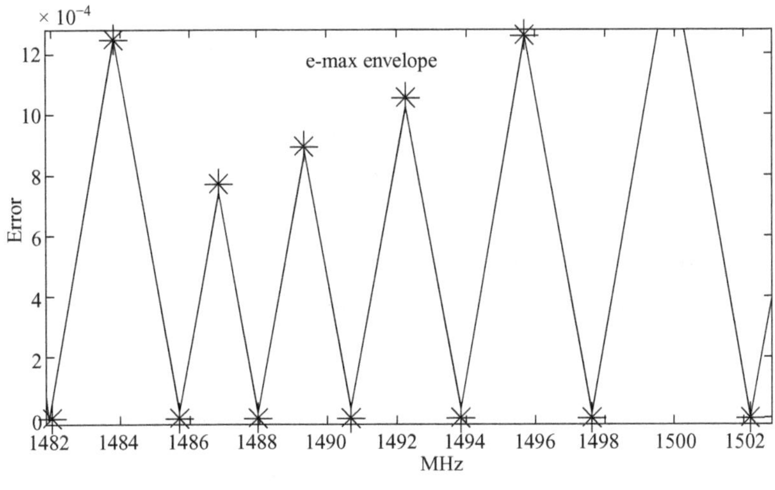

Fig. 4.B.4. The magnified plot of Fig. 4.B.3 (left) around 1492 MHz.

the three cases the maximum error agrees with what is predicted from Eq. 4.12: 0.01 = 1/(2*50), 0.005 = 1/(2*100), and 0.0025 = 1/(2*200).

APPENDIX 4.C: THE *SEED* AND *SET* IN INTEGER-FLYING-ADDER PLL

In Sections 4.9.2 and 4.9.3, the integer-flying-adder PLL was introduced. Equations 4.13 and 4.14 have been derived. They are repeated here as Eqs. 4.C.1 and 4.C.2, respectively.

$$f_{VCO} = \frac{F \cdot N}{K} f_r = (F \cdot N) \cdot \frac{f_r}{K} \tag{4.C.1, 4.13}$$

$$f_o = \frac{F_2 \cdot N}{F_1 \cdot M} f_r \tag{4.C.2, 4.14}$$

Compared to integer-N PLL, the number of available frequencies from these equations is greatly boosted. To better assist analysis, the concepts of *seed* and *set* are created as follow.

- **Frequency Seed:** A reference source with known frequency for generating other frequencies.
- **Frequency Set:** Based on a seed, the group of frequencies that can be generated from a particular frequency generator.

In integer-N PLL, the available frequencies are expressed as $f_o = N{\cdot}f_r$. Hence, f_r can be considered a seed. Those frequencies can be arranged as a straight line in the *f-N* plan, as shown in Fig. 4.C.1. This line is solely determined by seed f_r, and the frequency resolution is f_r. When the flying-adder divider is used inside the PLL, from Eq. 4.C.1, the resolution becomes either $F{\cdot}(f_r/K)$ when N varies or $N{\cdot}(f_r/K)$ when F varies. For a given divide ratio $N = N_x$, all the fractions $1/N_x$, $2/N_x$, . . . , $(N_x - 1)/N_x$ can be used in F because of the post divider fractional bit recovery (PDFR) introduced in Section 4.6. This can help make the frequency resolution to f_r/K as explained below. In Fig. 4.C.1, the line $F = K$ separates the whole area into two parts. The lower part is the area of $F < K$ and the upper area is $F > K$. The line $F = K$ corresponds to integer-N PLL (the FA divider is virtually bypassed). For a given N_x, we can draw a vertical line that represents the F change. When F moves

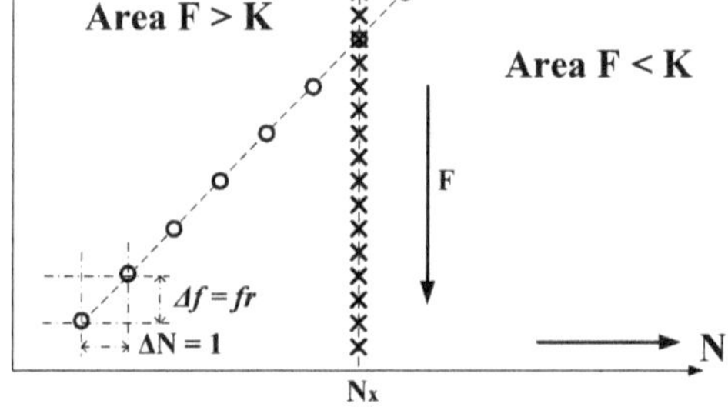

Fig. 4.C.1. The concept of frequency *seed* and frequency *set*.

along this line, for each integer F, it can include $N_x - 1$ fractions from $1/N_x$, to $(N_x - 1)/N_x$ (Fig. 4.C.1, not drawn to scale). As a result, frequency resolution becomes f_r/K. For each valid N_x, a line like this can be drawn. These lines form a *set* that origins from the *seed* f_r. The plots in Figs. 4.30 and 4.31 are examples of this concept of frequency set.

The frequency resolution from Eq. 4.C.2 is expected to be better than that of Eq. 4.C.1 since there are more variables in the equation and all of them are adjustable (in contrast, K is fixed in Eq. 4.C.1). Consequently, the analysis on frequency resolution is more involved. We consider first the case of $M = 1$. It results in

$$f_o = \frac{F_2 \cdot N}{F_1} f_r \tag{4.C.3, 4.15}$$

Based on previous discussions, we know that the resolution is f_r/F_1 when the PDFR is applied for the pair of (F_2 and N). Moreover, F_1 is a variable that can take any value from 2 to $2K$. Thus, for a given $N = N_x$, along the F_2 vertical line, we have frequency steps of $f_r/2, f_r/3, f_r/4, \ldots, f_r/(2K)$ with finest step of $f_r/(2K)$. Figure 4.C.2 is an example that illustrates this fact. In this example, $K = 8$. Therefore, F_1 can take any value from 2 to 16. For each given F_1, it divides the f_r into F_1 segments with each segment occupying the frequency band of f_r/F_1. When projected into the f axis, unlike the case in Eq. 4.C.1 where all the frequency points collapse, some of these frequency points interlace, resulting in a reduced band. In other words, it improves resolution. In this example, the finest step is $f_r/(2K)$, or $0.0625f_r$. However, a particular step could be smaller due to the interlacing. Figure 4.C.3 shows the scenario when F_1, F_2, and N are used together to form the set from a seed of f_r. By putting M back into consideration, the resolution could become even finer. From Eq. 4.C.2 and Fig. 4.C.1, the frequency resolution becomes $f_r/(F_1{*}M)$ when F_2 moves along the vertical line.

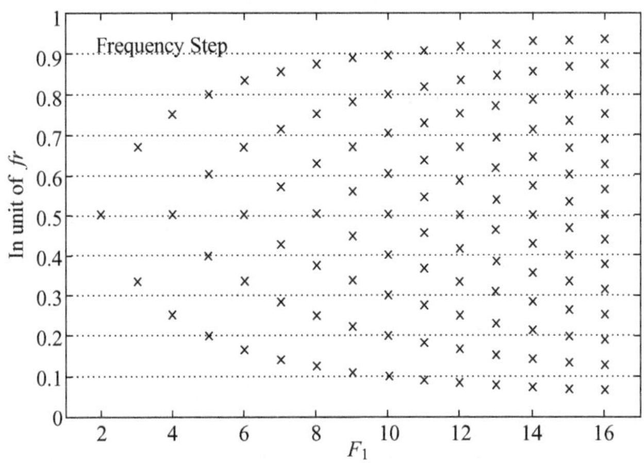

Fig. 4.C.2. The number of frequency points vs. F_1.

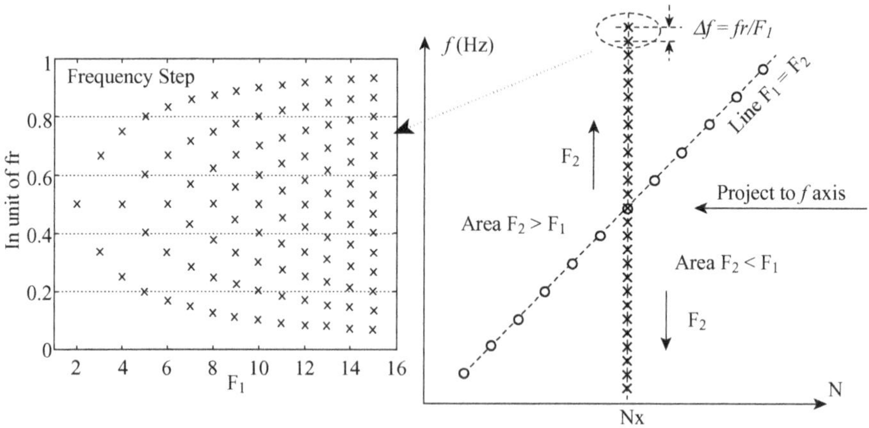

Fig. 4.C.3. The seed and set from integer-FAPLL.

Moreover, PDFR can be applied in the pair (F_I, M), which further improves the resolution. This fact can be appreciated from the curves in Fig. 4.33. The lower the output frequency f_o is, the finer the resolution would be.

APPENDIX 4.D: THE NUMBER OF CARRIES FROM AN XIU-ACCUMULATOR

Section 4.11 introduces a new type of accumulator, the XIU-accumulator, to be used in a flying-adder synthesizer. In this appendix, we will prove that the number of carries generated from this accumulator is the same as that produced from a conventional CON-accumulator. Assume that the flying-adder control word FREQ takes this value: $FREQ = I + r$, where r is the fraction $0 < r < 1$. Using base-b, r can be represented as Eq. 4.D.1 where an m-bits system is used. This number is fed into both the CON-accumulator and XIU-accumulator for "error" accumulation.

$$r = r_1 b^{-1} + r_2 b^{-2} + r_3 b^{-3} + \ldots + r_m b^{-m} \tag{4.D.1}$$

After b^m accumulations (b^m clock cycles), the sum from the CON-accumulator can be calculated as:

$$S_1 = b^m r = r_1 b^{m-1} + r_2 b^{m-2} + r_3 b^{m-3} + \ldots + r_m b \tag{4.D.2}$$

Equation 4.D.2 clearly shows that after b^m operations, all the fractional contents are propagated to the integer part. The total number of carries generated during these b^m operations is $b^m r$. At the same time, the mathematically equivalent r can be represented using Eq. 4.D.3 as well.

$$
\begin{aligned}
r = r_1 b^{-1} &+ 0b^{-2} + 0b^{-3} + \ldots + 0b^{-m} \\
&+ 0b^{-1} + r_2 b^{-2} + 0b^{-3} + \ldots + 0b^{-m} \\
&+ 0b^{-1} + 0b^{-2} + r_3 b^{-3} + \ldots + 0b^{-m} \\
&+ \ldots \\
&+ 0b^{-1} + 0b^{-2} + 0b^{-3} + \ldots + r_m b^{-m}
\end{aligned} \tag{4.D.3}
$$

Define:

$$R_1 \equiv r_1 b^{-1},\ R_2 \equiv r_2 b^{-2},\ R_3 \equiv r_3 b^{-3}, \ldots,\ R_m \equiv r_m b^{-m} \tag{4.D.4}$$

For each of these R_1, R_2, R_3 . . . , a 1-bit XIU-accumulator can be used to carry out the accumulation as depicted in Fig. 4.36. After b^m clock cycles, the result becomes

$$b^m \cdot R_1 = r_1 b^{m-1},\ b^m \cdot R_2 = r_2 b^{m-2},\ b^m \cdot R_3 = r_3 b^{m-3}, \ldots,\ b^m \cdot R_m = r_m b \tag{4.D.5}$$

Since the m 1-bit XIU-accumulators are serially connected as shown in Fig. 4.36, the carries generated at each stage are propagated forward progressively at each clock cycle. None of them is lost in this process. Therefore, after b^m operations, the result can be derived as

$$S_2 = b^m R_1 + b^m R_2 + b^m R_3 + \ldots + b^m R_m = r_1 b^{m-1} + r_2 b^{m-2} + r_3 b^{m-3} + \ldots + r_m b \ = S_1 \tag{4.D.6}$$

Equation 4.D.6 proves that, for every b^m cycles of accumulation, the results from the two accumulators equal. This fact is precisely what is needed for time-average-frequency-based flying-adder operations. Also worth being mentioned is that only b^x clock cycles are needed to satisfy the condition of $S_2 = S_1$ if r_x is the first (from LSB) nonzero bit in Eq. 4.D.1. As a matter of fact, the length-in-time that covers these b^x cycles (made of both T_A and T_B) is the fundamental period defined in Eq. 3.4.

APPENDIX 5.A: THE FLYING-ADDER STATE MACHINE MODEL (PERL)

```
#####################################################
#######################################
# This program is used to describe the flying-adder
operation as a finite state machine; please refer to
Section 5.4.
#####################################################
#######################################
```

```
#### Get information from user
#######################################################

print "\n\n*************************************************************************\n" ;
print "Please provide your answers to the following questions all as integers\n" ;
print "*******************************************************************\n" ;
print "\n\n\nWhat is the size of your flying-adder accumulator: n, in number of bits? ";
chop ( $n = <STDIN> ) ;
print "\nThe number you typed in is $n\n" ;

print "\nHow many inputs? In other words, what is the size of the MUX: m, in number of bits? ";
chop ( $m = <STDIN> ) ;
print "\nThe number you typed in is $m\n" ;

$w_max = 2**($n) ;
print "\nWhat is the frequency control word w? must be < $w_max ! ";
chop ( $w = <STDIN> ) ;
print "\nThe number you typed in is $w\n" ;
#### Done with user input
############################################################

print "###### Here is the summary ######\n" ;
$num_inputs = 2**($m) ;          # This is the number of FA inputs
$base_frac = 2**($n-$m) ;        # Any number smaller than this is a fraction
$G_n = GCD($w, $w_max) ;         # This is the greatest common divider between w and 2**n
$r = log($G_n)/log(2) ;          # a temporary variable
$K = $w_max/$G_n ;               # This is the period of state variables xk and yk
$G_nm = GCD($w, $base_frac) ;    # This is the greatest common divider between w and 2**(n-m)
$L = $base_frac/$G_nm ;          # This is the period of state variables xk and yk
```

```
$Ts = $w/$G_nm ;                      # This is the
                                      fundamental period on
                                      signal s and v (if
                                      L > 1)
$Txy = $K/$L*$Ts ;                    # This is the
                                      fundamental period on
                                      internal state xk and
                                      yk

#Our conventional frequency control word FREQ = I + r
#In a real FA circuit, there are two paths. Thus w
needs to be doubled
$FREQ = 2*$w/$base_frac ; $I = int($FREQ) ; $rr =
$FREQ - $I ;

print "\n\nThere are $n bits in the accumulator, $m
bits for MUX ($num_inputs input signals)\n" ;
print "In frequency control word w, the boundary
between integer and fraction is $base_frac\n" ;
print "The frequency control word is $w ---> the GCD
of $w and 2**$n is $G_n (r = $r)\n" ;
print "The period of state variables xk and yk is --->
K = $K\n" ;
print "The period of state variables dk and modified dk
is ---> L = $L\n" ;
print "\nThe frequency control word w can be converted
to FREQ as:\nFREQ = $FREQ, integer $I, fraction $rr\n"
;

for ($k=0; $k<2**($n-$r); $k++) { push(@xk, $k*$G_n)
;}
print "\n\nThe state variable xk will take value from
this set, and only from this set; each one is used
once\n";
foreach ( @xk ) { print "$_ " ; }
print "\n\n" ;

for ($k=0; $k<2**($n-$r); $k++) { $tmp =
int($k*$G_n/$base_frac); $yk{$tmp} = $tmp; }
print "The state variable yk will take value from this
set, and only from this set; each one is used at
least once\n";
foreach ( sort {$a <=> $b} keys %yk) { print "$yk{$_}
"; }
print "\n\n" ;
```

```
####### The FA operation starts from here
##################
open (F_TAF, ">TAF_waveform.txt") ;
$high_or_low = 1 ;
$xk_previous = 0 ;                  # initial value,
                                    could be any
                                    value < 2**n
$number_of_K_cycles = 1 ;           # you can change it
$stop = $K*$number_of_K_cycles ;
$yk_previous = 0;
for ($k=0; $k< $stop; $k++ ) {      #k is the index for
                                    advancing the
                                    discrete time
    $xk = ($xk_previous + $w ) % 2**$n ;
    $yk = int($xk/2**($n-$m)) ;
    $dk = ($yk - $yk_previous) % 2**$m ;
    if ( $dk == 0 ) { $m_dk = 2**$m ;} else { $m_dk =
    $dk ; }
    for ($i=0; $i<$m_dk; $i++) {
        if ($k < 2*$L ) { print F_TAF "$high_or_
        low\n"; $count++;}    #just do two L cycles
    }
    if ($high_or_low) { $high_or_low = 0 ;} else {
    $high_or_low = 1 ;}
    $xk_previous = $xk ;
    $yk_previous = $yk ;
    print "k = $k\t\tX = $xk\t\tY = $yk\t\tD = $dk\t\
tM_D = $m_dk\t\tTs(Tv) = $Ts\t\tTxy = $Txy\n" ;
}

#This part is for generating the 50% duty cycle ideal
waveform
open (F_IDEAL_x, ">IDEAL_waveform_x.txt") ;
open (F_IDEAL_y, ">IDEAL_waveform_y.txt") ;
#For two TV time frames, there are L cycles, 2*L high/
low levels.
$high_low_bound = $count/$L/2 ;      # boundary for
                                     high/low transition:
                                     $count/(2L)
print "Total data point count: $count, boundary:
$high_low_bound\n" ;
$factor = 100 ;
$new_bound = $high_low_bound*$factor ;
$value = 1.1 ;
$x_initial = 1.0 ;
```

```
for ($j=1; $j<=$count*$factor; $j++) {
    $y = $value ; print F_IDEAL_y "$y\n" ;
    $x = $j/$factor + $x_initial; print F_IDEAL_x
    "$x\n" ;
    if ( ($j % ($new_bound) ) == 0 ) { if ( $value ==
1.1 ) { $value = -0.1 } else { $value = 1.1} }
}

sub GCD {
    local($x, $y) = ($_[0], $_[1]) ;
    local($g) = ($y) ;
    while ( $x > 0 ) { $g = $x ; $x = $y % $x ; $y =
    $g ; }
    return $g ;
}
```

```
% Using Matlab to plot the waveform
>> TAF_waveform = load('TAF_waveform.txt') ;
>> Ideal_waveform_x = load('IDEAL_waveform_x.txt') ;
>> Ideal_waveform_y = load('IDEAL_waveform_y.txt') ;
>> plot(TAF_waveform)
>> hold on
>> plot(Ideal_waveform_x, Ideal_waveform_y)
```

APPENDIX 5.B: THE FLYING-ADDER WAVEFORM GENERATOR (PERL)

```
#####################################################
##################################
# This program is used to generate the flying-adder
time waveform. Please refer to Eqs. 5.1 and 5.3.
#####################################################
##################################

$fvco = 1.332 ;              #in GHz
$K = 8 ;                     #No. of flying-adder inputs
$delta = 1000/($K*$fvco) ;   #in ps

######################### Get Inputs
####################################################
print "\n\n\nWhat is the frequency control word? Input
as real number, ex: FREQ = 8.0\n";
chop ( $FREQ = <STDIN> ) ;
```

```
$FREQ =~ /(\S*)(\.\S*)/ ;
$F_integer = $1; $F_dec = $2;
print "\n\nThe number you typed can be split as:\n" ;
print "INTEGER = $F_integer, DECIMAL = $F_dec\n" ;

print "\n\n\nHow long do you want to do the simulation
(in unit of us)?\n";
chop ( $sim_time = <STDIN> ) ;
print "\n\nThe number you typed is $sim_time\n" ;
$sim_time = $sim_time*1e6/$delta ;
######################### All inputs are done
#########################################################

$Time = 0 ;                    #in units of delta
$accumulator_out = 0 ;              #Flying-adder
accumulator result
$F_current = $F_integer + $F_dec; ;

open (FX2, ">time.txt") ;
open (FY2, ">period.txt") ;
open (FY3, ">waveform.txt") ;

$FREQ_integer = int($F_current) ;
$FREQ_frac = $F_current - $FREQ_integer ;
$FA_period = $FREQ_integer ;
print FX2 "$Time\n" ;
print FY2 "$FA_period\n" ;

$delta_count = 0 ;
#Time is continuously running in units of delta until
the stop signal occurs
while ( $Time < $sim_time ) {
    $Time++ ;                  #in units of delta
    $delta_count++ ;

    if ( $delta_count == $FA_period ) { #hit flying-
    adder cycle's edge, finish one FA cycle
        $delta_count = 0 ;
        print FX2 "$Time\n" ;
        print FY2 "$FA_period\n" ;
        print "Time $Time, Period $FA_period\n" ;
        $accumulator_out += $FREQ_frac ;    #fractional
        accumulation
        $FA_period = $FREQ_integer + int($accumulator_
out) ;         #get it into integer
        $period_M{$Time} = $FA_period ;
```

```
        if ( $accumulator_out >= 1) { $accumulator_out
        -= 1 ; }    #then reset it
        waveform_gen ($FA_period) ;
    }
}


######## take care of post divider
############################
while ( $M != 0 ) {
print "\n\n\nWhat is post divider M? Input it as an
integer. "0" will exit.\n";
chop ( $M = <STDIN> ) ;
print "\n\nThe number you typed in is $M\n" ;

open (FX4, ">time_$M.txt") ;
open (FY4, ">period_$M.txt") ;

$index_M = 0 ;
$sum = 0 ;
foreach ( sort {$a <=> $b} keys %period_M) {
    $index_M++ ;
    $sum += $period_M{$_} ;
    if ( $index_M % $M == 0 ) { #hits the MCLK
    boundary
        $period_after_M = $sum;
        for ( $x=1; $x<=$M; $x++ ) { print FX4
"$_\n"; print FY4 "$period_after_M\n" ; }
        $sum = 0 ;    #reset
    }
    print "Index: $index_M, Period: $period_M{$_}, SUM:
$sum, AVG: $period_after_M\n" ;
}
} else { exit ;}


sub waveform_gen {
    $one_FA_cycle = $_[0] ;  #get the FA period
    for ($ii=1; $ii<=$one_FA_cycle; $ii++ ) {
        if ( $ii <= int($one_FA_cycle/2) ) { print FY3
        "0\n" }
        else { print FY3 "1\n" ; }
    }
}

# Please see Appendix 5.C for plotting the result
```

APPENDIX 5.C: THE FLYING-ADDER WAVEFORM GENERATOR WITH TRIANGULAR MODULATION (PERL)

```
############################################################
############################################################
####
# This program is used to add a triangular disturbance
to the flying-adder synthesizer control word FREQ.
Please refer to Section 5.7.
############################################################
############################################################
####

$fvco = 2 ;              #in GHz
$K = 8 ;                     #No. of flying-adder inputs
$delta = 1000/($K*$fvco) ;   #in ps

######################### Get Inputs
###########################################################
print "\n\n\nWhat is the central frequency control
word? Input as real number, ex: FREQ = 8.25\n";
chop ( $FREQ = <STDIN> ) ;
$FREQ =~ /(\S*)(\.\S*)/ ;
$F_integer = $1; $F_dec = $2;
print "\n\nThe number you typed can be split as:\n" ;
print "INTEGER = $F_integer, DECIMAL = $F_dec\n" ;
$F_central = $F_integer + $F_dec;

print "\n\n\nWhat is the magnitude of the spread?
Input as real number, ex: MAG = 0.125\n";
chop ( $MAG = <STDIN> ) ;
$MAG =~ /(\S*)(\.\S*)/ ;
$M_integer = $1; $M_dec = $2;
print "\n\nThe number you typed can be split as:\n" ;
print "INTEGER = $M_integer, DECIMAL = $M_dec\n" ;
$MAG = $M_integer + $M_dec ;

print "\n\n\nWhat is the step of the spread? Input as
real number, ex: STEP = 0.01\n";
chop ( $STEP = <STDIN> ) ;
$STEP =~ /(\S*)(\.\S*)/ ;
$S_integer = $1; $S_dec = $2;
print "\n\nThe number you typed can be split as:\n" ;
print "INTEGER = $S_integer, DECIMAL = $S_dec\n" ;
$STEP = $S_integer + $S_dec ;
```

```
print "\n\n\nWhat is the frequency of the SS CLK in
MHz? Input as real number, ex: SSCLK = 20.0 \n";
chop ( $SSCLK = <STDIN> ) ;
$SSCLK =~ /(\S*)(\.\S*)/ ;
$SSCLK_integer = $1; $SSCLK_dec = $2;
print "\n\nThe number you typed can be split as:\n" ;
print "INTEGER = $SSCLK_integer, DECIMAL = $SSCLK_
dec\n" ;
$SSCLK_MHz = $SSCLK_integer + $SSCLK_dec ;

$N1 = 1e6/$SSCLK_MHz/$delta;         #number_of_deltas_
in_one_SSCLK cycle
print "\n\nThere are $N1 deltas in each SSCLK cycle\n"
;
print "\nThe central FREQ: $F_central, The magnitude:
$MAG, The step: $STEP\n" ;
######################### All inputs are done
###################################################


$N2 = 2*$MAG/$STEP ;                #number of steps
needed for a half triangle

$number_of_triangular_cycle = 5 ;          #do the
analysis for this many of full triangles
$number_of_dir_change = 0 ;
$Time = 0 ;                     #in units of delta
$accumulator_out = 0 ;                #flying-adder
accumulator result
$F_up = $F_central + $MAG ;               #upper limit
$F_dn = $F_central - $MAG ;               #lower limit
$F_current = $F_dn ;                #Triangle starts from
bottom
$inc_direction = 0 ;                   #incremental
direction: 0 for increase, non-0 for decrease
$index_SSCLK_cycles = 0 ;                #this is the
index for SSCLK cycle

open (F1, ">notes.txt") ;
open (FX, ">x_asix.txt") ;
open (FY, ">FREQ.txt") ;
open (FX2, ">time.txt") ;
open (FY2, ">period.txt") ;
open (FY3, ">waveform.txt") ;

print FX "$Time\n" ; print FY "$F_current\n" ;
```

```
$FREQ_integer = int($F_current) ;
$FREQ_frac = $F_current - $FREQ_integer ;
$FA_period = $FREQ_integer ;
print FX2 "$Time\n" ; print FY2 "$FA_period\n" ;

#Time is continuously running in units of delta until
the stop signal occurs
while ( $number_of_dir_change < 2*$number_of_
triangular_cycle ) {        #this is for doing spread
    $Time++ ;                                #in units of
    delta

    if ( $Time % $FA_period == 0 )
    {                #hit flying-adder cycle's edge,
    finish one FA cycle
        $accumulator_out += $FREQ_frac
        ;             #fractional accumulation
        $FA_period = $FREQ_integer + int($accumulator_
        out) ;    #get it into integer
        print FX2 "$Time\n" ; print FY2 "$FA_period\n"
        ;
        $period_M{$Time} = $FA_period ;
        if ( $accumulator_out >= 1) { $accumulator_out
        -= 1 ; }    #then reset it
        waveform_gen ($FA_period) ;
    }

    if ( $Time % $N1 == 0 ) {                         #hit
SSCLK cycle's edge, take care of F update
        print F1 "Time: $Time, SSCLK index: $index_
SSCLK_cycles, Current FREQ: $F_current\n" ;
        if ( $inc_direction ) { $F_current -= $STEP ;
} else { $F_current += $STEP ; }

        $FREQ_integer = int($F_current) ;
        $FREQ_frac = $F_current - $FREQ_integer ;

        print FX "$Time\n" ; print FY "$F_current\n" ;

        $index_SSCLK_cycles++ ;
        if ( $index_SSCLK_cycles % $N2 == 0 ) { $inc_
direction = ~$inc_direction ; $number_of_dir_change++ ;
}
    }
}
```

```
######## take care of post divider
##############################
while ( $M != 0 ) {

print "\n\n\nWhat is post divider M? Input it as an
integer. "0" will exit.\n";
chop ( $M = <STDIN> ) ;
print "\n\nThe number you typed in is $M\n" ;

open (FX4, ">time_$M.txt") ;
open (FY4, ">period_$M.txt") ;

$index_M = 0 ;
$sum = 0 ;
foreach ( sort {$a <=> $b} keys %period_M) {
    $index_M++ ;
    $sum += $period_M{$_} ;
    if ( $index_M % $M == 0 ) { #hits the MCLK
    boundary
        $period_after_M = $sum;
        for ( $x=1; $x<=$M; $x++ ) { print FX4
"$_\n"; print FY4 "$period_after_M\n" ; }
        $sum = 0 ;    #reset
    }
}} else { exit; }


sub waveform_gen {
       $one_FA_cycle = $_[0]
       ;                          #get the FA period
         for ($ii=1; $ii<=$one_FA_cycle; $ii++ ) {
           if ( $ii <= int($one_FA_cycle/2) ) { print
           FY3 "0\n" }
           else { print FY3 "1\n" ; }
         }
}


%%% Plot the results in Matlab
%sample_rate = 8e9 ;    # 1 GHz@ 8 phase
%fvco = 1 GHz, 8 phases -> equivalent sample_rate 8
GHz, make it display in MHz
sample_rate = 8e3 ;

%125 ps, make it display as us
delta = 125e-6 ;
```

```
freq = load('FREQ.txt');
time_SSCLK_cycle = load('x_asix.txt');
plot(delta*time_SSCLK_cycle, freq)
title('Real time FREQ setting of flying-adder
synthesizer')
xlabel('Time (us)')

period = load('period.txt');
time_FA_cycle = load('time.txt');
plot(delta*time_FA_cycle, period)
title('Real-time flying-adder output period')
xlabel('Time (us)')

waveform = load('waveform.txt');
len = length(waveform) ;
NFFT = 2^nextpow2(len) ;
Y = fft(waveform, NFFT)/len ;
f = sample_rate/2*linspace(0,1,NFFT/2+1) ;
plot(f, 20*log10(abs(Y(1:NFFT/2+1))))
title('Flying-adder spread spectrum output')
xlabel('Frequency (MHz)')
```

APPENDIX 5.D: THE FLYING-ADDER WAVEFORM GENERATOR WITH RANDOM MODULATION (PERL)

```
#######################################################
#######################################################
#
# This program is used to add a random disturbance to
the flying-adder synthesizer control word FREQ. Please
refer to Section 5.5.
#######################################################
#######################################################
#

$fvco = 2 ;                #in GHz
$K = 8 ;                      #No. of flying-adder inputs
$delta = 1000/($K*$fvco) ;    #in ps

######################### Get Inputs
#######################################################
print "\n\n\nWhat is the central frequency control
word? Input as real number, ex: FREQ = 8.25\n";
```

```
chop ( $FREQ = <STDIN> ) ;
$FREQ =~ /(\S*)(\.\S*)/ ;
$F_integer = $1; $F_dec = $2;
print "\n\nThe number you typed can be split as:\n" ;
print "INTEGER = $F_integer, DECIMAL = $F_dec\n" ;
$F_central = $F_integer + $F_dec;

print "\n\n\nWhat is the magnitude of the random
number? Input as real number, ex: MAG = 0.2\n";
chop ( $MAG = <STDIN> ) ;
$MAG =~ /(\S*)(\.\S*)/ ;
$M_integer = $1; $M_dec = $2;
print "\n\nThe number you typed can be split as:\n" ;
print "INTEGER = $M_integer, DECIMAL = $M_dec\n" ;
$MAG = $M_integer + $M_dec ;

print "\n\n\nWhat is the frequency of the modulation
CLK in MHz? Input as real number, ex: MODCLK = 20.0
\n";
chop ( $MODCLK = <STDIN> ) ;
$MODCLK =~ /(\S*)(\.\S*)/ ;
$MODCLK_integer = $1; $MODCLK_dec = $2;
print "\n\nThe number you typed can be split as:\n" ;
print "INTEGER = $MODCLK_integer, DECIMAL = $MODCLK_
dec\n" ;
$MODCLK_MHz = $MODCLK_integer + $MODCLK_dec ;


$N1 = 1e6/$MODCLK_MHz/$delta;        #number_of_deltas_
in_one_MODCLK cycle
print "\n\nThere are $N1 deltas in each MODCLK
cycle\n" ;
print "\nThe central FREQ: $F_central, The magnitude:
$MAG, The step: $STEP\n" ;


print "\n\n\nHow long do you want to do the simulation
(in units of delta)?\n";
chop ( $sim_time = <STDIN> ) ;
print "\n\nThe number you typed is $sim_time\n" ;
########################## All inputs are done
#########################################################


$Time = 0 ;                  #in units of delta
```

```
$accumulator_out = 0 ;             #flying-adder
accumulator result
$F_current = $F_central ;
$index_MODCLK_cycles = 0 ;            #This is the
index for MODCLK cycle

open (F1, ">notes.txt") ;
open (FX, ">x_asix.txt") ;
open (FY, ">FREQ.txt") ;
open (FX2, ">time.txt") ;
open (FY2, ">period.txt") ;
open (FY3, ">waveform.txt") ;

print FX "$Time\n" ; print FY "$F_current\n" ;

$FREQ_integer = int($F_current) ;
$FREQ_frac = $F_current - $FREQ_integer ;
$FA_period = $FREQ_integer ;
print FX2 "$Time\n" ; print FY2 "$FA_period\n" ;

#Time is continually running in units of delta until
the stop signal occurs
while ( $Time <= $sim_time ) {
    $Time++ ;                            #in units of
                                         delta

    if ( $Time % $FA_period == 0 )
{                 #hit flying-adder cycle's edge, finish
one FA cycle
        $accumulator_out += $FREQ_frac
        ;             #fractional accumulation
        $FA_period = $FREQ_integer + int($accumulator_
        out) ;    #get it into integer
        print FX2 "$Time\n" ; print FY2 "$FA_period\n"
        ;
        $period_M{$Time} = $FA_period ;
        if ( $accumulator_out >= 1) { $accumulator_out
        -= 1 ; }    #then reset it
        waveform_gen ($FA_period) ;
    }

    if ( $Time % $N1 == 0 ) {                      #hit
MODCLK cycle's edge, take care of F update
        print F1 "Time: $Time, MODCLK index: $index_
MODCLK_cycles, Current FREQ: $F_current\n" ;
```

```
        $F_current = $F_central + rand($MAG) - $MAG/2
        ;

        $FREQ_integer = int($F_current) ;
        $FREQ_frac = $F_current - $FREQ_integer ;

        print FX "$Time\n" ; print FY "$F_current\n" ;

        $index_MODCLK_cycles++ ;
    }
}


######## take care of post divider
############################
while ( $M != 0 ) {

print "\n\n\nWhat is post divider M? Input it as an
integer. "0" will exit.\n";
chop ( $M = <STDIN> ) ;
print "\n\nThe number you typed in is $M\n" ;

open (FX4, ">time_$M.txt") ;
open (FY4, ">period_$M.txt") ;

$index_M = 0 ;
$sum = 0 ;
foreach ( sort {$a <=> $b} keys %period_M) {
    $index_M++ ;
    $sum += $period_M{$_} ;
    if ( $index_M % $M == 0 ) { #hits the MCLK
    boundary
        $period_after_M = $sum;
        for ( $x=1; $x<=$M; $x++ ) { print FX4
"$_\n"; print FY4 "$period_after_M\n" ; }
        $sum = 0 ;    #reset
    }
#    print "Index: $index_M, Period: $period_M{$_},
SUM: $sum, AVG: $period_after_M\n" ;
}} else { exit;}


sub waveform_gen {
      $one_FA_cycle = $_[0]
      ;                          #get the FA period
```

```
        for ($ii=1; $ii<=$one_FA_cycle; $ii++ ) {
          if ( $ii <= int($one_FA_cycle/2) ) { print
          FY3 "0\n" }
          else { print FY3 "1\n" ; }
        }
}

# Please see Appendix 5.C for plotting the result
```

APPENDIX 6.A: THE FA-DCXO TANGENT LINE AND LINEARITY MEASUREMENT

Please refer to the drawing on the right in Fig. 6.14. Assume that in general the flying-adder synthesizer's transfer function can be expressed as Eq. 6.A.1, where c is a constant. Its tangent line is Eq. 6.A.2, and a, and b are constants as well.

$$f_1(x) = \frac{c}{x} \tag{6.A.1}$$

$$f_2(x) = ax + b \tag{6.A.2}$$

Since at x_0, the two functions have the same value and the same first derivative, the following equations are valid:

$$a = f_1'(x_0) = -\frac{c}{x_0^2} \tag{6.A.3}$$

$$f_1(x_0) = \frac{c}{x_0} = f_2(x_0) = ax_0 + b \tag{6.A.4}$$

From Eqs. 6.A.3 and 6.A.4, the function of the tangent line can be deduced as

$$f_2(x) = -\frac{c}{x_0^2}x + \frac{2c}{x_0} \tag{6.A.5}$$

If we define Δy as the difference between the two functions at any given point x (please refer to the right-hand drawing in Fig. 6.14), then:

$$\Delta y = f_1(x) - f_2(x) = \frac{c}{x} + \frac{c}{x_0^2}x - \frac{2c}{x_0} \tag{6.A.6}$$

Since x can be expressed as $x = x_0 + \Delta x$, thus

$$\frac{1}{x}=\frac{1}{x_0+\Delta x}=\frac{1}{x_0(1+\Delta x/x_0)}=\frac{1}{x_0}\left(1-\frac{\Delta x}{x_0}+\left(\frac{\Delta x}{x_0}\right)^2-\left(\frac{\Delta x}{x_0}\right)^3+\dots\right) \quad (6.A.7)$$

If the accuracy is kept to the second order, we have

$$\frac{1}{x}\approx\frac{1}{x_0}\left(1-\frac{\Delta x}{x_0}+\left(\frac{\Delta x}{x_0}\right)^2\right)=\frac{1}{x_0}-\frac{\Delta x}{x_0^2}+\frac{\Delta x^2}{x_0^3} \quad (6.A.8)$$

$$\Delta y = f_1(x)-f_2(x)=\frac{c}{x_0}-\frac{c\Delta x}{x_0^2}+\frac{c\Delta x^2}{x_0^3}+\frac{c}{x_0^2}(x_0+\Delta x)-\frac{2c}{x_0}=\frac{c\Delta x^2}{x_0^3} \quad (6.A.9)$$

$$\frac{\Delta y}{y}=\frac{(f_1(x)-f_2(x))}{f_1(x)}=\frac{\frac{c\Delta x^2}{x_0^3}}{\frac{c}{x}}=\frac{\Delta x^2}{x_0^3}x=\frac{\Delta x^2}{x_0^3}(x_0+\Delta x)=\left(\frac{\Delta x}{x_0}\right)^2+\left(\frac{\Delta x}{x_0}\right)^3$$

$$\approx\left(\frac{\Delta x}{x_0}\right)^2 \quad (6.A.10)$$

Equation 6.A.10 is the error between the two functions and can be used as a linearity measurement in an FA-DCXO specification.

INDEX

Nanometer Frequency Synthesis Beyond the Phase-Locked Loop, First Edition. Liming Xiu.

Published 2012 by John Wiley & Sons, Inc.

www.ingramcontent.com/pod-product-compliance
Lightning Source LLC
LaVergne TN
LVHW011912200926
848484LV00002B/603
9781118162637